Nanomagnetism and Spintronics

Nanomagnetism and Spintronics

Second Edition

Edited by

Teruya Shinjo

Professor Emeritus,
Institute for Chemical Research,
Kyoto University,
Japan

ELSEVIER

AMSTERDAM • BOSTON • HEIDELBERG • LONDON • NEW YORK • OXFORD
PARIS • SAN DIEGO • SAN FRANCISCO • SINGAPORE • SYDNEY • TOKYO

Elsevier
32 Jamestown Road, London NW1 7BY, UK 225 Wyman Street, Waltham, MA
02451, USA

Second edition 2014

Notices
Knowledge and best practice in this field are constantly changing. As new research and
experience broaden our understanding, changes in research methods, professional practices,
or medical treatment may become necessary.

Practitioners and researchers must always rely on their own experience and knowledge in
evaluating and using any information, methods, compounds, or experiments described herein.
In using such information or methods they should be mindful of their own safety and the
safety of others, including parties for whom they have a professional responsibility.

To the fullest extent of the law, neither the Publisher nor the authors, contributors, or editors,
assume any liability for any injury and/or damage to persons or property as a matter of
products liability, negligence or otherwise, or from any use or operation of any methods,
products, instructions, or ideas contained in the material herein.

British Library Cataloguing-in-Publication Data
A catalogue record for this book is available from the British Library

Library of Congress Cataloging-in-Publication Data
A catalog record for this book is available from the Library of Congress

ISBN: 978-0-444-63279-1

For information on all Elsevier publications
visit our website at store.elsevier.com

This book has been manufactured using Print On Demand technology. Each copy is
produced to order and is limited to black ink. The online version of this book will show
color figures where appropriate.

Contents

3.1.1 Spin Injection   108
3.1.2 Spin-Transfer Torque   118
3.1.3 Field-Like Torque and Rashba Torque   125
3.1.4 Electric Field-Induced Anisotropy
      Change and Torque   126
3.2 Spin-Injection Magnetization Reversal   129
3.2.1 Amplification of the Precession   129
3.2.2 Linearized LLG Equation and Instability Current   133
3.2.3 Spin-Injection Magnetization Switching   136
3.2.4 Switching Time and Thermal Effects   139
3.2.5 High-Speed Measurements   143
3.3 High-Frequency Phenomena   148
3.3.1 Spin-Transfer Oscillation   148
3.3.2 Spin-Torque Diode Effect   154
3.3.3 Electric Field-Induced Dynamic Switching   159
3.4 From Spin-Transfer Torque RAM to Magnetic Logic   161
3.4.1 The Magnetic Random Access Memory   162
3.4.2 The Spin-Transfer Torque MRAM (or STT-RAM,
      or Spin-RAM)   167
3.4.3 Toward Magnetic Logic   167
Acknowledgments   169
References   169

4 **Dynamics of Magnetic Domain Walls in
  Nanomagnetic Systems**   **177**
Teruo Ono and Teruya Shinjo

4.1 Introduction   177
4.2 Field-Driven DW Motions   178
4.2.1 Detection of DW Propagation by Using GMR Effect   178
4.2.2 Ratchet Effect in DW Motions   181
4.2.3 DW Velocity Measurements   183
4.3 Current-Driven DW Motions   189
4.3.1 Concept of Current-Driven DW Motion   189
4.3.2 Magnetic Force Microscopy Direct Observations   190
4.3.3 Toward Applications of Current-Driven DW Motion   194
4.4 Topics on Nanodot Systems   197
4.4.1 MFM Studies on Magnetic Vortices in Dot Systems   197
4.4.2 Current-Driven Resonant Excitation of Magnetic Vortex   201
4.4.3 Switching a Vortex Core by Electric Current   203

Preface

Permanent magnets have been known to exist in nature since antiquity and their behavior has always been a matter of great interest. In the twentieth century, a great variety of magnetic materials have been found and magnetism has become a central feature in condensed matter physics and the subjects of various theoretical and experimental studies. At the same time, remarkable progress was achieved in developing industrial applications of magnetism, and many kinds of magnetic materials were utilized for practical purposes. A characteristic feature of magnetism is that theoretical and experimental studies are performed in tight collaboration. Another characteristic is that the gap between basic studies and the development of actual technical applications is rather small. The rapid development of magnetic recording technology can be cited as an example of the great success of the industrial application of magnetism. The modern hard-disk-drive system built in each computer, which is a typical magnetic device designed on the nanoscale, has been critical to the recent enhancement in computational capacity. Then, one might suppose that magnetism is already a too mature field to expect any more novel discoveries in the twenty-first century. However, this speculation is apparently wrong. If we look back at the progress in magnetism research, we see that many fruitful breakthroughs have appeared in a rather continuous manner. Hence, it is very probable that we will often meet something new in future studies on magnetism. A rapidly growing area in the study of magnetism is spintronics, which is the main subject of this book. State-of-the-art spintronics devices require nanoscale designs and fabrication techniques, thus making nanomagnetism an essential aspect of modern magnetism.

In the last quarter of the twentieth century, the most outstanding breakthrough in the field of magnetism was the discovery of giant magnetoresistance (GMR) effect In 1988, GMR effect was reported in Fe/Cr multilayers by Baibich et al. (Ref. 9 in Chapter 1), which was the first experiment to reveal that the electric conductance is significantly influenced by the structure of spin arrangement, parallel or antiparallel, even at room temperature. The discovery of GMR attracted great attention to the interaction between magnetism and transport phenomena and inspired many investigations into the role of spin in transport phenomena not only for the understanding of basic magnetism but also from the viewpoint of developing technical applications. By utilizing the GMR principle, magnetic recording heads were successfully fabricated rather soon after the discovery. The resistance change between spin parallel and antiparallel arrangements can also be observed by using a tunneling current through an insulating barrier. Studies on the tunneling magnetoresistance (TMR) effect have been advanced more recently and further improvement of recording heads was achieved by using the TMR. "Spintronics," which means the field of studies on spin-dependent transport phenomena, has been launched since the discovery of GMR, and nowadays rapidly developed. Owing to the great impact of the discovery of GMR effect, the 2007 Nobel prize in physics was awarded to the discoverers of GMR, Albert Fert (France) and Peter Grünberg (Germany).

"Spin" is always the leading actor in magnetism. In classical studies of magnetism, to be investigated was the behavior of spin at a fixed position. In modern magnetism, however, "spin current" is often a subject of attention. Electric current with polarized spin plays a crucial role in such phenomena as GMR and TMR. More recently, the concept of pure spin current has been established, which means the transport of spin without electric current, and novel extensive studies on the energy conversion between spin and electron, photon, phonon, etc. have started. The first edition of this book was issued in early 2009, which included the results reported by 2008. Since then, spintronics studies have accumulated significant progresses and exploitation of new branch has been reported every year. In order to cover the recent progresses in this field, the revised version of the book is prepared in electronic and printed forms simultaneously. This book consists of an overview in Chapter 1, followed by seven chapters by 15 coauthors describing the various aspects of spintronics. Each chapter begins with a short introduction and main content covers the latest developments up to 2012. I hope that this book will be useful to graduate students and those engaged in industrial research on nanomagnetism and spintronics.

Finally, I would like to express my sincere gratitude to all the coauthors for their laborious cooperation.

Teruya Shinjo
June 2013

List of Contributors

Claude Chappert Institut d'Electronique Fondamentale, Université Paris-Sud, Orsay cedex, France

Jun-ichiro Inoue* Department of Applied Physics, Nagoya University, Nagoya, Japan

Hiroshi Kohno Graduate School of Engineering Science, Osaka University, Toyonaka, Japan

Fumihiro Matsukura Center for Spintronics Integrated Systems, Tohoku University, Aoba-ku, Sendai, Japan; WPI-Advanced Institute for Materials Research, Aoba-ku, Sendai, Japan

Yoshinobu Nakatani Department of Computer Science, University of Electro-communications, Chofu, Tokyo, Japan

Hideo Ohno Center for Spintronics Integrated Systems, Tohoku University, Aoba-ku, Sendai, Japan; WPI-Advanced Institute for Materials Research, Aoba-ku, Sendai, Japan; Laboratory for Nanoelectronics and Spintronics, Research Institute of Electrical Communication, Tohoku University, Aoba-ku, Sendai, Japan

Teruo Ono Institute for Chemical Research, Kyoto University, Uji, Japan

Teruya Shinjo Institute for Chemical Research, Kyoto University, Uji, Japan

Youichi Shiota Department of Materials Engineering Science, Graduate School of Engineering Science, Toyonaka, Osaka, Japan

Yoshishige Suzuki Department of Materials Engineering Science, Graduate School of Engineering Science, Toyonaka, Osaka, Japan

Gen Tatara RIKEN Center for Emergent Matter Science (CEMS), Wako, Saitama, Japan

* Present address: Faculty of Pure and Applied Science, University of Tsukuba, Tsukuba, Japan

André Thiaville Laboratoire de Physique des Solides, Université Paris-Sud, Orsay, France

Ashwin A. Tulapurkar Department of Electrical Engineering, Indian Institute of Technology, Powai, Mumbai, India

1 Overview

Teruya Shinjo

Institute for Chemical Research, Kyoto University, Uji, Japan

Contents

1.1   Introduction

An electron has two attributes, "charge" and "spin." The main aim of condensed matter physics is to understand the behavior of electrons and for the most part, the subject is the charge of the electron. In contrast, magnetism originates from the other attribute, spin. Uncompensated electron spins are the reason why individual atoms possess local magnetic moments. If there is an exchange coupling between the magnetic moments of neighboring atoms, a magnetic order on a macroscopic scale may form at low temperatures. If the sign of the coupling is positive, the magnetic moments are aligned parallel to each other (i.e., ferromagnetism) and if negative, antiparallel to each other (i.e., antiferromagnetism). The critical temperature at which this magnetic order is lost becomes higher, if the coupling is stronger. The critical temperature of a ferromagnetic material is called the Curie temperature (T_C) and that of an antiferromagnetic material, the Něel temperature (T_N). Before the discovery of giant magnetoresistance (GMR), the investigations on the charges and spins of electrons were usually considered to be independent of each other and little attention was paid to the correlation between these two attributes, charge and spin.

Magnetoresistance (MR) is a term widely used to mean the change in the electric conductivity due to the presence of a magnetic field. A variety of MR effects are known and their characteristics depend on the material. Namely, MR effects in metallic, semiconducting, and insulating materials have different characteristics. Ferromagnetic materials with metallic conductance exhibit the anisotropic magneto-resistance (AMR) effect, that is, the dependence of conductance on the relative

Nanomagnetism and Spintronics. DOI: http://dx.doi.org/10.1016/B978-0-444-63279-1.00001-0

angle between the electric current and magnetization. Normally the resistance is smaller if the electric current flows in a direction perpendicular to the direction of magnetization than parallel. AMR is regarded to originate from spin-orbit interactions. The change of resistance (MR ratio) due to the AMR effect is fairly small, a few percent for $Ni_{80}Fe_{20}$ alloy (permalloy) at room temperature, but this phenomenon is very useful in technical applications, for instance in sensors. Before the discovery of GMR, the construction of read-out heads utilizing the AMR effect for magnetic storage devices had already been planned. The principle of magnetic recording is as follows: data are stored by nanoscale magnets in a recording medium (disk or tape) and the direction of individual magnetization in the small magnets corresponds to one bit. In order to read out the data, a sensor (i.e., read-out head) must detect very small magnetic fields straying on the surface of the recording medium. Compared with a conventional coil head, a head using the MR effect (i.e., MR head) can be much smaller and has the advantage of being able to convert magnetically stored data directly into electric signals. High-density recording can be realized by reducing the size of each memory region and by enhancing the sensitivity of the detecting head. For ultrahigh-density recording, a much larger MR ratio than that possible with the AMR effect is necessary but a search for new materials having a large MR ratio at room temperature appeared to be hopeless. Some magnetic semiconductors have been found to exhibit very large MR ratios but their Curie temperatures are lower than room temperature and they require excessively large magnetic fields, making them unsuitable for technical applications.

There have been a number of resistance measurements on ferromagnetic thin films and small resistance change was generally observed in the vicinity of the magnetization reversal field. In the process of magnetization reversal, domain walls are formed and the spin directions in the domain wall are deviated from the easy direction. Then, a change in resistance is expected owing to the AMR effect. On the other hand, a noncollinear spin structure that forms in the reversal process can serve as an electron scattering center and eventually the resistance is increased. In practice, an increase in resistance at the magnetization reversal is often observed in the case of ferromagnetic amorphous alloy films with perpendicular magnetization. From such results, it was recognized that the spin structure has an influence on conductance, but still not much attention was paid to these phenomena since the observed MR anomalies were not satisfactorily large. Velu et al. [1] studied the behavior of metallic sandwich systems with the structure, nonmagnetic/magnetic/nonmagnetic layers. The design of their sample was Au 30 nm/Co 0.3 nm/Au 30 nm. They observed an increase in resistance during magnetization reversal; 6% at 4 K and 1% at 300 K, respectively. The obtained MR ratio was not remarkably large. However, if the Co layer thickness is taken into account, which is only a few atomic layers and is much smaller than the total thickness of the Au layers, the contribution of the magnetic structure change to the total conductance is considerably large.

During 1980s, multilayers with artificial superstructures were actively investigated [2,3]. Because of the progress in thin film preparation techniques, it has become possible to deposit two or more elements alternately in order to construct artificially designed periodic structures with nanoscale wavelengths. Such artificial superstructured multilayers are new materials that do not exist in nature and can therefore be expected to possess novel physical properties. Actually multilayers were fabricated by combining various metallic elements and their superconducting, magnetic, and lattice dynamical properties have been investigated. Resistance measurements also were performed on magnetic multilayers, for example, Au/Co superlattices, but the observed MR effect was not significantly large [4]. This was because the role of interlayer coupling was not yet properly taken into consideration. In these experiments on multilayers, it was shown that noticeable enhancement in the MR effect was not induced by a superlattice effect or an interface effect.

1.2   Discovery of GMR

Grünberg and his group [5] were investigating the magnetic properties of Fe/Cr/Fe sandwich systems. They measured the magnetic behavior of the system with two Fe layers by changing the thickness of Cr spacer layers. Initially, the main aim of their experiment was to clarify the role of the Cr layer inserted in between Fe layers. If an ultrathin Cr layer has an antiferromagnetic spin arrangement analogous to that of bulk Cr, the relative spin directions of the two outermost atom layers should change from parallel to antiparallel, depending on the number of atomic layers in the Cr layer, (odd or even). As the number of atomic Cr layers is increased, the interlayer coupling between Fe layers should alternate in a layer-by-layer fashion. In other words, the sign of the interlayer coupling should oscillate between plus and minus, with every additional atomic Cr layer. However, the observed result was somewhat different from the naïve speculation. The magneto-optic Kerr effect and spin-polarized electron diffraction measurements suggested that there exists a rather strong antiferromagnetic exchange interaction between Fe layers separated by a Cr spacer layer when the Cr layer thickness is around 1 nm [6,7]. That is, the magnetizations in the two Fe layers are spontaneously oriented antiparallel to each other and can be aligned parallel if the external field is enough large. Binasch et al. [8] measured also the resistance of Fe/Cr/Fe sandwich films and found that the resistance in the antiparallel alignment is larger than that in the parallel alignment. This clearly evidences that the conductance is influenced by the magnetic structure and thus the physical principle of the GMR effect was demonstrated in such sandwich structures. However, the observed MR ratio (the difference of resistance between parallel and antiparallel magnetic alignments), about 1.5%, was not large enough to draw wide attention.

Really "giant" MR was first observed in Fe/Cr multilayers by the group of Fert in 1988 [9]. They were interested in the curious behavior of the interlayer coupling in the Fe/Cr/Fe structure found by Grünberg et al. [5] and intended to visualize the role of interlayer coupling in a multilayered structure. They have prepared epitaxial Fe(001)/Cr(001) multilayers with the typical structure [Fe(3 nm)/Cr(0.9 nm)] × 60 and systematically measured the magnetic properties including MR. The magnetization curves indicated that the remanent magnetization is zero and ferromagnetic saturation occurs at magnetic fields higher than 2 T. These features correspond to the existence of rather strong antiferromagnetic interlayer coupling. Surprising results were obtained in the measurements of resistance under external fields. The resistance decreased with an increase in the applied field and became almost to a half at the saturation field at 4 K (see Figure 2.10). The MR ratio was nearly 20% even at room temperature; a strikingly large value at that time for a metallic substance. This fantastic discovery was first reported very briefly at the International Conference on Magnetism (Paris, 1988) as an additional part of a paper. The new MR data were not originally intended results and therefore not yet mentioned in the rě sumě of the conference. A great discovery is often obtained as such an unexpected observation.

The results of this GMR measurement confirmed the existence of a strong antiferromagnetic interlayer coupling between Fe layers separated by a Cr spacer layer. The mechanism of the GMR was phenomenologically explained rather soon after the discovery by considering the spin-dependent scattering of conduction electrons. The scattering probability for conduction electrons at the interface of the ferromagnetic layer should depend on the spin direction, up or down. For instance, an up-spin electron is considered to penetrate without scattering from a Cr layer into an Fe layer with magnetization in the up-spin direction, while a down-spin electron is scattered. If the Fe layers have antiparallel magnetic structure, both up- and down-spin electrons soon meet an Fe layer having a magnetization in the opposite direction (within two Fe layers' distance) and accordingly the possibility of scattering is rather high for both types of electrons. In contrast, if all the Fe layers are aligned parallel, down-spin electrons are scattered at every Fe layer whereas up-spin electrons can move across long distance, without scattering. In other words, up-spin electrons will have a long mean free path but down-spin electrons have a very short mean free path. Total conductance of the system is the sum of that by up-spin electrons and by down-spin electrons. Because of the long mean free path of up-spin electrons, the total resistance is much smaller in the state with parallel magnetization than in the antiparallel state. A comprehensive explanation of the GMR effect is presented by Inoue in Chapter 2.

The GMR experiment brought two key issues to the fore; interlayer coupling and spin-dependent scattering. Although interlayer coupling was reported in the Fe/Cr/Fe sandwich system and later in Co/Cu multilayers by Cebollada et al. [10], before the discovery of GMR, it was hard to image a multilayered structure with antiparallel magnetizations, that is, "giant antiferromagnet." By applying an external field, the giant antiferromagnet can be converted into ferromagnetic. The GMR effect is the difference in conductance between these two states. In general, very large magnetic fields are necessary to change an intrinsic antiferromagnetic spin structure into ferromagnetic. In contrast, in the case of multilayers, the antiparallel structure (giant antiferromagnet) generated by interlayer coupling can be turned into a parallel structure (ferromagnetically saturated structure) by a moderate magnetic field. This is the key behind the discovery of GMR, which seems to be the first successful experiment to utilize spin structure manipulation. The antiparallel alignment of Fe layers' magnetizations at zero field and the reorientation into parallel alignment by an increase in the external field were confirmed by neutron diffraction technique for Fe/Cr multilayers [11]. A magnetic diffraction peak corresponding to the twice of the adjacent Fe layer distance was observed, which indicates that the direction of magnetization alternates at every adjacent Fe layer. This is clear evidence for the formation of a giant antiferromagnetic arrangement in an Fe/Cr multilayer. The origin of GMR is thus attributed to the change in the internal magnetic structure. This is apparently different from that of AMR, which is induced by a directional change of magnetization, while the ferromagnetic spin structure is invariable.

The behavior of Cr spacer layers sandwiched between ferromagnetic Fe layers has been extensively studied by Grünberg et al. [5] and also many other groups, using sandwich films and multilayers. The dependence of the interlayer coupling on the Cr layer thickness has been examined in detail. For a systematic experiment on thickness dependence, a sample with a wedge-shaped spacer layer is very useful [12]. A wedge layer is prepared by slowly sliding the shutter during the film deposition to effect a variation in thickness from zero to some 10 nm over a macroscopic length. Then, by applying Kerr rotation technique, the magnetic hysteresis curves at confined regions are measured. This method became very fashionable and was utilized not only for Fe/Cr/Fe structure but also for many metallic elements. Bulk Cr metal is known to have peculiar antiferromagnetic properties and the spin structure of ultrathin Cr layers is very complicated, being not satisfactorily understood even today. Although many studies have been performed on the interlayer coupling, the relation between the interlayer coupling and the intrinsic antiferromagnetism of Cr metal is not fully accounted for and the effect of this antiferromagnetism is usually neglected in discussions on the GMR properties of Fe/Cr systems. In the resistance measurements on Fe/Cr systems, no anomaly is noticed at the transition temperature (T_N) of Cr layer.

The discovery of GMR effect in Fe/Cr multilayers inspired various experiments on interlayer coupling in many other metals aiming to explore the nature of the MR effect in general. The existence of interlayer coupling was confirmed in many nonmagnetic metals making it clear that the interlayer coupling does not originate from the intrinsic magnetic properties the spacer layer. If the interlayer coupling is antiferromagnetic, the GMR effect is almost always observed, that is, the resistance in antiferromagnetic state is larger than that in ferromagnetic state. In the study of Co/Cu multilayers, a striking result was obtained: the interlayer coupling across the Cu layer oscillates with variations in its thickness [13,14]. Since the MR effect is caused by antiferromagnetic interlayer coupling, the MR measurement can be utilized as a tool to clarify that the sign of the interlayer coupling is negative. In the plot of the MR ratio as a function of Cu layer thickness, peaks of MR ratio were found to appear periodically with an interval of about 1 nm. Parkin et al. [15,16] prepared multilayers combining Co and various nonmagnetic metals, and found that the oscillation of interlayer coupling occurs rather generally with a wavelength of $1 \sim 1.5$ nm. The oscillation of the interlayer coupling was an amazing result and was the subject of many subsequent investigations. In the case of simple normal metals, the oscillatory feature was accounted for by considering the band structure and a relation with the quantum well state has been argued. Thus, through the studies on the oscillatory interlayer coupling behavior, our understanding of the electronic structure of thin metal film has been significantly advanced. About 10 years after the discovery, GMR witnessed a boom in studies on interlayer coupling but scientific progress in more recent years has not been remarkable. This book does not include a chapter on interlayer coupling. See Refs. [17] and [18] for review articles on interlayer coupling studies.

1.3 Development of GMR Studies

The GMR effect is caused by the change in the magnetic structure, between antiparallel and parallel alignments. In the cases of Fe/Cr and Co/C multilayers, the antiparallel configuration that originates from the antiferromagnetic interlayer exchange coupling is converted into ferromagnetic configuration by an externally applied field. The magnitude of the external field necessary for this conversion is determined by the strength of the interlayer coupling. Because of the strong interlayer coupling, the magnetic field required to induce the GMR effect in Fe/Cr multilayers is significantly large (about 2 T). In the case of Co/Cu system, the coupling is somewhat weaker and the necessary field smaller. Nevertheless, the saturation field value is too high for the MR effect to be exploited in technical applications, such as magnetic recording sensors.

Another type of GMR was demonstrated in 1990, by using noncoupled multilayer samples [19]. Multilayers comprising two magnetic elements were prepared by successively stacking NiFe(3 nm), Cu(5 nm),Co(3 nm), and Cu(5 nm) layers. Since the Cu spacer layer is not very thin, the interlayer coupling between the NiFe and Co layers is negligibly small and their magnetizations are independent. NiFe is a typical soft magnetic material but Co is magnetically rather hard. Owing to the small coercive force of the NiFe layer compared with that of the Co layer, the magnetization of the NiFe layer changes direction much earlier than that of the Co layer. Thus, an antiparallel alignment of magnetizations is realized when the external field is increasing (and also when it is decreasing). This is not due to interlayer coupling but because of the difference in coercive forces. A remarkable enhancement in resistance (i.e., GMR) was observed in the field region for this induced antiferromagnetic configuration. The experimental results are presented in the next chapter (Figure 2.12). The demonstration of noncoupled GMR confirms that the interlayer coupling has no direct influence on the MR phenomena. In other words, GMR and interlayer coupling are independent issues. For these noncoupled multilayers as well, the establishment of an antiparallel magnetic structure was confirmed by using the neutron diffraction method [20]. Noncoupled GMR multilayers can serve as a model system for fundamental research, with several advantages, for instance, the fact that the spin structure is easily manipulated [21]. A survey of the basic studies on noncoupled GMR multilayers is presented elsewhere [22]. A feature of noncoupled GMR, that is very important from a technical point of view, is the high sensitivity to external field. The resistance change occurs at weak fields if the soft magnetic component has a sufficiently small coercive force. Since NiFe behaves as a soft magnetic material, the MR effect in a multilayer including NiFe component can show a high sensitivity under fields on the order of 10 Oe.

The potential for the use of the GMR effect in technical applications was revealed in the result of studies on noncoupled multilayers. A practical application of GMR effect for magnetic recording heads was achieved by using noncoupled type sandwich films with only two magnetic components. At nearly the same time as the studies on noncoupled type GMR multilayers, Dieny et al. [23] published a paper on a noncoupled GMR sandwich system, that was named the "spin valve" [23]. The initial design of the spin-valve structure was NiFe(15 nm)/Cu(2.6 nm)/ NiFe(15 nm)/FeMn(10 nm). There are two ferromagnetic NiFe layers and an anti-ferromagnetic FeMn layer is attached to one of the NiFe layers to increase the required coercive force via the exchange anisotropy. The other NiFe layer behaves freely as a soft magnet. Therefore, the two NiFe layers are called the "pinned" and "free" layers, respectively. Because of the ease in controlling the magnetic proper-ties, the spin-valve system was adopted for commercial magnetic recording heads. Although the initial spin-valve structure was very simple, various kinds of improve-ments were attempted promptly soon after. To enhance the coercive force of the pinned layer, a simple antiferromagnetic layer (FeMn) used originally was replaced by a complicated structure combined with an antiferromagnet (MnPt) and a syn-thetic antiferromagnetic layer. An example of a synthetic antiferromagnet is FeCo/ Ru/FeCo, which acts as a powerful magnetic anchor due to the strong interlayer coupling across the Ru layer. Because the large surface magnetic moments are essentially important for spin-dependent scattering, surfaces of both free and pinned layers were covered by ultrathin FeCo layers with a few atom layers thick, which are supposed to have a large magnetic moment. Concerning the material for the spacer layer, Cu seems to be the best choice and has always been used. At the beginning, sandwich systems did not show such large MR values as multilayer sys-tems. However, remarkable improvements were achieved within a short time and fairly large MR ratios were realized in refined spin-valve systems. Perhaps the improvement in quality from a crystallographic point of view was one of the keys to this success. There are many ideas for further progress; the introduction of reflective layers (ultrathin oxide layers) on each surface, which will reflect the con-duction electrons without energy loss, and the insertion of a nano-oxide layer with many microscopic holes in the spacer layer, which may be useful to collimate the electron path. A number of industrial research groups joined in the competition for the GMR head business and consequently various trials were performed [24].

Eventually the MR ratio of the spin-valve system has been increased satisfacto-rily for commercial purposes. Within 10 years from the discovery, the GMR princi-ple has been successfully exploited in commercial magnetic recording technology. The commercial products called spin valve or GMR head have greatly contributed to the progress of magnetic recording technology as shown in Figure 1.1. The progress of recording technology is typically expressed by the increase in recording density. The GMR head was integral to the recent increase from 10 to

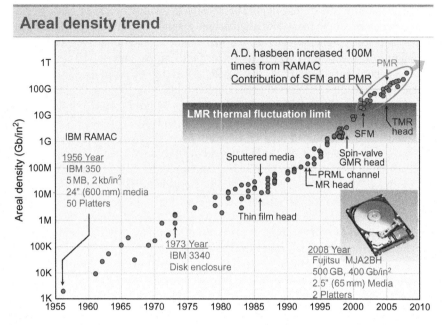

Figure 1.1 Progress of magnetic recording technology: density of recording (bit per square inch) versus year (by courtesy of Fujitsu ltd.). LMR, SFM, and PMR mean longitudinal magnetic recording, synthetic antiferromagnetism, and perpendicular magnetic recording, respectively. Thermal fluctuation limit indicates the highest attainable boundary for recording density, due to superparamagnetism, supposed before the appearance of GMR, TMR, SFM, and PMR.

100 Mbit/in^2. The industrial application of the new GMR phenomenon was realized in such a short interval because the application of AMR effect in a similar manner was just in progress. It is interesting to note that although interlayer coupling and multilayer structure were key conditions for the discovery of the GMR effect, commercial spin-valve heads have neither a periodic multilayered structure nor antiferromagnetic interlayer coupling through a spacer layer. As a matter of fact, a strong antiferromagnetic interlayer coupling through Ru layer is utilized in the structure of the pinned layer but the magnetic coupling between pinned and free magnetic layers through a Cu spacer layer is negligibly small. On the other hand, initially the spin-valve structure started with only a few layers but the sophisticated structure of improved spin valve was actually a multilayer consisting of more than 10 layers.

The magnetic recording technology consists of two principal nanomagnetic systems: One is the magnetic heads (recording and reading), as mentioned already, and the other is recording media. The element in recording media is nanoscale magnetic particles. For instance, a surface layer of a disk comprises nanoscale Co clusters dispersed in a nonmagnetic metallic matrix and the magnetizations in nanoscale areas express the memorized signal of one bit. The magnetization directions in recording media was traditionally in plane but nowadays the perpendicular magnetic recording method has been adopted and the capacity of magnetic recording has been greatly enhanced. The perpendicular recording method was initially proposed by Iwasaki [25] and recently commercial products using perpendicular recording have also been performed. An ultimate material for magnetic recording, which will be realized in near future, must be "patterned" media. Using nanofabrication techniques, nanoscale magnetic dots are prepared on a disk and each single nanodot corresponds to one bit of memory. As well as heads, magnetic recording media are very interesting subjects for nanomagnetism studies but because of the limit of space, no more description is presented in this book.

1.4   Recent Progress in MR Experiments

Further enhancement of the MR effect is an attractive challenge for scientists in fundamental physics and also in industries. The GMR effect has been observed first in multilayers and then in sandwich samples in many combinations of magnetic and nonmagnetic metallic elements but concerning the magnitude of MR ratio, eventually the combinations of Fe/Cr and Co/Cu seem to be the optimum selections among metallic elements. There can be several strategies to search larger GMR effects as the following: (i) taking the CPP geometry, (ii) using the tunneling current, (iii) using half metal as the magnetic constituent, and (iv) using the ballistic current. Usually resistance measurements for thin metallic specimens are carried out in a conventional geometry to use an electric current flowing in the film plane. Such configuration is called the current-in-plane (CIP) geometry. In contrast, resistance measurements in the other geometry, the current perpendicular to plane (CPP), are very inconvenient for thin metallic films. An enhancement of MR ratio is however expected in the CPP geometry compared with the CIP geometry because the GMR effect is a phenomenon for the electrons passing through interfaces. Before the discovery of GMR, it was not expected that any remarkable MR effect may happen in the CIP geometry. Fortunately, this speculation was not correct and significantly large MR effect has been observed in the CIP geometry, even at room temperature. However, if measurements in the CPP geometry are possible, further enhancement of MR ratio is obtainable. The first measurement on extremely small resistance of GMR systems in the CPP geometry has been attempted by Pratt Jr. et al. [26] using superconducting electrodes, and an apparent increase of MR ratio at low temperatures was observed. In order to avoid the inconvenience in the measurements on a too small resistance in the CPP geometry, the application of nanofabrication technique to the thin film samples is worthwhile for metallic GMR systems. Gijs et al. [27] have prepared micro-columnar samples of GMR system for the first time and confirmed the enhancement of MR ratio in the CPP geometry at room temperature. Experiments in the CPP geometry are important not only for the purpose to enhance the MR ratio but also to investigate the mechanism of spin-dependent scattering. In the case of the CPP geometry, the electric current is regarded to be constant in the sample, while the current in the CIP geometry is not homogeneous and the estimation of current density distribution is a hard job. It is therefore difficult to argue quantitatively the spin-dependent scattering probability from CIP experimental results. CPP-GMR has a definite potential for the enhancement of the MR ratio and actually rather large MR ratios at room temperature, about 75%, have recently been achieved using Heusler alloy electrodes [28]. The merit of Heusler alloy will be described soon later. CPP-GMR may have a technical potential to use as a sensor with low resistance. The experiments for the study of CPP-MR were the first evidence that the nanoscale fabrication techniques are very crucial to explore magnetic devices. Nowadays, nanofabrication techniques are being used rather routinely to prepare samples for spintronic devices and the properties of nanomagnetic systems are regarded as the subjects of great interest both from fundamental and technical viewpoints.

Remarkable advance has been achieved in MR experiments using tunneling current (tunneling magnetoresistance, TMR) Basically, the sample structure for TMR measurements is very simple; two magnetic electrodes are separated by an insulating barrier and the difference of tunneling conductance in the states of parallel and antiparallel magnetizations is measured. Since the TMR is essentially a phenomenon for the electrons passing through the barrier, the geometry of measurement is equal to CPP-GMR. Trials to use a tunneling current were already initiated by Julliere [29] and by Maekawa and Gäfvert [30], and were followed by several groups. But observation of perceivable MR effect was very difficult and the reproducibility was rather poor, because at that time it was difficult to prepare ultrathin tunneling barriers without pinhole. The preparation techniques for thin oxide films have progressed in the 1990s, in relation with the flourishing of high T_C superconducting oxide research. Inspired by the success of GMR measurements, attempts for TMR have revived and outstanding breakthrough was obtained in 1996 [31, 32]. Miyazaki and Tezuka prepared three-layer junctions, $Fe/Al_2O_3/Fe$, and observed MR ratio of 30% at 4 K and 18% at 300 K. Afterward many groups joined in active research on TMR. In recent experiments on TMR, the sample sizes are very small, being prepared by nanoscale fabrication, and such samples with very limited area have an advantage that the possibility of pinhole is relatively less. Thus it has become rather easy to obtain large MR ratio at room temperature reproducibly. More recently a remarkable progress was achieved by using MgO layer with nanoscale thickness as the tunneling barrier instead of Al_2O_3[33, 34]. Yuasa et al. [33] prepared FeCo/MgO/FeCo junctions using epitaxially grown MgO layers with a crystallographically high quality as tunneling barriers, and succeeded in observing a much larger MR ratio at room temperature. The possible mechanism of the enhancement of MR effect is supposed to be the coherent tunneling current through MgO barrier, which will be discussed in the next chapter. The application of TMR effect with very large MR ratios into commercial read-out heads has been carried out and TMR heads have become the successor of GMR heads. In the case of TMR also, the initial sample structure was a simple three-layer structure but the actual structure of recent TMR heads is a sophisticated multilayer including more than 10 different layers, similar to that of spin-valve GMR heads. The largest MR ratio so far reported in the TMR junction of CoFeB/MgO/CoFeB is 604% at 300 K (1144% at 5 K) [35].

The theoretical background of TMR phenomena is given in Chapter 2. The geometry of TMR is analogous to CPP-GMR and the conductance is determined by the spin polarization of two electrodes. It is an attractive approach to utilize a half metal as the electrodes in a TMR junction, because a ferromagnetic metal electrode with a larger polarization can make a larger MR ratio. The definition of half metal is that only one kind of spin exists at the Fermi level owing to a big spin splitting of the energy band, and only up spins participate in the electric conduction. From band calculation, certain metallic compounds such as Heusler alloys are regarded as examples of half metal. If two ideal half metals are used as the electrodes in a TMR junction and if the magnetizations are antiparallel with each other, 100% spin-polarized electrons are emitted from one electrode, while the other electrode cannot accept any electron without spin flipping. Then the resistance of an antiparallel arrangement becomes extremely high and eventually an infinitively large MR ratio may be realized. On the other hand, although trials to adopt Heusler alloy layers in TMR junctions have been practiced, a large MR ratio as theoretically expected was not obtained easily. Very recently, an enormously large MR ratio at low temperature has been reported using a TMR junction including Heusler layers, with the structure of $Co_2MnSi/MgO/Co_2MnSi$; 1995% at 4 K [36]. Such a great MR ratio certifies that the Heusler alloy layer is indeed working as a half metal and therefore regarded as an appropriate component in a TMR junction. However, although the MR ratio at 4 K is extremely large, that at 300 K, 354%, is not a satisfactorily large value, if we take into account that the Curie temperature of bulk Co_2MnSi is considerably high (985 K). The experimentally observed temperature dependence of the MR effect is much steeper than that of magnetization. It is therefore suggested that the MR ratio depends not on the bulk magnetic behavior but on that of interface atom layers. If the temperature dependence of interface magnetization is steeper than the bulk, the effective polarization of the tunneling current may be seriously decreased with the increase of temperature. Therefore, in order to obtain large MR ratios at room temperature, it is necessary to improve the quality of interface. Although the knowledge on magnetic properties of interfaces are crucial to understand the TMR behaviors, experimental means for the study of interface magnetism are very limited and our understanding is not enough. For the further exploration of spintronics, it is required to establish the technique to create a current with a full spin polarization (i.e., an ideal spin current source). Although "interface magnetism" has been a subject of great interest for a long time [37], our understanding is still very primitive with regard to the relation between interface magnetic behavior and spin current across the interface. Intensive studies on interface magnetism, including nanoscale crystallographic observations, are indispensable. Before the appearance of spintronics, interface magnetism was regarded as a subject in purely basic physics but now it is a crucial issue from the viewpoint of applications.

Already GMR and TMR phenomena have been successfully utilized in commercial products as read-out heads in magnetic storage systems and also as the elements of magnetic random access memory (MRAM) systems. If the MR ratio at room temperature is able to be enhanced, up to more than 1000%, novel technical applications of TMR as sensitive detectors for weak magnetic field, for instance working in biological systems, will become possible and more extensive industrial applications may be developed, since the handling of TMR sensor is much easier than the conventional superconducting quantum interference device (SQUID). An example of future target for industrial application is a sensing system to analyze the magnetic field variations in human brains [38].

1.5 The Scope of This Book

The first edition of this book was published in early 2009. This second edition is organized with the same structure of the last version, having seven chapters following this overview. The fundamental knowledge on up-to-date topics relating to nanomagnetism and spintronics is presented here. The authors for the seven chapters are Japanese, French, and Indian who are actively involved in the current investigations. Chapter 2 described by Inoue is an introduction to spin-dependent transport in ferromagnetic metallic systems and the theoretical backgrounds for GMR, TMR, and other MR effects are explained. Recent developments in the studies on spin Hall effect and spin caloritronics also are briefly mentioned. This chapter will be useful as a text for students who begin to study physics on magnetotransport phenomena in ferromagnetic metallic materials. The main subject of Chapter 3 by Suzuki et al. is the spin injection of which studies are recently progressing remarkably. Novel phenomena induced by the spin torque transferred by electric current, such as current-induced magnetization switching and spin-torque diode effect, in GMR and TMR junctions are described. As a recent topic in the field of spintronics, the dependence of magnetism on an electric field is argued. It is introduced that spin switching can be induced by applying an electric field, due to the electric field effect on interface anisotropy. Basic physical concepts and feasibility for application are argued. In Chapter 4, Ono and Shinjo explain experimental results on magnetic domain wall motion in ferromagnetic nanowires. Dynamical properties of magnetic vortex core in ferromagnetic nanodot systems are also introduced. Theoretical aspects of domain wall motion induced by electric current are discussed by Kohno and Tatara in Chapter 5. Studies on dynamical behavior of magnetic domain wall with micromagnetic simulation are presented in Chapter 6 by Thiaville and Nakatani. Finally, in Chapter 8, Matsukura and Ohno survey recent developments on ferromagnetic III-V compound semiconductors, typically Mn-substituted GaAs. Their electric and magnetic properties are described and novel phenomena relating to spintronics, such as current-induced domain wall motion and electric field control of ferromagnetic phase are introduced.

Although the title of this book is nanomagnetism, there is no specific section for the traditional issues on nanoscale magnetic clusters (or magnetic ultrafine particles). From a long time ago, magnetic properties of clusters (with limited number of magnetic atoms) have been of great interests from theoretical and experimental points of view, but the progress in recent years is not remarkable. In industrial applications, on the other hand, such as magnetic recording technology, the size of magnetic elements becomes smaller and smaller, down to the scale of a few nanometers. Therefore, it is very crucial to understand the influence of interface atom layer and size reduction on local magnetic moment, anisotropy, and dynamical characteristics. Comprehensive studies using large-scale computers will give us useful guidance for further development of spintronic studies. As an example of computational simulation for nanoscale ferromagnetic clusters, a paper was reported by Entel et al. [39]. Industrial applications of spintronics also are not included in this book, although technologies for hard-disk drive systems in magnetic recording technology and MRAM systems have been established and novel spintronics devices are regarded to be promising. Here is no chapter dealing with spintronic properties of compounds, such as perovskite oxides [40], carbon nanotubes and graphenes, and organic molecules, although they may become key players for future spintronic devices.

This book is not able to cover whole relevant areas of nanomagnetism and spintronics. However, the author hopes that this book will be useful for the readers to recognize the significance of this field. It is certain that the field, nanomagnetism and spintronics, will continue to grow.

In this chapter, the author introduced a part of his investigation carried out at Institute for Chemical Research, Kyoto University, where he had served for 36 years. He would like to express his gratitude for the collaborators.

References

[1] Velu E, Dupas C, Renard D, Renard JP, Seiden J. Phys Rev 1988;B37:668.
[2] Shinjo T, Hosoito N, Kawaguchi K, Takada T, Endoh Y, Friedt JM. J Phys Soc Jpn 1983;52:3154.
[3] An example of book is Shinjo T, Takada T, editors. Metallic superlattices. Amsterdam: Elsevier; 1987
[4] Takahata T, Araki S, Shinjo T. J Magn Magn Mater 1989;82:287.
[5] Grünberg P, Schreiber R, Pang Y, Brodsky MB, Sowers H. Phys Rev Lett 1986;57:2442.
[6] Saurenbach F, Walz U, Hinchey L, Grünberg P, Zinn W. J Appl Phys 1986;63:3473.
[7] Carbone C, Alvarado SF. Phys Rev 1987;B39:2433.
[8] Binasch G, Grünberg P, Saurenbach F, Zinn W. Phys Rev 1989;B39:4828.
[9] Baibich MN, Broto JM, Fert A, Nguyen Van Dau F, Etienne P, Creuzet G, et al. Phys Rev Lett 1988;61:2472.
[10] Cebollada A, Martinez JL, Gallego JM, de Miguel JJ, Miranda R, Ferrer S, et al. Phys Rev 1989;B39:9726.
[11] Hosoito N, Araki S, Mibu K, Shinjo T. J Phys Soc Jpn 1990;59:1925.
[12] Ungaris J, Celotta RJ, Pierce DT. Phys Rev Lett 1991;67:140.
[13] Mosca DH, Petroff F, Fert A, Schroeder P, Pratt Jr. WP, Loloee R. J Magn Magn Mater 1991;94:1.
[14] Parkin SSP, Bhadra R, Roche KP. Phys Rev Lett 1991;66:2152.

[15] Parkin SSP, More N, Roche KP. Phys Rev Lett 1990;64:2304.
[16] Parkin SSP. Phys Rev Lett 1991;67:3598.
[17] Hartmann U, editor. Magnetic multilayers and giant magnetoresistance. Berlin: Springer; 1999.
[18] Mills DL, Bland JAC, editors. Nanomagnetism, ultrathin films, multilayers and nanostructures. New York, NY: Elsevier; 2006.
[19] Shinjo T, Yamamoto H. J Phys Soc Jpn 1990;59:3061.
[20] Hosoito N, Ono T, Yamamoto H, Shinjo T, Endoh Y. J Phys Soc Jpn 1995;64:581.
[21] Okuyama T, Yamamoto H, Shinjo T. J Magn Magn Mater 1992;113:79.
[22] Maekawa S, Shinjo T, editors. Spin transport in magnetic nanostructures. London: Taylor & Francis; 2002.
[23] Dieny B, Speriosu VS, Parkin SSP, Gurney BA, Wilhoit DR, Mauri D. Phys Rev 1991;B43:1297.
[24] A survey on applicational aspects of GMR phenomena, for instance, is Hirota E, Sakakima H, Inomata K, editors. Giant magneto-resistance devices. Berlin: Springer; 2002
[25] Iwasaki S. IEEE Trans Magn 1980;16:71.
[26] Pratt Jr. WP, Lee S, Slaughter FJMR, Loloee R, Schroeder PA, Bass J. Phys Rev Lett 1991;66:3060.
[27] Gijs MA, Lenczowski MSKJ, Giesbers JB. Phys Rev Lett 1993;70:3343.
[28] Sato J, Oogane M, Naganuma H, Ando Y. Appl Phys Exp 2011;4:113005.
[29] Julliere M. Phys Lett 1975;54A:225.
[30] Maekawa S, Gäfvert U. IEEE Trans Magn 1982;18:707.
[31] Tezuka N, Miyazaki T. J Appl Phys 1996;79:6262.
[32] Moodera JS, Kinder LB. J Appl Phys 1996;79:4724.
[33] Yuasa S, Nagahama T, Fukushima A, Suzuki Y, Ando K. Nat Mater 2004;3:868.
[34] Parkin SSP, Kaiser C, Panchula A, Rice PM, Hughes B, Samant B, et al. Nat Mater 2004;3:862.
[35] Ikeda S, Hayakawa J, Ashizawa Y, Lee YM, Miura K, Hasegawa H, et al. Appl Phys Lett 2008;93:082508.
[36] Liu H-X, Honda Y, Taira T, Matsuda K, Arita M, Uemura J, et al. Appl Phys Lett 2012;101:132418.
[37] Shinjo T. Surf Sci Rep 1991;12:49.
[38] Fujiwara K, Oogane M, Yokota S, Nishikawa T, Naganuma H, Ando Y. J Appl Phys 2012;111:07C71.
[39] Entel P, Grunner ME, Rollmann G, Hucht A, Sahoo S, Zayak AT, et al. Phil Mag 2008;88:2725.
[40] A review article on perovskite manganites, for instance, is Tokura Y. Rep Prog Phys 2006;69:797.

2 GMR, TMR, BMR, and Related Phenomena

Jun-ichiro Inoue[†]

Department of Applied Physics, Nagoya University, Nagoya, Japan

Contents

[†] Present address: Faculty of Pure and Applied Science, University of Tsukuba, Tsukuba, Japan

Nanomagnetism and Spintronics. DOI: http://dx.doi.org/10.1016/B978-0-444-63279-1.00002-2

2.1 Introduction

The magnetism of materials [1] is carried by electron spin, while electrical transport is caused by the motion of electron charge. While these two fundamental properties of solids have been well known for many centuries, the electron and spin were not discovered until the beginning of the twentieth century [2,3]. The fields of magnetism and electrical transport have developed almost independently. However, as the fabrication techniques of micro- and nanoscale samples have progressed rapidly, the field of spin electronics or spintronics has been developed, where the coupling of electron spin and charge plays an important role. In paramagnets, the number of up- and down-spin electrons is the same and no effect of spin appears in the electrical transport. However, the difference in the number of up- and down-spin electrons in ferromagnets causes complex properties in which magnetism effects electrical transport and vice versa. For example, the control of spins by an electric field and the control of electrical current by a magnetic field are fundamental issues in the field of spintronics.

The fundamental properties of spintronics are closely related to the length scale L characteristic of samples and to the motion of electrons in metals. There are several length scales that characterize the properties of electrons in metals.

The z-component of spin s_z takes one of two values $\pm 1/2$ and is not necessarily conserved, that is, it is time-dependent due to such effects as the spin−orbit interaction (SOI) and interactions between electrons. Therefore, the length for which the spin of an electron is conserved is finite. This length is called the spin-flip mean free path and typically takes values in the range 10^2 nm $- 10^1 \mu$m. Due to scattering of electrons, the length an electron travels with a fixed spin direction is much shorter than the spin-flip mean free path. This length is called the spin-diffusion length λ_{spin}. To find the spin-polarized current in nonmagnetic metals, it is necessary that the system length L be much shorter than λ_{spin}.

In ferromagnetic metals, due to the imbalance between the number of electrons with up and down spins, the current may be spin polarized. Because the electrical resistivity is governed by the mean free path ℓ, which characterizes the scattering process of electrons, it is necessary that $\ell \ll \lambda_{\text{spin}}$ in order that the spin polarization of the current be meaningful. When this condition is satisfied, the spin polarization of the current is well defined and the up- and down-spin electrons may be treated independently. This is called Mott's two-current model [4]. When the condition is satisfied, the two-current model holds even in systems for which $L \gg \lambda_{spin}$.

Another important length scale is the Fermi wavelength λ_F, which characterizes the electronic states. In general, $\ell \gg \lambda_F$. This length scale becomes important when interference occurs between wave functions of electrons. The velocity of electrons on the Fermi surface is given by the Fermi velocity v_F and hence the time scale for an electron with v_F traveling a distance ℓ is given by $\tau = \ell / v_F$, the relaxation time.

As mentioned above, progress in nanofabrication techniques has made it possible to create artificial structures such as magnetic multilayers and nanocontacts, the characteristic scale length L of which can be shorter than λ_{spin} or ℓ and can even be close to λ_F. In these cases, novel transport phenomena occur; giant magnetoresistance (GMR), tunnel magnetoresistance (TMR), and ballistic magnetoresistance (BMR) are typical examples. GMR occurs when the layer thickness of magnetic multilayers is close to or shorter than ℓ. BMR occurs when the scale of the contact region of two ferromagnets is close to λ_F. TMR is a phenomenon in which the overlap of wave functions of electrons in two separated ferromagnetic metals becomes small.

In this chapter, we first review the spin dependence of electrical resistivity in metals and alloys and explain the phenomena of GMR, TMR, and BMR. Theoretical methods to calculate the conductivity or conductance will be presented in Section 2.3, though the reader may skip this section and move directly to the section on magnetoresistive properties.

GMR, TMR, and BMR appear in multilayers, tunnel junctions, and magnetic nanocontact, respectively. The magnetoresistive phenomena also appear in bulk systems. Typical examples are normal MR in normal metals and semiconductors, anisotropic MR (AMR) in transition metals (TMs) and alloys, and colossal MR (CMR) in manganites. To clarify the essential difference between these MRs, we will give a brief explanation of normal MR, AMR, and CMR in Section 2.7.

SOI, which is responsible to AMR, gives rise to other interesting transport properties such as anomalous Hall effect (AHE) and spin Hall effect (SHE) which recently attract much interests in both technological aspect and fundamental physics. Since SOI is a coupling of spin and orbital motion of electrons, current control of spin and magnetic control of charge via SOI are possible. This is the reason that SOI attracts much interest in the technological aspect. Therefore, we introduce AHE and SHE in Section 2.8, in addition to a spin accumulation caused by SOI in the nonequilibrium state. Finally, we give brief introduction of developing field of spin caloritronics, in addition to some aspect of spin current and spin transfer torque (STT).

Other aspects on the spin-dependent transport may be found in several textbooks and review articles [5−16]. Electronic and magnetic properties of solids may be found, for example, in Harrison's [17] and Chikazumi's textbook [18].

2.2 Spin-Dependent Transport in Ferromagnetic Metals

One of the most important requirements for MR in nanoscale ferromagnets is spin dependence of the electrical resistivity. In this section, we review spin-dependent resistivity (or conductivity) in ferromagnetic bulk metals and alloys, emphasizing the role of the electronic states on the resistivity at low temperatures.

2.2.1   Electronic States and Magnetism in TMs and Alloys

Few ferromagnetic materials are composed of a single element. The exceptions are the TMs, such as Fe, Co, and Ni, and rare earth metals. This is in marked contrast to superconductivity, which appears in many pure metals. In rare earth metals, electrons responsible for transport and magnetism can be distinguished. However, this distinction is not clear in TMs, that is, both s- and d-electrons contribute to transport and magnetism. A high Curie temperature is another characteristic of TM ferromagnets.

The electronic structure of TMs consists of mainly s- and d-orbitals. The relative position of the Fermi level E_F to the s- and d-states depends on the material, i.e., the number of s + d electrons per atom. Figure 2.1 shows the schematic density of states (DOS) of the typical TMs Cr, Fe, and Co, and the DOS of Cu. The electronic states are composed of wide s-bands and narrow d-bands. The d-part of the DOS is high because the d-states are almost localized near atoms. The s- and d-states hybridize to form complicated electronic states.

The electronic structures shown in Figure 2.1 for TMs give rise to the characteristic features of both magnetism and electrical transport. A typical example of the former is the Slater−Pauling curve of the magnetization of TM alloys, as shown in Figure 2.2[19−22]. The linear part of the slope with 45° may be easily understood by changing the filling of the DOS with electrons. The branches deviating from the main curves can be explained only by introducing changes in the DOS due to random impurity potentials.

As mentioned above, the two-current model for electrical transport holds well in TMs and their alloys. Hence, the electrical resistivity depends on spin in ferromagnetic metals and alloys. The spin dependence of the resistivity is governed by the spin dependence of the electronic states near the Fermi level, and by spin-dependent impurity potentials in ferromagnetic alloys. We will review the spin-dependent resistivity in detail in the next section.

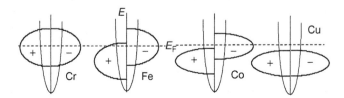

Figure 2.1 Schematic DOS of Cr, Fe, Co, and Cu. + and − indicate majority and minority spin states, respectively, identical to up (↑) and down (↓) spin, respectively, in uniformly magnetized materials.

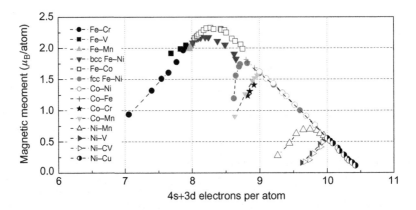

Figure 2.2 Slater–Pauling curve [22].

2.2.2 α-*Parameter*

The simplest formula for the electrical conductivity σ is given by the Drude formula:

$$\sigma = \frac{e^2 n \tau}{m} \tag{2.1}$$

where e, n, τ, and m are the electrical charge, carrier density, lifetime and effective mass of carrier electrons, respectively. For ferromagnets, the spin dependence of these quantities must be taken into account in the Drude formula, since the electronic states of ferromagnets are spin polarized due to the number of up ($\uparrow$) and down ($\downarrow$) spin electrons not being compensated. Basically, n, m, and τ are all spin dependent. Most important is the spin dependence of the lifetime, since it affects electron scattering very strongly.

The lifetime is related to the mean free path ℓ via the relation $\ell = v_F \tau$, where v_F is the Fermi velocity. For typical ferromagnetic metals, ℓ is much shorter that the spin-diffusion length λ_{spin}, and therefore the spins of the carrier electrons are well conserved in the time scale τ. In this case, $\uparrow$ and $\downarrow$ spin electrons can be treated independently in evaluating the electrical conductivity, that is, $\sigma = \sum_s \sigma_s$ with $s = \uparrow$ or $\downarrow$. This assumption is the Mott's two-current model.

Although Mott's two-current model explains the experimental results of electrical resistivity in ferromagnetic metals, it is rather difficult to confirm the model directly by experiment, since $\sigma_\uparrow$ and $\sigma_\downarrow$ cannot be separated independently from the σ data. However, Fert and Campbell [23,24] have approached the problem by measuring the residual resistivity and temperature dependence for various binary and ternary alloys and succeeded in deducing the ratio $\rho_\downarrow/\rho_\uparrow (= \sigma_\uparrow/\sigma_\downarrow)$ for diluted alloys of Fe, Co, and Ni metals.

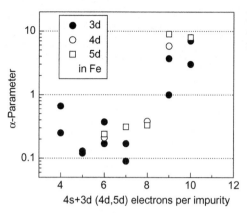

Figure 2.3 Experimental values of α-parameters for 3d, 4d, and 5d TM impurities in Fe [23,24].

The ratio is referred to as the α-parameter. α-Parameters for TM impurities in Fe are presented in Figure 2.3. We can see that α-parameter strongly depends on the species of the impurity atoms. In the next sections, we show how the material dependence of the α-parameter is related to the electronic states of ferromagnets.

2.2.3 Spin-Dependent Resistivity in TM Alloys

The spin dependence of τ caused by impurity scattering of electrons in ferromagnetic metals may be evaluated by using the formula:

$$\tau_s^{-1} = (2\pi/\hbar)N_i V_s^2 D_s(E_F) \tag{2.2}$$

which is given by the Born approximation, where N_i, V_s, and $D_s(E_F)$ are the impurity density, scattering potential, and DOS at the Fermi energy E_F, respectively. Here, both V_s and $D_s(E)$ are spin ($s = \uparrow$ or $\downarrow$) dependent. Equation (2.2) indicates that the lifetime becomes short as the scattering potential becomes large and the number of final states of the scattering process increases.

Let us consider TM impurities in Fe. The impurities give rise to a spin-dependent potential V_s in Fe even when the impurity is nonmagnetic, since the DOS $D_s(E)$ of Fe is spin dependent. Since $D_\uparrow(E_F) \sim D_\downarrow(E_F)$ for ferromagnetic Fe, the spin dependence of the lifetime is caused mainly by V_s.

The magnitude of V_s may be evaluated crudely by assuming that the DOS of TM impurities are unchanged from the bulk case and that the number of d-electrons and magnetic moment impurities are also unchanged from those of the bulk state. The latter assumption may be validated from the charge neutrality condition and from neutron diffraction measurements of local moments in ferromagnetic alloys. On the other hand, the former assumption is believed to be truly crude.

Under these assumptions, V_s is given by the relative shift of the d-level of impurities with respect to that of Fe, since the Fermi level (or the chemical potential) for TM impurities and Fe metal should coincide. The values of $\Delta V_{\xi s} = V_{\xi s} - V_{Fe0}$

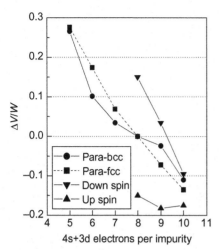

Figure 2.4 Calculated spin-dependent impurity potentials in Fe. The potentials are measured from the potential of paramagnetic Fe. W indicates the effective bandwidth of the 3d-bands [25].

thus determined are shown in Figure 2.4[25]. Here ξ indicates the atomic species of the impurities and V_{Fe0} is the d-level of paramagnetic Fe.

From Figure 2.4, we can see that $|V_{Fe\downarrow}| \simeq |V_{Cr}|$ and $|V_{Fe\uparrow}| \gg |V_{Cr}|$, where the spin suffix of V_{Cr} is omitted since Cr is assumed to be nonmagnetic in Fe. The results indicate that the band matching between Fe and Cr is quite good for the $\downarrow$ spin state, while it is rather poor for the $\uparrow$ spin state. Schematic shapes of the DOS for Cr, Fe, Co, and Cu with a common Fermi level are shown in Figure 2.1. The results deduced above may be easily understood from the relative positions of the d-DOS.

Since $\Delta V_{\xi s}$ is simply V_s in the Drude formula, we find $\rho_\uparrow \gg \rho_\downarrow$ for Cr impurities in Fe metal. This is in good agreement with the α-parameters shown in Figure 2.3. The present crude estimate of V_s may be validated by first-principles calculations, which gives the same results for the spin-dependent resistivity for Cr impurities in Fe. A detailed study of the residual resistivity in the first-principles method has been presented by Mertig [26]. The study reproduces the experimental trends of the spin-dependent residual resistivity in Fe, Co, and Ni.

2.2.4 Spin-Dependent Resistivity due to Ferromagnetic Impurities in Novel Metals

The residual resistivity due to TM impurities in metals is well described by the Anderson model [27]. The lifetime in this model is given as:

$$\tau_s^{-1} = 5(2\pi/\hbar)N_i V_{sd}^2 D_{ds}(E_F) \tag{2.3}$$

where V_{sd} represents s−d mixing between the conduction state and localized d-states of impurities, and $D_{ds}(E_F)$ is the DOS of impurities at E_F with spin s. The factor 5 comes from the degeneracy of the d-states of TM impurities.

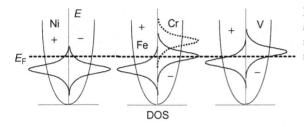

Figure 2.5 DOS of Ni, Fe, Cr, and V impurities in Cu. The + spin state of Cr impurities is the same as that of Fe impurities.

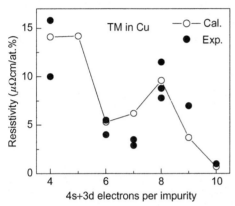

Figure 2.6 Calculated (open circles) and experimental (filled circles) values of residual resistivity caused by TM impurities in Cu [29].

Equation (2.3) is similar to Eq. (2.2) with V_s and D_s replaced by V_{sd} and D_{ds}. This is to be expected since the conduction electrons (s-electrons) are scattered into d-states via s−d mixing. It should be noted however that $D_{ds}(E)$ is not a bare DOS of the impurity d-states, but rather is a renormalized DOS broadened due to s−d mixing. Figure 2.5 shows the schematic shape of the DOS of V, Cr, Fe, and Ni impurities in a free-electron band. First-principles band calculations also show an electronic structure of TM impurities similar to those shown in Figure 2.5[28].

Despite its simplicity, the Anderson model satisfactorily explains the tendency of the residual resistivity caused by TM impurities, in Cu for example. Experimental and theoretical results are shown in Figure 2.6[29]. The horizontal axis of this figure is the number of 4s + 3d electrons n per atom, where $n = 5$, 6, 8, and 10 correspond to V, Cr, Fe, and Ni, respectively. Since the DOS of Ni impurities is almost occupied, $D_{ds}(E_F)$ is too low to be exchange split. Therefore, the residual resistivity is spin-independent and small for Ni impurities. Fe impurities, on the other hand, are magnetized and the DOS is exchange split, as shown in Figure 2.5. Because E_F is located near the peak of $D_{d\downarrow}(E)$, the residual resistivity becomes large. For Cr impurities, $D_{d\downarrow}(E)$ shifts to higher energy, while $D_{d\uparrow}(E)$ remains unshifted, E_F is located in a low $D_{d\downarrow}(E)$ region. As a result, the resistivity due to Cr impurities is smaller than that for Fe impurities. The resistivity becomes large again for V impurities, since $D_{d\downarrow}(E)$ also shifts to higher energy.

The interpretation of the residual resistivity for TM impurities in Cu gives $\rho_\downarrow/\rho_\uparrow \gg 1$ for Fe impurities and $\rho_\downarrow/\rho_\uparrow \ll 1$ for V impurities. The results are also consistent with the material dependence of the α-parameter.

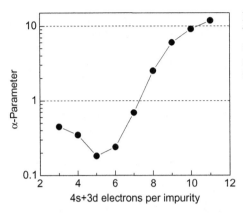

Figure 2.7 Calculated α-parameters for TM and noble metal impurities in Fe (as an example host metal) as a function of the number of 4s + 3d electrons of the impurities [30].

2.2.5 Two-Band Model

The material dependence of the α-parameter $\rho_\downarrow/\rho_\uparrow$ given by experiments (Figure 2.3) and that estimated theoretically for TM impurities in Fe and in Cu may be understood by adopting a two-band model [30]. The model consists of a broad s-like band and narrow d-like band with a mixing between the s- and d-bands.

Taking Fe as the host metal for example, the electronic state of TM−Fe alloys may be given by a random distribution of d-levels ε_{TM} and ε_{Fe} of the TM and Fe atoms, respectively. By applying the coherent potential approximation (CPA) [31−34] to the random distribution of d-levels, one may calculate the DOS and electrical resistivity of TM−Fe alloys.

The spin dependence of the residual resistivity $\rho_\downarrow/\rho_\uparrow$ thus calculated is shown in Figure 2.7 as a function of the number of 4s + 3d electrons per atom. The results reproduce the experimental tendency rather well. The calculated results can be easily understood in terms of matching/mismatching of the d-electronic states between impurities and host atoms.

For Ag impurities ($n = 11$) in Fe, the matching of the $\uparrow$ spin state is good, resulting in $\rho_\downarrow/\rho_\uparrow \gg 1$, while for Cr impurities band matching is better for $\downarrow$ spin bands, and therefore $\rho_\downarrow/\rho_\uparrow \ll 1$, as shown in Figure 2.7.

2.3 Microscopic Theory of Electrical Conductivity: Linear Response Theory

In this section, we describe linear response theory and its application to layered structures. In the theory, the conductivity is given as a current−current correlation function, since the conductivity is the response of a current to an external electric field which drives the motion of the electrons. The correlation function is calculated for electronic states in the equilibrium state. That is, the fluctuation−dissipation relation in the equilibrium state determines the response of the electrical charge to the external field. In the following, we give formulations for conductivity with current parallel and perpendicular to layer planes. Readers not concerned with the details of the theoretical framework may skip this section and jump to the sections on GMR, TMR, and BMR.

In this section, we assume that Mott's two-current model holds and omit the spin suffixes. The basic model used is the tight-binding model with isotropic δ-function-type impurity potentials. General formalism of the Green's function used below, and theory of the electrical transport, may be found in several textbooks [35,36].

2.3.1 Kubo Formula

Applying the one-electron approximation to the general expression of the linear response theory [37], we obtain the so-called Kubo−Greenwood formula for electrical conductivity:

$$\sigma = \frac{\pi\hbar}{\Omega} \text{Tr}\left[\mathscr{J}\delta(E_F - \mathscr{H})\mathscr{J}\delta(E_F - \mathscr{H})\right] \tag{2.4}$$

where $\mathscr{H}$ is the Hamiltonian and $\mathscr{J}$ is the corresponding current operator. The two δ-functions in the equation represent current conservation and the response of the electrons on the Fermi surface to the electric field. Using the Green's function $\mathscr{G}^{R(A)}(E)$, defined as

$$\mathscr{G}^{R(A)}(E) = [E + (-)i\eta - \mathscr{H}]^{-1} \tag{2.5}$$

$$= P[E - \mathscr{H}]^{-1} - (+)i\pi\delta(E - \mathscr{H}) \tag{2.6}$$

the δ-functions are expressed in terms of the imaginary part of the Green's function, where R and A indicates the retarded and advanced Green's functions, respectively. The conductivity is thus expressed as:

$$\sigma = \frac{\hbar}{\pi\Omega} \text{Tr}\left[\mathscr{J}\{\mathscr{G}^A - \mathscr{G}^R\}\mathscr{J}\{\mathscr{G}^A - \mathscr{G}^R\}\right] \tag{2.7}$$

where E_F in the Green's function is omitted.

There are two methods for practical calculations of the conductivity using the expression given above. One is to adopt suitable approximations in the calculation of the conductivity and the other is to simulate the conductivity numerically for finite size systems with leads. In the following, we demonstrate the methods of calculation of the conductivity or conductance of multilayers adopting a Hamiltonian with random potentials:

$$\mathscr{H} = \mathscr{H}_0 + V \tag{2.8}$$

where V indicates the random potentials.

2.3.2 Current Parallel to Planes

When the current flows parallel to the planes of the multilayers, the electrical conductivity can be calculated semianalytically. This is because the system exhibits translational invariance parallel to the planes and momentum conservation holds along this direction. In the formulation, the Green's function should first be evaluated, averaged statistically over the impurity distribution. We denote the average of the quantity A as $\langle A \rangle$. As a result of averaging, the self-energy due to electron scattering by the impurity potential V is introduced. The averaged Green's functions are expressed as:

$$\tilde{\mathscr{G}}^{R(A)}(E) = \langle (E + (-)i\eta - \mathscr{H})^{-1} \rangle \tag{2.9}$$

$$= (E + (-)i\eta - \mathscr{H}_0 - \Sigma^{R(A)})^{-1} \tag{2.10}$$

The self-energy Σ depends on both energy and momentum, in general. The simplest method to evaluate the self-energy is the Born approximation. In this approximation, it is given as $\Sigma^{R(A)} = \langle V \mathscr{G}_0^{R(A)} V \rangle$, where $\mathscr{G}_0^{R(A)} = (E + (-)i\eta - \mathscr{H}_0)^{-1}$, with an infinitesimally small real number η. When the magnitude of the impurity potential is large, the CPA is useful.

The statistical average for the electrical conductivity should be taken such that

$$\sigma = \frac{\hbar}{\pi\Omega} \text{Tr}\langle \mathscr{J}\{\mathscr{G}^A - \mathscr{G}^R\} \mathscr{J}\{\mathscr{G}^A - \mathscr{G}^R\}\rangle \tag{2.11}$$

We note that $\langle \mathscr{J}\mathscr{G}\mathscr{J}\mathscr{G} \rangle = \mathscr{J}\langle \mathscr{G}\mathscr{J}\mathscr{G} \rangle \neq \mathscr{J}\tilde{\mathscr{G}}\mathscr{J}\tilde{\mathscr{G}}$. Since the current operator does not account for the randomness caused by impurity potentials, the first equality holds. However, the average of the product of two Green's functions is not a product of two averaged Green's functions. Therefore, we must evaluate $\mathscr{J}\langle \mathscr{G}\mathscr{J}\mathscr{G} \rangle - \mathscr{J}\tilde{\mathscr{G}}\mathscr{J}\tilde{\mathscr{G}}$ correctly. We call this correction the "vertex correction" to the conductivity. Current conservation is satisfied when the vertex correction is evaluated self-consistently in the determination of the self-energy Σ. Since the imaginary part of the self-energy corresponds to the lifetime, the conductivity may diverge when Green's functions are used without the self-energy.

In the following, we consider a simple cubic lattice with isotropic δ-function-type impurity potentials. In this case, the vertex correction vanishes since the correct operator is odd in the momentum space. The current operator is a product of the electrical charge e and electron velocity along the current direction, that is, $\mathscr{J}_x = ev_x$. Since the velocity vector is the momentum derivative of the energy, we obtain

$$\mathscr{J}_x = e\frac{1}{\hbar}\frac{\partial E(\mathbf{k}_\parallel)}{\partial k_x} = \frac{e}{\hbar} 2ta \sin k_x a \tag{2.12}$$

for the energy eigenvalues $E(\mathbf{k}) = 2t(\cos k_x a + \cos k_y a + \cos k_z a)$ of a single orbital tight-binding model on the simple cubic lattice, where a and t are the lattice constant and transfer integral between the nearest-neighbor sites, respectively.

For multilayers with a finite number of layers, which can be considered pseudo-two-dimensional systems, the electrical conductivity at zero temperature may be evaluated as follows. Since there is no translational invariance along the direction perpendicular to the layers, it is convenient to express the coordination of the sites as (ℓ, i), where ℓ is the layer index and i is the position of the site within a layer. After recovering the translational invariance by taking the statistical average over the impurity distribution of a layer, the wave vector $k_{\parallel}$ along a layer plane can be defined. Therefore, a Fourier transformation can be performed and the representation $(\ell, k_{\parallel})$ can be used, where $k_{\parallel} = (k_x, k_y)$ for stacking along the z-axis.

In this representation, the current operator along the x-direction is given as $\mathscr{J}_x(k_{\parallel}) = e v_x(k_{\parallel})$ and the vertex correction vanishes, as described above. Therefore, the electrical conductivity is given as:

$$\sigma_{xx} = \frac{e^2 \hbar}{\Omega \pi} \sum_{s\ell m k_{\parallel}} v_x(k_{\parallel}) v_x(k_{\parallel}) \mathrm{Im} \tilde{\mathscr{G}}^{\mathrm{R}}_{\ell m s}(k_{\parallel}) \mathrm{Im} \tilde{\mathscr{G}}^{\mathrm{R}}_{m \ell s}(k_{\parallel}) \tag{2.13}$$

where s denotes spin and ℓ and m are the layer suffixes. The Green's function is expressed as a matrix with a size determined by the number of layers.

2.3.3 Current Perpendicular to Layer Planes

We now consider multilayers with two leads attached to the top and bottom of the multilayers with current flowing perpendicular to the layer planes. Because there is no translational invariance along the direction of current flow, the multilayers become scattered and produce electrical resistivity even if the multilayers are perfect with no defects, impurities, etc. Since the system size is finite, it is convenient to consider the conductance defined as $\Gamma = (\sigma/L)S$ instead of the conductivity σ. Here, S and L are the cross section and length of the sample, respectively. From Eq. (2.4), the conductance is given as:

$$\Gamma = \frac{\pi \hbar}{L^2} \mathrm{Tr} \left[\mathscr{J} \delta(E_{\mathrm{F}} - \mathscr{H}) \mathscr{J} \delta(E_{\mathrm{F}} - \mathscr{H}) \right] \tag{2.14}$$

As the width of the multilayers is much larger than the thickness, we may regard the width as being nearly infinite and can express the electronic states using a mixed representation of layer number and wave vector parallel to the layer planes, that is, $(\ell, k_{\parallel})$ when no defects, impurities, etc. are included. When the multilayers include impurities or when the shape of the sample is complex, we must adopt a real space representation using (ℓ, i). Here, we define "sample" as being the region that exhibits electrical resistivity. For multilayers with leads, the region of the multilayers is the sample. For complex structures, we may choose the sample arbitrarily, even including the leads. For point contacts, the sample region may include only a few atoms. In tunnel junctions, the sample may be the insulating barrier region since the resistivity is governed by the insulating materials.

We rewrite Eq. (2.14) in a form applicable to numerical calculations. When electron hopping between layers, given by t, is nonzero only between nearest-neighbor sites, the conductance is given by:

$$\Gamma = \frac{e^2 t^2}{2h} \mathrm{Tr}\left[\overline{G}_{\ell,\ell+1}\overline{G}_{\ell+1,\ell} + \overline{G}_{\ell+1,\ell}\overline{G}_{\ell+1,\ell} - \overline{G}_{\ell,\ell}\overline{G}_{\ell+1,\ell+1} - \overline{G}_{\ell+1,\ell+1}\overline{G}_{\ell,\ell}\right] \quad (2.15)$$

where $\overline{G} = \mathscr{G}^{\mathrm{A}} - \mathscr{G}^{\mathrm{R}}$ and the trace indicates a sum over spins and sites in the layer. We have used the fact that the electrical current is independent of the position of the layers. When the multilayers are clean and the wave vector $\boldsymbol{k}_{\|}$ is well defined, the Green's functions are functions of $\boldsymbol{k}_{\|}$. However, when the sample has a complex structure, we must adopt a real space representation, and the Green's functions are matrices with a size determined by the sample width. Expression (2.15) has been given by Lee and Fisher [38].

2.3.4 Recursive Green's Function Method

The Green's function may be calculated once the Hamiltonian of the whole system is given. We here present a simple example to treat the Green's function using a one-dimensional model, in which the hopping integral between the nearest-neighbor sites is given by t and the atomic potentials are v_i. The Green's function is given as:

$$\mathscr{G}(z) = \begin{bmatrix} \cdot & t & 0 & 0 & 0 \\ t & z-v_{\ell-1} & t & 0 & 0 \\ 0 & t & z-v_\ell & t & 0 \\ 0 & 0 & t & z-v_{\ell+1} & t \\ 0 & 0 & 0 & t & \cdot \end{bmatrix}^{-1} \quad (2.16)$$

where $z = E \pm i\eta$. We divide the system into left and right semi-infinite parts. When the ℓth site is an edge atom of the left part, $\mathscr{G}_{\ell\ell}$ is given as:

$$\mathscr{G}_{\ell\ell}(z) = \frac{1}{z - v_\ell - t^2\mathscr{G}_{\ell-1\ell-1}(z)} \quad (2.17)$$

which connects the Green's function of the $(\ell-1)$th atom with the ℓth atom. When the ℓth atom is far from the edge, we set $\mathscr{G}_{\ell\ell}(z) = \mathscr{G}_{\ell-1\ell-1}(z)$, since these atoms are equivalent. Solving this equality gives $\mathscr{G}_{\ell\ell}(z)$. The Green's functions for the right part are calculated using the same procedure. The Green's functions of the left and right parts are connected by the relation

$$\mathscr{G}(z) = \begin{bmatrix} \mathscr{G}_{\ell\ell}^{-1}(z) & t \\ t & \mathscr{G}_{\ell+1\ell+1}^{-1}(z) \end{bmatrix}^{-1} \quad (2.18)$$

from which we obtain $\mathcal{G}_{\ell\ell}(z)$ and $\mathcal{G}_{\ell\ell+1}(z)$. This procedure to determine the layer Green's function is called the recursive Green's function method.

This method may be applied to systems which are much smaller than real systems in order to evaluate the conductivity of a real system. In this case, we first calculate the conductance and evaluate the conductivity as follows. Since the conductivity in the linear response regime is characterized by Ohm's law, where the resistivity increases linearly with the sample length, and the conductivity and conductance for finite size samples with width S and length L is given as $1/\Gamma = 1/\Gamma_c + (L/\sigma S)$, we calculate the conductance as a function of L and estimate the value of σ from the slope of the $1/\Gamma$ versus L curve. Here, $1/\Gamma_c$ is the contact resistance of the contact between the sample and the leads. Furthermore, the conductance is calculated for many samples with different distributions of impurities by taking a statistical average of the conductance. This procedure is necessary since we are dealing with conductivity in the diffusive regime, in which electrical scattering is important for the conductivity.

2.3.5 Conductance Quantization and Landauer Formula

The current of a one-dimensional chain subject to a voltage is given as:

$$I = 2e \int_{E_F}^{E_F + eV} D(E)v(E)dE \tag{2.19}$$

using the DOS $D(E)$ and the velocity of electrons $v(E)$. Using the relations for free electrons,

$$D(E) = \frac{1}{2\pi\hbar}\sqrt{\frac{m}{2E}} \tag{2.20}$$

and $v(E) = p/m = \sqrt{2E/m}$, we get:

$$I = \frac{2e^2}{h}V \tag{2.21}$$

Therefore, the conductance is given as:

$$\Gamma = 2e^2/h \tag{2.22}$$

in terms of physical constants. The factor 2 indicates spin degeneracy. Because only two propagating states exist in a one-dimensional chain, that is, left going and right going waves, there is only one conducting path (called a channel) where electrons flow for each spin. The channel gives a conductance of $2e^2/h$, which is called the quantum conductance. This value is 3.4×10^7 cm/s and is equal to $(26 \text{ k}\Omega)^{-1}$. The conductance of multipass systems is given by integer multiples of $2e^2/h$.

Even if the conductance is ballistic, the conductance for systems which have no translational invariance along the current direction is given as:

$$\Gamma = \frac{2e^2}{h} T(E_F) \tag{2.23}$$

using the transmission coefficient $T(E_F)$ at the Fermi energy. When there are many channels, as in three-dimensional systems, and the wave vector parallel to planes is well defined, the total conductance is written as:

$$\Gamma = \frac{2e^2}{h} \sum_{k_{\parallel}} T(E_F, k_{\parallel}) \tag{2.24}$$

For a finite voltage, the expression is generalized to be:

$$\Gamma = \frac{2e^2}{h} \sum_{k_{\parallel}} \int dE \{ f(E) - f(E + eV) \} T(E_F, k_{\parallel}) \tag{2.25}$$

which is called the Landauer formula [39]. The result is also obtained in the non-equilibrium Green's function method [40].

2.4   Giant Magnetoresistance

Magnetic multilayers are composed of an alternating stack of thin magnetic and nonmagnetic layers. The thickness of each layer is a few nanometers. Trilayers, where a nonmagnetic layer is sandwiched by two magnetic layers, can also be considered to be multilayers. Some magnetic multilayers show large MR. When the nonmagnetic layers are metals, the MR is called giant MR (GMR) and when the nonmagnetic layer in a trilayer is an insulator, the MR is called tunnel MR (TMR).

Magnetic multilayers have the following two important characteristics:

1. The alignment of the magnetization of the magnetic layers is easily controlled by an external magnetic field, since the coupling between the magnetization of the magnetic layers is weakened by the presence of the nonmagnetic layer between them.
2. Each layer is thin enough for carrier electrons to feel a change in the magnetization direction of the magnetic layers.

GMR and TMR depend strongly on the type of magnetic and nonmagnetic layers, and their combination. In this section, we first explain how the experimental GMR results may be understood in terms of spin-dependent resistivity described in Section 2.2, and that the material dependence of GMR is strongly related to the electronic structure of the constituent metals of the multilayers.

2.4.1 Magnetic Multilayers

The basic structure of magnetic multilayers composed of a ferromagnetic A metal and a nonmagnetic B metal is shown in Figure 2.8. The thickness of each layer is 1−10 nm and the number of layers ranges from 3 (for trilayers) to about 100. Fe, Co, Ni, and their alloys are frequently used for the ferromagnetic A layers, while nonmagnetic TMs such as Cr and Ru or noble metals Cu, Ag, and Au are used for the nonmagnetic B layers.

To fabricate high-quality magnetic multilayers, matching of the lattice constants of the constituent metals is important. Figure 2.9 shows the distance between nearest-neighbor atoms and the lattice structure of 3d, 4d, and 5d TMs. We find that the matching of the atomic distance and lattice structure of Fe with Cr and Co with Cu are sufficiently good.

Figure 2.8 Schematic figure of magnetic multilayers with ferromagnetic A and nonmagnetic B layers. d and d' indicate the layer thickness.

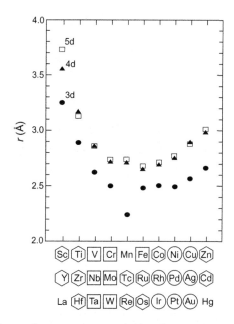

Figure 2.9 Atomic distance between nearest-neighbor sites and lattice structures in 3d, 4d, and 5d TMs. Circles, squares, and hexagons under the figure show that the crystal structures of metals are fcc, bcc, and hcp, respectively.

2.4.2 Experiments on GMR

2.4.2.1 GMR and Exchange Coupling

The first observation of antiparallel (AP) coupling between magnetic layers was reported by Grünberg *et al.* [41] for Fe/Cr trilayers. They also observed negative MR, that is, a resistivity reduction under an external magnetic field. The magnitude of the MR of Fe/Cr trilayers was observed to be a few percent. Two years later, Fert's group [42] reported MR as large as 40% for Fe/Cr multilayers. This MR was the largest so far observed for magnetic metal films and was called giant MR (GMR). After the discovery of GMR, many experimental works have been performed [43–48].

Figure 2.10 shows the experimental results for Fe/Cr multilayers [42]. The resistivity decreases with increasing magnetic field due to a change in the alignment of the magnetization of the Fe layers. The resistivity is high when the alignment is AP and is low when the alignment is parallel (P).

The magnitude of the MR is expressed by the so-called MR ratio, defined as:

$$MR = \frac{\rho_{AP} - \rho_{P}}{\rho_{AP}} \tag{2.26}$$

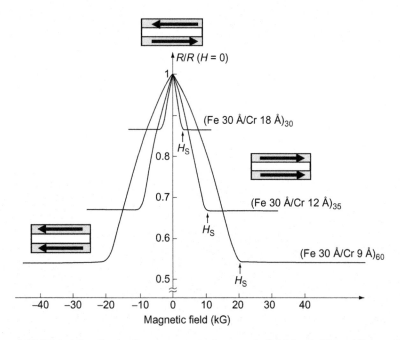

Figure 2.10 Resistivity change due to an external magnetic field for Fe/Cr multilayers [42].

or

$$MR = \frac{\rho_{AP} - \rho_P}{\rho_P} \tag{2.27}$$

where ρ_{AP} and ρ_P are the resistivity in AP and P alignment of the magnetization of the magnetic layers. Since usually $\rho_P < \rho_{AP}$, the definitions (2.26) and (2.27) are called as pessimistic and optimistic definitions, respectively. In experiments, the optimistic definition, Eq. (2.27), is usually used, while we use pessimistic definition, Eq. (2.26), for the theoretical results in this chapter. Tables 2.1 and 2.2 show MR ratios observed for several combinations of magnetic and nonmagnetic metals. We see that the MR ratio depends on the combination of metals. Fert' group pointed out in their first paper on GMR that the spin-dependent resistivity could be responsible for GMR; however, no detailed discussion on the material dependence was presented. (The many-body effect was ruled out since no anomalous dependence of ρ on temperature has been observed [45]. The residual resistivity, however, was found to be very high.) Thus, an issue to be clarified is the material dependence of GMR and the relation between the MR ratio and the electronic structures of the constituent metals of the magnetic multilayers.

Table 2.1 MR Ratios Measured for Various Magnetic Multilayers for Current Parallel to the Layer Planes

Multilayer	$\Delta\rho/\rho_P$ (%)	$\Delta\rho/\rho_{AP}$ (%)
Fe/Cr	108	52
Co/Cu	115	53
NiFe/Cu/Co	50	33
FeCo/Cu	80	44
NiFeCo/Cu	35	26
Ni/Ag	26	21
Co/Au	18	15
Fe/Mn	0.8	
Fe/Mo	2	
Co/Ru	7	
Co/Cr	2.6	
Fe/Cu	12	

Table 2.2 MR Ratios Measured for Selected Magnetic Multilayers for Current Perpendicular to the Planes

Multilayer	$\Delta\rho/\rho_P$ (%)	$\Delta\rho/\rho_{AP}$ (%)
Ag/Fe	42	30
Fe/Cr	108	52
Co/Cu	170	63

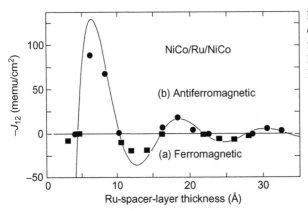

Figure 2.11 Oscillation of coupling energy as a function of nonmagnetic layer thickness in NiCo/Ru/NiCo multilayers [51].

GMR appears when the AP alignment of the magnetization is changed to P alignment by an external magnetic field. Therefore, AP alignment of the magnetization is a prerequisite for GMR. A detailed study of the alignment of magnetization has found that the coupling of magnetization in magnetic layers changes as a function of the nonmagnetic layer thickness [49]. The coupling between magnetic layers is called interlayer exchange coupling [50−54]. Figure 2.11 shows an experimentally determined oscillation of coupling energy as a function of layer thickness [51]. The period of the oscillation is rather long and the magnitude decays as the thickness of the nonmagnetic layer increases. The long period of oscillation has been confirmed in various experiments [55−57]. The features are very similar to those of the so-called Ruderman-Kittel-Kasuya-Yosida (RKKY) interaction between magnetic impurities in metals [58]. The period of the oscillation of the interlayer exchange coupling is determined by the Fermi wave vector k_F, as in the RKKY interaction [59−64]. In the present case, however, the thickness of the nonmagnetic layer changes discretely and therefore $(\pi/a - k_F)$ can also be the period of oscillation, where a is the lattice distance. Since k_F in Cu, for example, is close to π/a, the period of oscillation of the exchange coupling becomes long. The decay of the magnitude for multilayers is proportional to L^2, where L is the nonmagnetic layer thickness, in contrast to r^3 for the RKKY interaction.

2.4.2.2 Noncoupling Type of GMR

Magnetic layers in multilayers are usually coupled magnetically (interlayer exchange coupling). The interlayer exchange coupling in Fe/Cr multilayers is rather strong to be controlled by the magnetic field. Multilayers with thicker nonmagnetic layers have nearly zero exchange coupling; however, the magnetization direction of the magnetic layers may be controlled by using the difference in the coercive force between magnetic layers of different metals. An example is Co/Cu/NiFe multilayers shown in Figure 2.12, in which NiFe is a soft magnet with a magnetization easily changed by a weak external magnetic field [65,66].

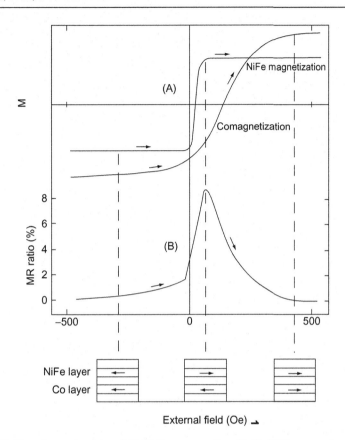

Figure 2.12 Experimental results of GMR for noncoupling Co/Cu/NiFe magnetic multilayers [65].

2.4.2.3   Spin Valve

Technological applications of GMR, for example, sensors, require a sharp response of the magnetization direction to the external magnetic field within a few Oersteds. To achieve such sensitivity, a trilayer structure with an attached antiferromagnet has been designed. The magnetization of the magnetic layer adjacent to the antiferromagnetic layer is pinned by the antiferromagnetism and only the other magnetic layer responds to the external magnetic field. This kind of trilayer is called a spin valve [67−69]. PtMn and FeMn are typical antiferromagnets used in spin valves. An example of GMR in a spin-valve-type trilayer is shown in Figure 2.13.

2.4.2.4   CPP-GMR

The experiments presented so far have used a geometry with the current flowing parallel to the layer planes. GMR with this geometry is called current-in-plane

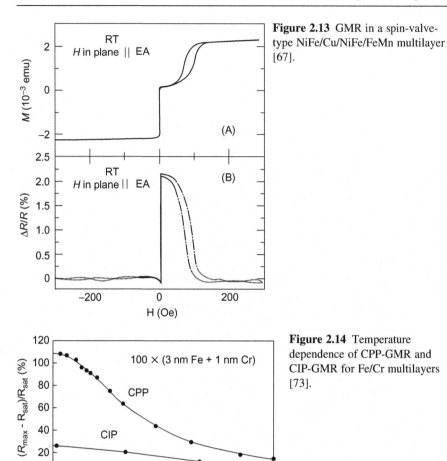

Figure 2.13 GMR in a spin-valve-type NiFe/Cu/NiFe/FeMn multilayer [67].

Figure 2.14 Temperature dependence of CPP-GMR and CIP-GMR for Fe/Cr multilayers [73].

GMR (CIP-GMR). GMR with a geometry with current flowing perpendicular to the planes is called CPP-GMR.

In CPP-GMR, the resistivity of a sample is too small to be detected, since the layer thickness is usually less than micrometers and the resistivity of the leads is overwhelming. To make a measurement of sample resistivity possible, several methods have been adopted. One is to utilize superconducting leads [70−72], the second one is to microfabricate the samples [73,74], and to fabricate multilayered nanowire formed by electrodeposition into nanometer-sized pores of a template polymer membrane [75]. In the second case, the resistivity of the sample becomes as large as that of the leads because the resistivity of such systems is governed by the narrow region of the system.

The temperature dependence of CPP-GMR for a microfabricated sample of Fe/Cr multilayers is compared with that of CIP-GMR in Figure 2.14. We see that

CPP-GMR is much larger than CIP-GMR, which is a general trend for GMR. To interpret the results, the geometry of multilayers should be taken into account, in addition to the spin-dependent resistivity of ferromagnetic metals. Shinjo's group fabricated a zigzag structure of multilayers in which the current flows in angle to planes, and measured a rather high MR ratio [76,77].

2.4.2.5 Granular GMR

GMR has been observed not only in magnetic multilayers but also in granular films in which magnetized metallic grains of Co or Fe, for example, are distributed in nonmagnetic metals, typically Cu or Ag [78−84]. The size of the grains is of nanoscale and the magnetic moments of the grains are nearly isolated from each other.

When an external magnetic field is applied, the random orientation of the magnetic moments of the grains are forced to be parallel, resulting in a decrease of the resistivity, as for magnetic multilayers. Example experimental results are shown in Figure 2.15. The dependence of the resistivity on the magnetic field is strongly affected by the annealing temperature of the sample. When the annealing temperature is low, the MR does not saturate even at high magnetic field. This may be because isolated magnetic atoms and/or clusters still remain for low annealing temperature and they continue to respond to high magnetic fields.

2.4.3 Phenomenological Theory of GMR

The typical length scale of a multilayer is the thickness L of each layer, which is of the order of 1 nm. Since the length scale is shorter than the mean free path and much shorter than the spin-diffusion length, Mott's two-current model is applicable to GMR in multilayers as a first approximation. The model is also applicable to granular GMR because the length scale of the sample is the diameter of the magnetic grains. Detailed experiments have shown that the spin-diffusion length in the multilayer, especially in CPP-GMR-type multilayers, may be different from that in

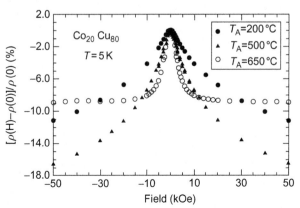

Figure 2.15 GMR in granular magnetic metals. Here, T_A is the annealing temperature of the samples [79].

metals and alloys [15,85−89]. In the following, however, we adopt the simplest picture to explain the effect of GMR [25,90].

In applying the two-current model to magnetic multilayers, the direction of the spin axis, up (↑) or down (↓), should be defined to deal with the resistivity. Because the magnetization of the magnetic layers is reversed by the magnetic field and the magnetization alignment can be either P or AP, ↑ and ↓ spin states should be distinguished from the majority (+) and minority (−) spin states of each magnetic layer. We henceforth use the notation ↑ and ↓ spin states as the global spin axes and + and − spin states to express the electronic states of the magnetic metals. For P alignment, ↑ and ↓ spin states coincide with + and − spin states, respectively; however, they do not coincide for AP alignment. In the following, we adopt Eq. (2.26) for the MR ratio.

For simplicity, we consider a case where the current traverses a trilayer composed of a nonmagnetic layer sandwiched by two ferromagnetic layers. Let ρ_+, ρ_-, and ρ_0 be the majority and minority spin resistivities in the ferromagnetic layers and the resistivity in the nonmagnetic layer, respectively. The total resistivity for P and AP alignments may be easily obtained by referring to the equivalent circuits shown in Figure 2.16A and B, respectively. The resultant MR ratio is given as:

$$MR = \left(\frac{\rho_+ - \rho_-}{\rho_+ + \rho_- + 2\rho_0}\right)^2 \tag{2.28}$$

The MR ratio increases as the difference between the spin-dependent resistivities $\rho_+ - \rho_-$ increases.

When $\rho_+ + \rho_0$ and $\rho_- + \rho_0$ are rewritten as ρ_+ and ρ_-, the expression above is written as:

$$MR = \left(\frac{1-\alpha}{1+\alpha}\right)^2 \tag{2.29}$$

using the α-parameter $\alpha = \rho_-/\rho_+$. Thus, a combination of materials which gives a large value of the α-parameter gives rise to a large GMR. The combination of

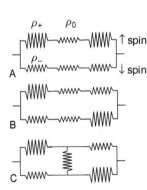

Figure 2.16 Equivalent circuit for the resistivity for (A) P alignment and (B) AP alignment of magnetization. (C) Equivalent circuit for the resistivity in multilayers which include spin-flip scattering.

Fe and Cr gives a large GMR. In addition, the lattice matching between Fe and Cr layers is good.

2.4.4 Mechanism of GMR

2.4.4.1 Electronic States and Spin-Dependent Resistivity in Multilayers

The spin-dependent resistivity in ferromagnetic metals is closely related to the electronic structure of the metal and impurities introduced into the metal. The electronic structures of the constituent metals of the magnetic multilayers also govern the spin-dependent resistivity (or resistance) in multilayers. In this section, we describe possible sources of spin-dependent resistivity in multilayers from the viewpoint of electronic states:

- *Spin-dependent resistivity caused by interfacial roughness.* As described, the origin of the spin-dependent resistivity in metals and alloys is the spin dependence of the scattering potentials caused by roughness. The roughness due to random arrangement of atoms also exists in multilayers. In molecular beam epitaxy (MBE) and sputtering fabrication methods, it is impossible to avoid intermixing of atoms at interfaces. The intermixing of magnetic A atoms and nonmagnetic B atoms at an A/B interface gives rise to spin-dependent random potentials near the interface. The situation is similar to that in ferromagnetic alloys.
- *Band matching/mismatching at interfaces.* The essence of the origin of electrical resistivity is the absence of translational invariance along the current direction, because the momentum of electrons need not be conserved in this case. When the interfaces are clean, translational invariance parallel to the layer planes is satisfied and there is no electrical resistivity. Even in this case, however, there is electrical resistivity perpendicular to the layer planes, since there is no translational invariance along this direction for thin multilayers. In this case, the difference between the electronic structure of the constituent metals of the multilayers acts as a source of spin-dependent electrical resistivity and gives rise to CPP-GMR. In Co/Cu multilayers, for example, band matching between $+$ spin states is much better than between $-$ spin states, as schematically shown in Figure 2.1. Therefore, $\rho_+ \ll \rho_-$ is realized. In Fe/Cr multilayers, the opposite relation, $\rho_- \ll \rho_+$, is realized. The spin dependence of the resistivity is the same as that obtained by spin-dependent random potentials.

Thus, both random potentials and matching/mismatching of the electronic structure cause the same spin dependence of the resistivity [25,90−94]. For current flowing perpendicular to the planes, both are crucial for the spin dependence of the resistivity, while the spin-dependent random potential is likely to be more greatly responsible for GMR for current flowing parallel to the planes. The scattering at interfaces and layer-resolved GMR effect have also been studied for CIP-GMR [95,96].

2.4.4.2 Estimate of GMR Using Spin-Dependent Potentials

We now estimate the MR ratio considering spin-dependent random potentials at interfaces. As mentioned, the material dependence of these random potentials is similar to that of the α-parameter. However, we can evaluate these random potentials more precisely by calculating the electronic states at interfaces for various multilayers and evaluate the spin-dependent resistivity and then calculate the MR ratio using Eq. (2.29). The calculated MR ratios are shown in Figure 2.17 for Fe/TM and Co/TM multilayers [91].

Figure 2.17 shows that the MR ratio has a peak for Fe/Cr multilayers and Co/Ru multilayers. In Fe/Cr multilayers, Cr atoms dissolved in the Fe layers and Fe atoms dissolved in the Cr layers give rise to spin-dependent random potentials. Because $V_{\text{Fe}-} \sim V_{\text{Cr}}$, while $V_{\text{Fe}+} \ll V_{\text{Cr}}$, both the dissolved Cr and Fe atoms result in $|V_+| \gg |V_-|$, which produces spin-dependent resistivity in such a way that $\rho_+ \gg \rho_-$. Therefore, the MR ratio of Fe/Cr multilayers becomes high. A similar situation may occur also in Fe/Pd multilayers, resulting in $\rho_+ \ll \rho_-$. In Fe/Pd multilayers, however, GMR has not been observed. This is because the Pd metal produces long-range spin polarization when it is in contact with ferromagnetic metals, and as a result, the interlayer exchange coupling between the Fe layers can be ferromagnetic irrespective of the thickness of the Pd layers. In Co/TM multilayers, the atomic potential of Co is lower than that of Fe (see Figure 2.4) and the relations $V_{\text{Co}-} \sim V_{\text{TM}}$ and $V_{\text{Co}+} \ll V_{\text{TM}}$ occur for TM = Rh and Ru. Therefore, the MR ratios for Co/Rh and Co/Ru become high.

The results reproduce the trend of the material dependence of GMR. However, the calculated magnitude is much larger than the experimental results. This may be attributed to the simplified treatment of the random potential for the resistivity and to the fact that the spin-independent resistivity caused by phonons, lattice imperfections due to lattice mismatch, etc. is not accounted for. The high MR ratio of Fe/Cr multilayers may also be related to good lattice matching between the Fe and Cr layers, in addition to the fact that they possess the same bcc structure.

For Co/Cu multilayers, Co atoms dissolved in the Cu layers may govern the spin-dependent resistivity, since the conductivity of Cu is much higher than that of Co metal. As shown above, the α-parameter of Co impurities in Cu is larger than 1, because the Co + spin state is similar to that of Cu, while the Co − spin state is

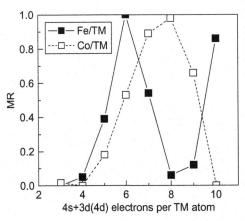

Figure 2.17 Theoretical estimate of MR ratio for Fe/TM and Co/TM multilayers for various TMs which are characterized by 4s + 3d(4d) electrons per atom [91].

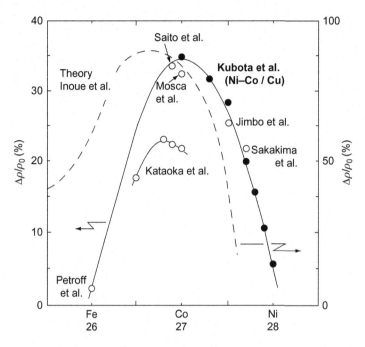

Figure 2.18 Theoretical estimate (broken curve) and experimental values (symbols) of MR ratio for TM/Cu multilayers as a function of the number of d-electrons [97].

quite different from that of Cu. Therefore, we expect $\rho_+ \ll \rho_-$ for Co/Cu multilayers. The spin-dependent resistivity due to magnetic TM impurities in Cu may be estimated by applying the Anderson model. The MR ratio thus calculated is shown in Figure 2.18 as a function of the number of d-electrons of the TM impurities [90]. The results are compared with the experimental values in Figure 2.18 [97–102].

We see that the MR ratio becomes high for Co/Cu multilayers, consistent with experimental results. The calculated results indicate that Ni/Cu shows no MR, in contradiction to the experimental results. Since the DOS of the TM impurities is given by a Lorentzian shape in the present model and the Ni impurities carry no magnetic moments, no MR occurs. It has been confirmed, however, that a more realistic model does give MR for Ni/Cu multilayers [103,104]. As for Fe/Cu multilayers, the theory gives a rather high MR ratio, whereas the MR ratios obtained in experiment are usually very small. The contradiction may be caused by the fact that the lattice structure is not taken into account in the theory; because Fe is bcc and Cu is fcc, the lattice mismatch should reduce the MR ratio.

2.4.4.3 Microscopic Theory of GMR

The discussion in the previous subsection does not take into account the geometry of the multilayers. Therefore, they are not able to distinguish between CIP-GMR

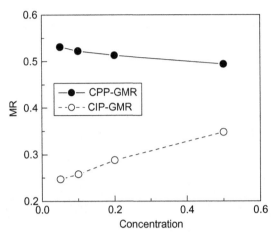

Figure 2.19 Calculated MR ratios for CIP- and CPP-GMR in a single-band tight-binding model, where the MR ratio is defined in Eq. (2.26) [105].

and CPP-GMR. To account for the effects of geometry, microscopic theory should be applied to the resistivity, which may be done by adopting the Kubo formula explained in the previous section.

Figure 2.19 shows the calculated MR ratios for CIP-GMR and CPP-GMR in a simple tight-banding model for Fe/Cr multilayers with roughness at interfaces [105]. A supercell method is adopted in the calculations. The horizontal axis indicates the concentration of random atoms dissolved into the first layer of both the magnetic and nonmagnetic layers. We find that CPP-GMR is much larger than CIP-GMR and that the former decreases with an increase in the degree of roughness, while the latter increases. Similar results have been obtained for the recursive Green's function method [106,107].

The reason that CPP-GMR is much larger than CIP-GMR is that the current in CPP geometry always traverses the layer and both the spin-dependent potential and the effect of matching/mismatching of DOS contribute to the MR. On the other hand, in CIP geometry, the effect of matching/mismatching of the DOS is less effective and the MR becomes smaller. Furthermore, the spin-independent contribution to the resistivity from the nonmagnetic layers decreases the MR. The decrease of the MR ratio in CPP-GMR with increasing roughness shown in Figure 2.19 may be understood as follows: without roughness (zero concentration), the matching and mismatching of ↓ spin and ↑ spin DOS gives rise to strong spin dependence of the resistivity, resulting in a high MR ratio. In contrast, the spin dependence of the resistivity becomes weak with increasing roughness.

The MR ratio including a realistic electronic structure has been calculated for Fe/Cr multilayers and is shown in Figure 2.20 as a function of nonmagnetic layer thickness. The electronic structure was calculated using a supercell method and a Linear Muffin Tin Orbital (LMTO) method without roughness [109,110]. The symbols [100] and [110] in Figure 2.20 indicate the stacking direction of the layer planes. Note that the definition of the MR ratio used is $MR = (\Gamma_P - \Gamma_{AP})/\Gamma_{AP} \times 100$ (%), where Γ is the conductance calculated by the

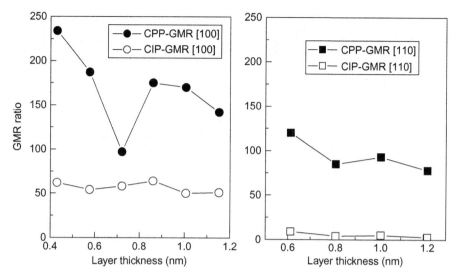

Figure 2.20 Calculated MR ratios for CIP-GMR and CPP-GMR in first-principles band calculations for Fe/Cr multilayers [108].

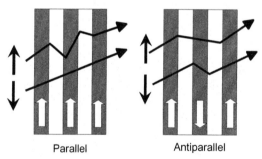

Parallel   Antiparallel

Figure 2.21 Schematic representation of spin-dependent resistivity in multilayers.

Landauer formula. We can see that the qualitative features are the same as those obtained in the simple model.

The microscopic theory has also been applied to granular GMR. The results show that the MR ratio increases with decreasing size of the grains, consistent with the experimental results.

2.4.4.4 Simple Picture of GMR

We have presented the mechanism of GMR in terms of spin-dependent resistivity and band matching/mismatching between magnetic and nonmagnetic layers. These origins are closely related to each other and present a simple picture for the mechanism of GMR in magnetic multilayers, as shown in Figure 2.21. In P alignment of the magnetization, one of the two spin channels has a low resistance (or resistivity) while the other has a high resistance. In AP alignment, both spin channels have high resistance (or resistivity), since both ↑ and ↓ spin electrons are scattered at some interfaces. As a result, the resistance in AP alignment is larger than that in P alignment.

2.4.5 Effects of Spin-Flip Scattering

So far, we have not taken spin-flip or spin-diffusion into account. It is easy to do so in phenomenological theory. Adopting the equivalent circuit model shown in Figure 2.16C, the resistivity in P and AP alignment may be given as:

$$\rho_P = \frac{2\rho_+\rho_-}{\rho_+ + \rho_-} \tag{2.30}$$

$$\rho_{AP} = \frac{2\rho_+\rho_- + \rho_{sf}(\rho_+ + \rho_-)}{\rho_+ + \rho_- + 2\rho_{sf}} \tag{2.31}$$

where ρ_{sf} is the contribution to the resistivity by spin-flip scattering. ρ_P does not include ρ_{sf}, since the voltage drop at the middle of the circuit is the same for ↑ spin and ↓ spin channels. In AP alignment, taking $\rho_{sf} \to 0$ gives $\rho_{AP} = \rho_P$ and the MR disappears because the current is shorted. For an infinite ρ_{sf}, there is no spin-flip effect and the MR ratio decreases with decreasing ρ_{sf}. (Note that $2\rho_+ \equiv \rho_\uparrow$ and $2\rho_- \equiv \rho_\downarrow$ according to Fert et al. [24].)

Valet and Fert [111] have incorporated the effects of spin-diffusion and layered structures by extending Boltzmann theory to include the spatial dependence of the distribution function in the nonequilibrium state. Under a finite voltage, they took into account spin accumulation and spin diffusion near the interfaces. They obtained general expressions for the spin-dependent resistivity, from which the following expressions are derived for simple cases:

$$J_s = \frac{\sigma_s}{e}\frac{\partial \bar{\mu}_s}{\partial x} \tag{2.32}$$

$$\frac{e}{\sigma_s}\frac{\partial \bar{\mu}_s}{\partial x} = \frac{\bar{\mu}_s - \bar{\mu}_{-s}}{\lambda_{spin}^2} \tag{2.33}$$

with

$$\bar{\mu}_s = \mu_s - eV(x) \tag{2.34}$$

where J_s, σ_s, and μ_s are the current, conductivity, and chemical potential of the spin s state, respectively, and λ_{spin} is the spin-diffusion length.

Equation (2.32) is simply Ohm's law and Eq. (2.33) indicates a balance between the spin accumulation and spin-flip rate. Using the parameters given in

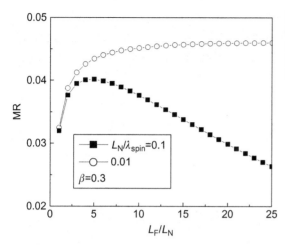

Figure 2.22 Calculated MR ratio in the Valet−Fert model. L_F and L_N are the thicknesses of the ferromagnetic and nonmagnetic layers, respectively.

Figure 2.22, the MR ratio is calculated for an F/N/F trilayer as a function of L_F/L_N, where L_F and L_N are the thicknesses of the F and N layers, respectively. We can see that the MR ratio becomes small when the spin-diffusion length is short. Since we are considering a trilayer, the MR ratio defined by Eq. (2.26) is very small. A nonzero value of $\bar{\mu}_s - \bar{\mu}_{-s}$ means that the chemical potential becomes spin dependent near the interface, due to the difference between the spin-dependent resistivity in the F and N layers, and indicates the existence of spin accumulation near the interface.

2.5   Tunnel Magnetoresistance

TMR in ferromagnetic tunnel junctions (FTJs) was reported prior to the discovery of GMR [112,113]. The observed MR ratios, however, were rather small at the time. A large TMR at room temperature was reported for Fe/Al−O/Fe in 1995 [114,115], and it has attracted considerable attention due to its wide potential application in sensors and memory storage devices in the near future. In the same year, tunnel-type MR was reported for metal-oxide ferromagnetic granular films [116,117].

Many theoretical studies have so far investigated tunnel resistance and TMR [11,118−122]. Since TMR is caused by tunneling current through an insulating barrier, it is sensitive to physical factors such as temperature, voltage, and the thickness and energy height of the barrier; it is also affected by scattering mechanisms. It is therefore vital to clarify the effects of each factor on TMR in order to develop technological applications of TMR.

In this section, we first overview the experimental results of TMR together with a phenomenological theory, and then describe the results obtained using microscopic theories. We also touch upon some recent experiments.

2.5.1 Ferromagnetic Tunnel Junctions

FTJs are made of a thin (about 1-nm thick) nonmagnetic insulator sandwiched between two ferromagnetic electrodes. A schematic diagram of such a junction is shown in Figure 2.23. The ferromagnetic metals used are predominantly Fe, Co, and their alloys, while amorphous Al_2O_3 is one of the most stable materials for the insulating barrier. Recently, a single-crystal MgO layer has been used as the barrier in order to generate high TMR ratios.

The resistivity of the FTJ is reduced when the magnetization alignment of two ferromagnets is changed from AP to P alignment. This is similar to GMR, and the MR ratio is defined as:

$$MR = \frac{\rho_{AP} - \rho_P}{\rho_{AP}} \qquad (2.35)$$

An alternative definition is often used in experimental studies; in this definition ρ_{AP} in the denominator is replaced with ρ_P.

2.5.2 Experiments for TMR

Figure 2.24 shows the experimental results for the resistance in an Fe/Al−O/Fe tunnel junction as a function of the external magnetic field [114]. The resistance is

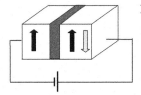

Figure 2.23 A schematic representation of the FTJ.

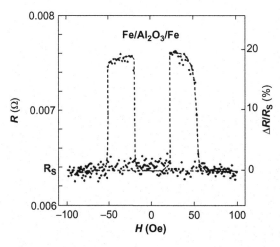

Figure 2.24 Experimental results for the resistance in an Fe/Al−O/Fe tunnel junction [114].

high when the magnetization of the two ferromagnetic electrodes is AP, and it is low when the magnetization is parallel. One of the characteristics of TMR is that the external magnetic field required to rotate the magnetization is sufficiently low. This is because there is almost no coupling between the magnetizations of the two electrodes as a result of the insulating barrier inserted between them. The AP and P alignments of the magnetization are realized by using a small difference in the coercive force between the two ferromagnets. The current flows perpendicular to the layer planes, which is similar to CPP-GMR. The resistance in the FTJs is much higher than that in CPP-GMR. This makes it possible to measure the junction resistance without microfabricating the samples. This could be considered to be another characteristic of TMR.

2.5.3 A Phenomenological Theory of TMR

2.5.3.1 MR Ratio and Spin Polarization

The experimental results for TMR can be understood phenomenologically as follows. Let us denote the left and right electrodes as L and R, respectively. When the tunneling process is independent of the wave vectors of tunneling electrons, the tunnel conductance Γ is proportional to the product of the densities of states of the L and R electrodes, and is given by $\Gamma \propto \sum_s D_{Ls}(E_F) D_{Rs}(E_F)$, where s denotes the spin [112,123]. The proportionality constant includes the transmission coefficient of electrons through the insulating barrier. Henceforth, the Fermi energy E_F is omitted for simplicity.

By using this expression, the conductance for P magnetization alignment is given by $\Gamma_P \propto D_{L+}D_{R+} + D_{L-}D_{R-}$, and that for AP magnetization alignment is given by $\Gamma_{AP} \propto D_{L+}D_{R-} + D_{L+}D_{R-}$. Since the resistivity $\rho_{P(AP)}$ corresponds to $\Gamma_{P(AP)}^{-1}$, the MR ratio is given by:

$$\text{MR} = \frac{\Gamma_{AP}^{-1} - \Gamma_P^{-1}}{\Gamma_{AP}^{-1}} = \frac{2P_L P_R}{1 + P_L P_R} \tag{2.36}$$

where $P_{L(R)}$ is the spin polarization of L(R) electrodes and is defined by:

$$P_{L(R)} = \frac{D_{L(R)+} - D_{L(R)-}}{D_{L(R)+} + D_{L(R)-}} \tag{2.37}$$

Although the transmission coefficient governs the magnitude of the tunnel conductance, it does not appear in the expression for the MR ratio.

An intuitive picture of the tunneling process explained above is shown in Figure 2.25, where in P alignment, majority and minority spin electrons in the L electrode tunnel through the barrier into the majority and minority spin states in the R electrode, respectively. In AP alignment, however, the majority and minority spin electrons in L electrode tunnel into the minority and majority spin states in R

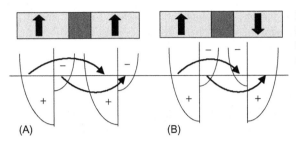

Figure 2.25 A schematic representation to show the tunneling process in FTJs for (A) P and (B) AP alignments.

Table 2.3 Spin Polarizations Observed in Tunnel Junctions and Point Contacts for Various Metals [124−128]

Materials	Tunnel Junctions	Point Contacts
Fe	+ 0.40	0.42
Co	+ 0.35	0.37, 0.42
Ni	+ 0.23	0.32, 0.43
$Ni_{0.8}Fe_{0.2}$		0.37
Cu		0.0
NiMnSb		0.58
$LaSrMnO_4$	+ 0.70	0.78
CrO_2		0.90

electrode, respectively. The difference between the conductances of the P and AP alignments gives rise to TMR.

2.5.3.2 Spin Polarization

Equation (2.36) indicates that the spin polarization of the electrodes govern the MR ratio. The spin polarization has been experimentally determined by using junctions of ferromagnet/Al/superconductor, or by analyzing the tunneling spectrum obtained using point contacts [124−130]. Measured values of the spin polarization are shown in Table 2.3. Using these values, the experimentally measured MR ratios are explained rather well. For example, the MR ratio measured for Fe/Al−O/Fe junctions is about 0.3 and this value is close to the theoretical value calculated by using the experimental value of P for Fe. However, the MR ratio observed for a single-crystal Fe/MgO/Fe junction is 0.7−0.8, which cannot be explained in terms of the spin polarization of Fe.

It is also difficult to explain the spin polarization P by using a first-principles calculation. The ratios of P deduced from the height of the DOS at the Fermi energy are inconsistent with experimentally measured ones, for example, theoretical values for P are negative for Co and Ni, while experimental ones are positive. Bulk Fe has a positive value of P, while surface Fe has a negative value. In order to explain TMR, one should study the spin polarization of the tunnel conductance itself.

2.5.4 Free-Electron Model

To investigate the tunneling process of electrons through a barrier, we first consider a one-dimensional free-electron model. Let us calculate the transmission probability for an electron injected from the left electrode with an energy E to tunnel through the barrier to the right electrode. Since the injection energy E is measured from the bottom of the energy band, it is spin dependent when the electrodes are ferromagnetic, and is expressed by E_s with $s = +$ (majority spin) or $-$ (minority spin).

Let Φ be the barrier potential determined from the Fermi energy and d be the barrier thickness, respectively, the decay rate of the wave function within the barrier is given by:

$$\kappa = \sqrt{2m\Phi}/\hbar \tag{2.38}$$

in a free-electron model, where m is the effective mass of tunneling electrons. The transmission coefficient may be given by:

$$T_s \simeq \frac{16E_s\Phi}{(E_s+\Phi)^2}\,\mathrm{e}^{-2\kappa d} \tag{2.39}$$

for $\kappa d \gg 1$. The spin dependence of the transmission coefficient appears in the prefactor of $\mathrm{e}^{-2\kappa d}$. Values of T_s are plotted as functions of Φ/E_+ for a given value of E_+/E_- in Figure 2.26, which shows that the sign of the spin polarization of the transmission coefficient depends on the value of Φ/E_+.

To evaluate the difference between the transmission coefficients for P and AP alignment of the magnetization, a generalized calculation should be performed [131,132]. By performing such calculations, the conductances Γ_P and Γ_{AP} for P and AP alignments of the magnetization respectively, can be obtained as [132]:

$$\Gamma_{P(AP)} \propto 1 + (-)P^2$$

$$P = \left| \frac{k_{F+} - k_{F-}}{k_{F+} - k_{F-}} \frac{\kappa^2 - k_{F+}k_{F-}}{\kappa^2 + k_{F+}k_{F-}} \right| \tag{2.40}$$

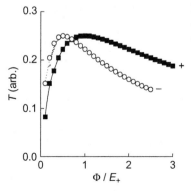

Figure 2.26 Calculated results for the spin-dependent transmission coefficient in the free-electron model for a ferromagnetic junction.

where k_{Fs} is the Fermi wave number for s-spin of the electrodes. Here, we have assumed that $\kappa d \gg 1$. The MR ratio defined as $MR = (\Gamma_P - \Gamma_{AP})/\Gamma_P$ gives $2P^2/(1 + P^2)$ for $P_L = P_R = P$.

It is possible to evaluate the value of P by calculating k_{Fs} in a first-principles calculation; however, the value of P is quite small (only a few percent) compared with experimentally measured values due to the factor $|\kappa^2 - k_{F+}k_{F-}|$. To make a quantitative comparison between the experimental and theoretical values of P, a more realistic formulation or numerical calculation of P is desired, one which employs realistic electronic structures and that incorporates the effects of disorder at, for example, junction interfaces.

2.5.5 Ingredients for TMR

In the simple theories described above, the origin of TMR is either the spin dependence of the DOS or that of the transmission coefficient. However, as shown below, realistic electronic structures and disorder at interfaces exert a larger effect on determining the magnitude of TMR [133–138].

2.5.5.1 Role of the Transmission Coefficient

In the free-electron model, the spin dependence of the transmission coefficient resides in the prefactor of $\exp(-2\kappa d)$, which is generated by the matching conditions of the wave functions of tunneling electrons at the interfaces. Although $\exp(-2\kappa d)$ is spin independent, it gives rise to a strong dependence on spin in the following way.

Equation (2.38) is the result obtained using a one-dimensional model, and it should be generalized to a three-dimensional case in such a way that

$$\kappa = \sqrt{2m(U - E(k_z))}/\hbar \tag{2.41}$$

with

$$E_F = E(k_\parallel) + E(k_z) \tag{2.42}$$

where U is the height of the barrier potential. We see that when $\kappa d \gg 1$, the electronic state which maximizes $E(k_z)$ (that is, minimizes $E(k_\parallel)$) makes the largest contribution to the tunneling conductance. Usually, such a state has $k_\parallel = (0,0)$, and the tunneling conductance calculated by taking into account this state only agrees with the result obtained using the one-dimensional model. For junctions with thin tunnel barriers, however, the contribution from states with wave numbers besides $k_\parallel = (0,0)$ becomes large. Thus, we find that the electrons injected perpendicular to the barrier planes make the biggest contribution to tunneling. This is merely the filter effect for the momentum of a tunneling electron.

Since the electronic states of ferromagnets depend on spin, the tunneling probability becomes spin dependent due to this momentum filter effect, and the decay rate may be written as $\kappa_s = \sqrt{2m(U - E_s(k_z))}/\hbar$. The spin-dependent κ_s can pro-duce high TMR ratios.

2.5.5.2   Effects of Fermi Surface

When a junction has no disorder and is translationally invariant along layer planes, the component of the wave vector $k_\parallel$ parallel to layer planes is conserved. This kind of tunneling process is referred to as specular tunneling. On the other hand, when $k_\parallel$ is not conserved, the tunneling process is referred to as diffusive tunneling. In this section, we consider specular tunneling.

Tunnel junctions are usually composed of an insulating barrier sandwiched between two different ferromagnetic metals. Therefore, the Fermi surfaces of the L and R electrodes are generally different. Since $k_\parallel$ is conserved in specular tunnel-ing, only states on the Fermi surface with the same $k_\parallel$ may contribute to tunneling.

In the free-electron model, the state with $k_\parallel = (0,0)$ is always included in the Fermi surface. However, this state may not be included in the complicated Fermi surfaces of, for example, TMs. In particular, ferromagnets have spin-dependent Fermi surfaces, and thus the state with $k_\parallel = (0,0)$ may be included in one spin state, but not in the other spin state. In this case, the tunneling conduc-tance is strongly spin dependent. Furthermore, the tunnel conductance in AP alignment becomes very small because the $k_\parallel = (0,0)$ state may not contribute to tunneling.

2.5.5.3   Symmetry of the Wave Function

In specular tunneling, the difference between the symmetries of the wave functions of tunneling electrons in L and R electrodes exerts a strong influence on TMR. The wave function of each state in metals has a specific symmetry, and electrons with a certain symmetry are not able to transfer into a state with different symmetry. Therefore, electrons on the L electrode can tunnel through the barrier into the R electrode only when the states specified by $k_\parallel$ in the L and R electrodes have the same symmetry; but this becomes impossible when the states are different. When the L and R electrodes are composed of different materials, the symmetry of the wave functions of tunneling electrons play an important role in tunnel conductance. In ferromagnets, the + and − electronic states are different in general. Therefore, when the magnetizations of the two electrodes are AP, the states on the Fermi sur-faces of the L and R electrodes can have different symmetries. In this case, the electrons in these states are not able to tunnel through the barrier. The situation makes the tunnel conductance for AP alignment much smaller than that for P align-ment, resulting in a large MR ratio.

Figure 2.27 shows the energy-momentum relation along the (001) direction for Fe and MgO [138,139]. It can be seen from this figure that the symmetry of the + spin band is Δ_1 on E_F for Fe, however, the symmetry does not appear in the − spin band. This situation is realized in Fe/MgO/Fe junctions.

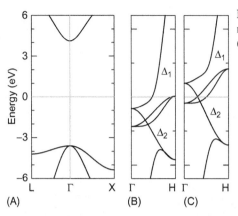

Figure 2.27 Electronic structure of Fe ((A) majority and (B) minority spin states) and (C) MgO along the (001) direction [138].

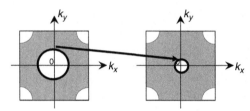

Figure 2.28 Projected Fermi surface on $k_\parallel = (k_x, k_y)$ plane of left and right electrodes.

2.5.5.4 Effect of Interfacial States

When a semiconductor (e.g., GaAs) is used as the barrier, interfacial states, called Shockley states, appear within the energy band gap. Since interfacial states are localized near the interface, they do not contribute to electron transport. For a thin semiconductor barrier, the states at one interface extend up to a few atomic layers inside the barrier and might overlap with those at the other interface. When the Fermi level is located within the energy region where the interfacial states exist, the tunnel conductance may be strongly enhanced by the interfacial states, thus effectively reducing the thickness of the barrier [140].

2.5.5.5 Effect of Electron Scattering

Let us now consider diffusive tunneling, which occurs when some disorder is present in junctions, and the wave vector $k_\parallel$ is not conserved. Interfacial roughness and amorphous-like insulators break translational invariance parallel to layer planes. Therefore, the parallel component of the wave vector no longer has to be conserved; that is, the wave vector $k_\parallel$ of incident electrons need not coincide with that $k'_\parallel$ of the transmitted electrons. In this section, we consider a situation in which the Fermi surfaces of the L electrode differ from that of the R electrode, as shown in Figure 2.28. In the specular tunneling case, states with Fermi wave vectors $k_{\parallel L} > k_\parallel > k_{\parallel R}$ cannot contribute to tunneling. In the diffusive tunneling case, on the other hand, this restriction is removed, and therefore those states having wave

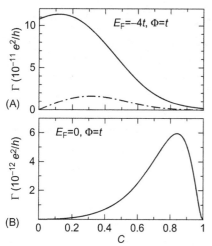

Figure 2.29 Calculated results of tunnel conductance as a function of roughness at interfaces of a tunnel junction for (A) $E_F = -4t$ and (B) $E_F = 0$, where c denotes the concentration of disordered atoms at the interface [138].

vectors $k_\parallel$ between $k_{\parallel L}$ and $k_{\parallel R}$ may contribute, which increases the tunnel conductance. Even when the Fermi surfaces of the L and R electrodes are the same and do not include the $k_\parallel = (0,0)$ state, tunneling electrons take the $k_\parallel = (0,0)$ state virtually in the tunneling process. The virtual process reduces the decay rate κ and consequentially increases tunnel conductance [136–138].

Let us consider some practical cases in which disorder exists between the electrodes and barrier layers. Here we adopt a single-band tight-binding model (with a hopping integral t) and calculated the tunnel conductance by varying the magnitude of the disorder by the following procedure. The disorder is introduced in such a way that the atomic layer at each interface includes a random distribution of two types of atoms that constitute the electrodes and barrier. The tunnel conductance Γ is plotted as a function of the concentration c of the barrier atom in Figure 2.29 for two different Fermi levels. The barrier is 10 atomic layers thick when $c = 0$.

When the Fermi level is located at $E_F = -4t$, Γ increases slightly at first and then decreases with increasing c. When $E_F = 0$, Γ increases considerably, forming a peak, and then becomes much smaller than the value at $c = 0$. Note that Γ at $c = 1$ is one order of magnitude smaller than that at $c = 0$. The chained curve in Figure 2.29A shows the contribution from diffusive tunneling. When $E_F = 0$, the diffusive tunneling is dominant and the chained curve is almost the same as the solid curve. These results show that the conductance increases with increasing roughness. This is in strong contrast with metallic transport in which electrical conductivity decreases with increasing roughness.

Once we have calculated the tunnel conductance, it is straightforward to evaluate the MR ratio. Figure 2.30 shows the calculated results of the MR ratio as functions of the number of atomic layers in the barrier for $E_F = 0$ (squares) and $-5t$ (circles). The position of the Fermi level of the former and the latter is close to the middle and bottom of the band, respectively. Filled symbols show the results without roughness, and open ones show those with roughness.

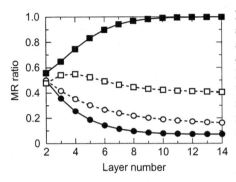

Figure 2.30 Calculated results of MR ratio as functions of barrier thickness (number of atomic layers) for $E_F = -5t$ (filled and open circles) and $E_F = 0$ (filled and open squares). Filled and open symbols denote the results without ($c = 0$) and with ($c = 0.5$) randomness, respectively. Results with randomness are plotted as functions of the average thickness of the barrier [137,138].

We see from the figure that the MR ratio depends strongly on the position of E_F. For $E_F = 0$, the state with $k_\parallel = (0,0)$ is occupied by only minority $(-)$ spin electrons. Therefore, the tunnel conductance for$-$spin electrons becomes large, and the MR ratio is close to the maximum value of 1. With increasing roughness, majority $(+)$ spin electrons in the state with $k_\parallel = (0,0)$ begin to contribute to tunneling, and the MR ratio decreases considerably. For $E_F = -5t$, the state with $k_\parallel = (0,0)$ is occupied by both $+$ and $-$ spin electrons, and the effect of roughness is smaller in this case.

2.5.5.6 Spin-Flip Tunneling

The results presented thus far were obtained by assuming that the spin of tunneling electrons is conserved. When there exist impurity spins within the barrier or near the interfaces, the tunneling electrons may interact with these impurity spins and reverse their spins in the tunneling process. This kind of spin-flip tunneling may reduce the MR ratio. Interaction between conduction electrons and localized spins in electrodes may also produce similar effects.

In the case when tunneling electrons interact with localized spins and flip their spins, the expression for TMR is easily obtained by using a simple model [141,142]. When the magnetizations of the two electrodes are canted by an angle θ, the tunnel conductance is given by:

$$\Gamma = \Gamma_0\{1 + \gamma\langle m^2\rangle\}(1 + P_L P_R\cos\theta) + \gamma\langle \ell_+^2\rangle(1 - P_L P_R\cos\theta)F(\beta\Delta)$$

where

$$F(\beta\Delta) = \frac{\beta\Delta}{1 - e^{-\beta\Delta}} \tag{2.44}$$

$$\langle m^2\rangle = S(S+1) + \langle m\rangle\coth(\beta\Delta/2) \tag{2.45}$$

$$\langle \ell_+^2 \rangle = \{\coth(\beta\Delta/2) - 1\}\langle m \rangle \tag{2.46}$$

$$\Delta = 2k_B T_C \langle m \rangle / S(S + 1) \tag{2.47}$$

where $\beta = 1/k_B T$ and S is the spin of the localized electrons. T_C is an effective Curie temperature of the localized spins, and $\langle m \rangle = SB_S(\beta\Delta)$ with the Brillouin function B_S. γ denotes the degree of spin-flip tunneling. When $\gamma = 0$, the expression above gives the usual expression for the MR ratio.

Numerical results for the temperature dependence of the MR ratio calculated by accounting for the spin-flip tunneling are presented later.

2.5.5.7   Voltage Dependence

The voltage drop in the tunnel junctions occurs at the insulating barrier since the resistance is highest at the barrier. When a voltage V is applied, the chemical potential shifts between the left and the right electrodes in such a way that $|\mu_L - \mu_R| = eV$. Since electrons between μ_L and μ_R contribute to tunneling in this case, a higher-order effect of V on the tunneling conductance occurs, and the spin dependence of the electronic states of the ferromagnets is averaged. As a result, the MR ratio is usually reduced. When the left and right electrodes are made from different materials, the voltage dependence of the tunnel conductance becomes asymmetric.

When tunneling electrons interact with elementary excitations such as phonons, magnons, they tunnel through the barrier being accompanied by these excitations. Because the tunneling electrons lose their energy by interacting with these excitations, a characteristic feature appears in the current–voltage relation, which gives useful information on the excitation [143].

2.5.6   TMR in Various Systems

2.5.6.1   Fe/MgO/Fe

Calculations of MR ratios for disorder-free Fe/MgO/Fe FTJs have been performed by using a first-principles method and a realistic tight-binding model [134,135]. The calculated results show that an extremely high MR ratio may be realized in clean samples. Being inspired by the results, single-crystal Fe/MgO/Fe samples were fabricated [144,145] and high TMR ratios have been observed [146–148]. Figure 2.31 shows one of the experimental results for the MR ratio.

The experimentally observed MR ratio is much higher than that predicted by the phenomenological model, and that elucidated using the principles mentioned in the previous subsection. First, the symmetry of the conduction bands of Fe plays an important role. The conduction band of + spin states of Fe has Δ_1 symmetry, which is composed of $s, p_z, d_{3z^2-r^2}$ orbitals, and contains the state with $k_\parallel = (0,0)$.

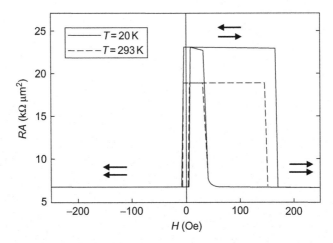

Figure 2.31 Experimental results for tunnel resistance as a function of external magnetic field [146].

The Δ_1 band hybridizes with the s and p_z atomic orbitals of MgO. On the other hand, the conduction band of $-$ spin states of Fe has Δ_2 symmetry composed of $d_{x^2-y^2}$ and does not contain the state with $k_{\parallel} = (0,0)$, nor hybridize with the s and p_z atomic orbitals of MgO. Therefore, the decay rate of the wave function of the $-$ spin electrons is much larger than that of $+$ spin electrons, and as a result, the transmission coefficient of $+$ spin electrons is much larger than that of $-$ spin electrons. This gives rise to a large MR ratio. The symmetry of the bands and the existence of $k_{\parallel} = (0,0)$ state in the conduction bands exert a strong influence on the MR ratio. When the MgO layer is thin, the state with $k_{\parallel} \neq (0,0)$ also contributes to tunneling, and the dependence of the MR ratio on MgO thickness is not simple in general.

When there is roughness in the junction, the momentum $k_{\parallel}$ need not be conserved. Then tunneling via the $k_{\parallel} = (0,0)$ state becomes possible for $-$ spin electrons, and the tunnel conductance of $-$ spin electrons increases resulting in a reduction in the spin asymmetry of the tunnel conductance, which causes the MR ratio to decrease. MR ratios calculated using the full tight-binding model that includes the effect of roughness are shown in Figure 2.32. It can be seen that the MR ratio decreases with increasing roughness. The MR ratio ~ 0.8 in Figure 2.32 corresponds to a measured MR ratio that is about 400% in the optimistic definition of the MR ratio [138,139]. Recently, similar high MR ratios have been reported.

Thus we understand how a high MR ratio is realized in clean Fe/MgO/Fe junctions. However, high MR ratios also occur in disordered ferromagnetic electrodes such as FeCoB, which has an amorphous structure [149]. Possible reasons for this may be that B atoms may reside on regular sites in the fcc structure after annealing, and the electronic states of B atoms have energies that are distant from the Fermi energy.

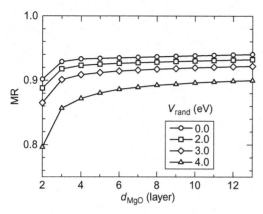

Figure 2.32 MR ratio calculated for Fe/MgO/Fe junctions with disorder as a function of MgO thickness [138].

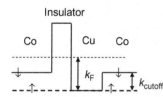

Figure 2.33 Schematic figure of the potential profile for a Co/insulator/Cu/Co junction in an AP alignment of the magnetization.

2.5.6.2 Oscillation of TMR

When a nonmagnetic layer is inserted between the barrier layer and one of the ferromagnetic electrodes, the TMR decreases rapidly with increasing thickness of the nonmagnetic layer [150−152]. The rapid decrease in the TMR ratio has been explained by taking into account the change in the electronic states near the interface [153]. However, an oscillation of TMR has been reported for samples in which an nonmagnetic layer is inserted between the insulator and one of the two ferromagnets [154,155]. The composition of the junction layers is NiFe/Al−O/Cu/Co, in which the Al−O layer is amorphous, but the Co and Cu layers are single crystals. Yuasa *et al.* [154] observed an oscillation in the MR ratio as a function of the thickness of the Cu layer. The results showed that the oscillation period is given by the Fermi wavelength of Cu, and that the MR ratio oscillates around zero.

These observations seem to be reasonable, but they contradict theoretical predictions [156,157]. Let us consider a Co/Al−O/Cu/Co junction for simplicity. A schematic representation of the potential profile is shown in Figure 2.33. A quantum well is formed in the Cu ↓ spin state when the magnetization of the right Co layer is AP to that of the left Co layer. In Figure 2.33, the left-hand-side level indicates the + spin potential of Co, and the right-hand-side level indicates the − spin potential of Co. Since the Cu potential matches the potential of the Co + spin state, the situation shown in Figure 2.33 is realized.

Now, the theoretical predictions for clean junctions in which $k_{\parallel}$ is conserved in the tunneling process are as follows:

1. $k_{\parallel}$ states on common Fermi surfaces of the L and R electrodes contribute to tunneling.

2. Since $k_\parallel = (0,0)$ contributes to tunneling in Co, its contribution is the largest.

3. The filtering effect of $k_\parallel = (0,0)$ state gives rise to an oscillation in the conductance with a period of k_F of Cu.

4. An oscillation due to the cutoff wave vector k_{cutoff} shown in Figure 2.33 appears in addition to the oscillation due to k_F.

Summing up these results, theory predicts that more than two periods appear in TMR oscillations and that the MR ratio oscillates about a finite value.

These theoretical results do not agree with experimental observations. These results, however, are modified when roughness is present, and become as follows [158]:

• Since $k_\parallel$ need not be conserved, in principle all states of the Fermi surface projected on $k_\parallel$ plane contribute to tunneling, and the conductance increases. In particular, tunneling via quantum-well states becomes possible. Therefore, tunnel conductance in AP alignment of the magnetization Γ_{AP} increases, which makes the MR ratio small. The important point is not just the existence of a quantum-well state, but how the quantum-well state is formed. As shown in Figure 2.33, the quantum-well state is formed by changing the magnetization alignment of the Co layers, and it is formed in the Cu layer inserted between the insulating barrier and the right Co layers.

• Due to the appearance of tunneling paths via quantum-well states, the contribution to the oscillation period by k_{cutoff} decreases. This makes the oscillation period determined by k_F dominant.

With increasing roughness, the average value of the MR ratio gradually approaches zero, and its period tends to the value determined by k_F. Figure 2.34 shows results calculated using a simple model for MR ratios for the cases with and without roughness. We see that the multiperiod oscillation becomes a single long period, and that the MR ratio oscillates about zero when disorder is introduced to the system. The calculated results show how specular tunneling crosses over to diffusive tunneling. In addition, the intensity of the MR ratio decays according to the following equation:

$$MR = \frac{1}{L_{NM}} \exp\left(-\frac{\lambda_{NM}}{L_{NM}}\right) \tag{2.48}$$

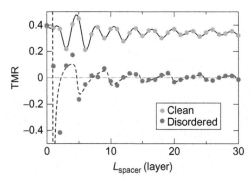

Figure 2.34 Calculated results of the MR ratio as functions of the spacer thickness obtained for clean (open circles) and disordered (filled circles) junctions. The solid curve is a visual guide, while the dashed curve is the result obtained using the stationary phase approximation [158].

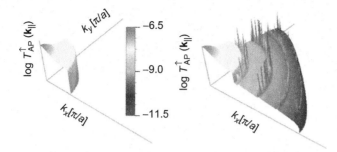

Figure 2.35 Momentum resolved transmission probabilities in the AP magnetization configuration for (A) clean and (B) disordered junctions [158].

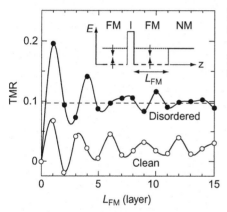

Figure 2.36 Calculated results for the TMR ratio for a tunnel junction of FM/I/FM/NM, the potential profile of which is shown in the inset. The MR ratio of the disordered junction is larger than that of the clean junction [159].

where L_{NM} is the thickness of the nonmagnetic layer (Cu) and λ_{NM} is the mean free path of electrons [158]. Figure 2.35 shows the new tunneling paths appears via the quantum-well states. Spikes shown in Figure 2.35B indicates the transmission coefficients appeared due to the disordered effects.

The theoretical result mentioned above shows that the roughness reduces the MR ratio due to additional tunneling paths via quantum-well states in AP alignment. It should be noted that this reduction in the MR ratio is not a general result, since an increase in the MR ratio can be realized depending on the type of the quantum well [138].

Let us consider the quantum well shown in the inset of Figure 2.36, where the quantum well is formed in the left ferromagnet between the insulator barrier and the nonmagnetic electrodes. In P alignment of the magnetization, the quantum well is formed in the ↑ spin state, while it is formed in the ↓ spin state when the magnetization of the right ferromagnet is reversed. Since, in P alignment, there is poor matching of the ↑ spin Fermi surfaces between the left FM and right NM electrodes, roughness increases the ↑ spin conductance due to tunneling channels via quantum-well states. In AP alignment, the matching of the ↓ spin Fermi surface of the left electrode and the right NM electrode is already good, and thus opening new channels via quantum-well states is ineffective. As a result, roughness increases the tunnel conductance in P alignment giving rise to an increase in the MR ratio. Figure 2.36 shows the calculated results for the MR ratio with and without roughness for the quantum-well state depicted in the inset. It shows that the MR ratio increases when roughness is present.

2.5.6.3 Tunnel Junctions with Half-Metals

The DOS of metallic ferromagnets (MFs) are usually spin dependent. When the DOS of either a $\uparrow$ or $\downarrow$ spin state is zero at the Fermi level, and one of two spin states is metallic and the other is insulating, the metals are referred to as half-metals. The spin polarization P of these half-metals is 100%, and therefore half-metals have potential applicability as magnetoresistive devices.

Many oxides, including CrO_2, Fe_3O_4, and perovskite $LaSrMnO_3$, have been shown to be half-metallic by using first-principles band calculations [160−162]. The first theoretical prediction for half-metallicity was done for Heusler alloys [160], which contain TM elements. Recently, it has been shown using the first-principles method that diluted magnetic semiconductors (DMSCs), such as (GaMn)As, may also be half-metallic. In experiments in which point contacts and tunnel junctions were used to measure spin polarization, lower values than 100% were obtained for P (e.g., 90% for CrO_2, 70−85% for $LaSrMnO_3$, and 60% for Heusler alloys) [163−165]. Recently, relatively high MR ratios have been observed in FTJs with Heusler alloys, suggesting that the value of P is about 86% [166,167].

In the following, we briefly review the electronic states of half-metals, and give some experimental results for TMR in FTJs with half-metals.

FTJs with Manganites

Perovskite manganites $(La-Sr)MnO_3$ and $(La-Ca)MnO_3$ show metallic ferromagnetism when La^{3+} ions in $LaMnO_3$, which is an antiferromagnet, are replaced with Sr^{2+} or Ca^{2+} ions [168,169]. Let us first understand the origin of metallic ferromagnetism in the oxides [170,171]. Since oxygen ions are always divalent, Mn ions are Mn^{3+} in $LaMnO_3$, and Mn^{4+} ions appear on replacing La ions with Sr or Ca. The Mn ions in the perovskite structure are surrounded by six oxygen ions which form a cubic octahedron, and therefore the degenerate d-levels are split into a doubly degenerate e_g state and a triply degenerate t_{2g} state (crystal field splitting) as shown in Figure 2.37A.

Four electrons in a Mn^{3+} ion occupy the d-level, as shown in Figure 2.37A to satisfy strong Hund's rule coupling. (The Hund's rule coupling energy is stronger than the energy splitting of the d-levels caused by the crystal field.) Since the orbitals in the t_{2g} state are d_{xy}-, d_{yz}-, and d_{zx}-orbitals, and they overlap weakly with oxygen p-orbitals, three electrons in the t_{2g} state behave as localized spins with $S = 3/2$ (Figure 2.36B). One electron in the e_g state cannot be itinerant because the Coulomb interaction is strong in $LaMnO_3$. That is, $LaMnO_3$ is a Mott insulator. When La ions are replaced with Sr or Ca ions and Mn^{4+} ions appear, the

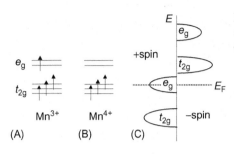

Figure 2.37 Electronic states of (A) Mn^{3+} and (B) Mn^{4+}, and schematic representation of the d-DOS manganites.

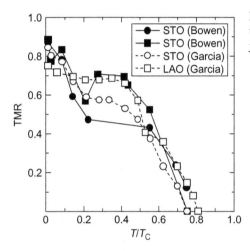

Figure 2.38 Temperature dependence of the MR ratio for manganite tunnel junctions [172–174].

e_g electrons on Mn^{3+} ions may hop onto vacant e_g states in Mn^{4+}. In this case, the spin of the e_g electron should be parallel to the t_{2g} localized spin on Mn^{4+} ion, because of the strong Hund's rule coupling. Since hopping of e_g electrons lowers their kinetic energy, a parallel alignment of the localized spins of neighboring Mn ions is favorable. This is a simple picture of the so-called double exchange interaction.

Because the ferromagnetism of perovskite manganites occurs due to strong Hund's rule coupling between e_g electrons and t_{2g} localized spins, only up-spin electrons exist in the e_g state, and down-spin states appear in the high-energy region. As a result, a half-metallic state is realized in the e_g state as shown in Figure 2.37C. The value of P measured for $(LaSr)MnO_4$ from the Andreev reflection is about 70%, the MR ratio of a tunnel junction made of $(LaSr)MnO_4$ is 1800% at low temperatures (using the optimistic definition), indicating that $(LaSr)MnO_4$ is almost half-metallic [172].

The TMR ratio of manganite tunnel junctions, however, decreases rapidly with increasing temperature and becomes almost zero near room temperature [172–177]. Some of the experimental results are shown in Figure 2.38, in which the pessimistic definition of the MR ratio is used. Here, the barrier materials are

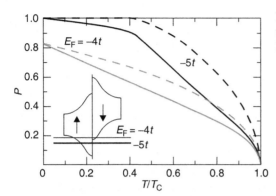

Figure 2.39 Theoretical results of the temperature dependence of the MR ratio [178].

either SrTiO$_3$ (STO) or LaAlO$_3$ (LAO). A dip of the MR ratio near $T/T_C \sim 0.2$ may be due to structural change in the barrier materials. The result that the MR ratio becomes zero at temperatures below T_C is another characteristic of manganite tunnel junctions.

Figure 2.39 shows the theoretical results for the temperature dependence of the MR ratio calculated using a simple model which takes into account the double exchange interaction [178]. Hund's rule coupling has been expressed as a ferromagnetic exchange interaction $K\mathbf{s}_i \cdot \mathbf{S}_i$ on site i with an interaction constant K. The spin of the tunneling electron may flip because of the exchange interaction, resulting in effective spin-flip tunneling. This reduces the MR ratio with increasing temperature, that is, increasing spin-fluctuations of the localized spins. This decrement, however, is not so strong as that observed at low temperatures. The rapid decrease in the MR ratio observed at low temperature may be due to a change in the electronic and magnetic states of manganites at the interfaces.

Hybrid tunnel junctions have been made using Co and manganite as ferromagnetic electrodes. In this case, the sign of the MR ratio depends on the barrier materials. Junctions with Al$_2$O$_3$ exhibit a negative MR, which is usual; however, those with SrTiO$_3$ have positive MR [179,180]. This may be attributed to the difference between the symmetry of the wave functions in Al$_2$O$_3$ and SrTiO$_3$, which changes the spin-dependent decay rate.

Tunnel Junctions with Heusler Alloys

The half-metallic characteristic of the electronic states in MFs was first predicted for NiMnSb in a first-principles band calculation. Later, it was shown in a first-principles band calculation that many Heusler alloys have half-metallic DOS. There are two types of structures in Heusler alloys, namely half-Heusler with an XYZ lattice and full-Heusler alloys with an X$_2$YZ lattice, where X = Ni, Co, Pt; Y = Cr, Mn, Fe; and Z = Sb, Ge, Al, etc. The basic lattice structure is composed of two nested bcc lattices. The half-Heusler lattice has vacant sites, which are occupied by X atoms in full-Heusler alloys.

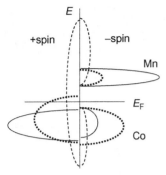

Figure 2.40 A schematic DOS of a full-Heusler alloy.

Let us look at the global features of the DOS of a full-Heusler alloy, for example, Co_2MnAl. The up- and down-spin states of Mn atoms are strongly exchange split, and the majority spin state is below the Fermi level. Although the minority spin band is located above the Fermi level, bonding bands appear in the low energy region due to strong pd hybridization. This strong pd hybridization produces a hybridization gap at the Fermi energy. Since the d-electron number of Co atoms exceeds that of Mn atoms, the majority spin state has unoccupied states. The strong pd hybridization also makes a gap in the local DOS of Co. A schematic figure of the half-metallic DOS is shown in Figure 2.40.

Thus, the origin of the half-metallicity of the Heusler alloys is the strong exchange splitting and pd hybridization. In experiments, however, the order of degree of ordered alloys is usually less than 1, and the value of P measured so far is much less than 100%. Recently, a tunnel junction made of full-Heusler alloys was produced and it exhibits a relatively high MR ratio at low temperature, indicating a substantial improvement in the quality of the sample. Figure 2.41 shows one of the experimental results [166,167,181]. Nevertheless, the temperature dependence of the MR ratios is rather strong, similar to that in manganite tunnel junctions.

There may be several reasons for the strong temperature dependence of the MR ratio. First, the electronic and magnetic states of Heusler alloys at the interface can differ from those in the bulk state [182]. Secondly, spin-flip tunneling might occur as in manganite tunnel junctions [142,183]. Figure 2.42 shows calculated results for the temperature dependence of the MR ratio in a phenomenological model that takes spin-flip tunneling into account. These results are rather similar to those shown in Figure 2.39; however, they are not sufficient to explain the strong temperature dependence at low temperatures observed in experiments.

Recently, quite high MR ratios have been observed in tunnel junctions using CoFeB ferromagnetic alloys and Heusler alloys [184–189]. These MR ratios at room temperature (RT) and helium temperature (LT) have been summarized in Table 2.4.

In addition, it should be noted that an oscillation of the TMR ratio observed in Ref. [148] has been confirmed by Matsumoto *et al.* [190] and Ishikawa *et al.* [189].

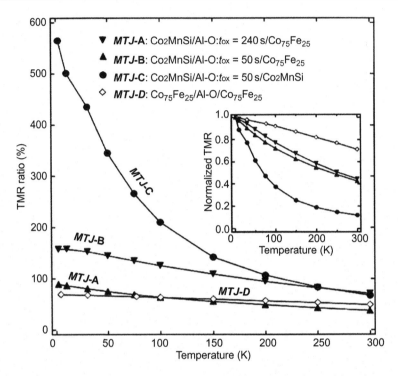

Figure 2.41 Experimental results for the temperature dependence of tunnel junctions made of Heusler alloys [167].

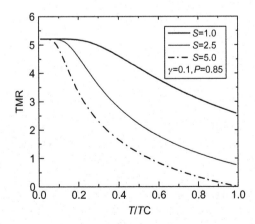

Figure 2.42 Calculated results for the temperature dependence of the MR ratio in a model that takes spin-flip tunneling into account.

Table 2.4 TMR Ratios Observed Using MgO Barrier with CoFeB and/or Heusler Alloys

Junctions	MR Ratios (%)		References
	LT	RT	
CoFeB/MgO/CoFeB	1010	500	[185,186]
$Co_2Cr_{0.6}Fe_{0.4}$/MgO/$Co_{50}Fe_{50}$	317	109	[187]
$Co_2FeAl_{0.5}Si_{0.5}$/MgO/$Co_2FeAl_{0.5}Si_{0.5}$	390	220	[188]
Co_2MnGe/MgO/$Co_{50}Fe_{50}$	376	160	[189]

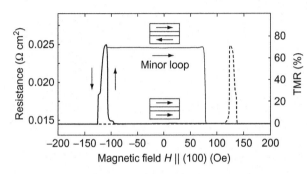

Figure 2.43 Experimental results for the TMR effect in tunnel junctions made of (GaMn)As [199].

Tunnel Junctions with DMSCs

Semiconductors are the essential materials for silicon-based technologies. Recently, the field of semiconductor spintronics has been developed; it is concerned with ways to control the spin degree of freedom of electrons in semiconductors [191]. One of the most direct methods to introduce spin degrees of freedom in semiconductors is to introduce magnetic ions into semiconductors. Such semiconductors are referred to as magnetic semiconductors, or as DMSCs when the magnetic ions are diluted.

Basic research into DMSCs of II−VI compounds has been performed for decades [192]. However, their Curie temperatures are not sufficiently high for technological applications and they are not metallic. In the late 1990s, DMSCs of III−V based compounds were successfully fabricated by introducing Mn ions and by using a low-temperature MBE technique [193−195]; using this technique, it is possible to grow crystals in a nonequilibrium state, preventing Mn ions from precipitation. The highest T_C achieved at that time was 110 K for 5.5% Mn doped (GaMn) As [196]. Recent fabrication techniques have succeeded in producing DMSCs of II−VI compounds that have T_C as high as approximately 200 K [197,198].

First-principles band calculations have shown that the electronic structure of (GaMn)As is half-metallic. Thus, a high MR ratio might be expected for tunnel junctions having electrodes made from (GaMn)As and a barrier of AlAs. The experimental results for the resistivity as function of the external magnetic field strength are shown in Figure 2.43[199]. Recently, MR ratios as high as 290% have been reported [200].

A different mechanism of the TMR, however, has recently been proposed. Gould *et al.* [201] have observed a spin-valve-like TMR in (GaMn)As/Al−O/Au junctions in which only one magnetic layer exists. The result has been attributed to a change in the DOS at the top of the valence band due to the magnetic anisotropy of (GaMn)As, and the phenomenon is called tunneling anisotropic magnetoresistance (TAMR). TAMR has been confirmed also for (GaMn)As/GaAs/(GaMn)As junctions [202,203]. Therefore, the previous results observed in DMSC junctions might be caused by a change in the relative direction between magnetization and crystal orientation. However, Saito *et al.* [204] have reported that the intrinsic TMR exists in (GaMn)As/ZnSe/(GaMn)As junctions and the contribution by TAMR can be one-tenth of the total MR.

2.5.7 Coulomb Blockade and TMR

2.5.7.1 TMR in Granular Magnets

A fairly large MR was observed in thin films in which magnetic grains of Co are imbedded in oxides such as Si−O [116,117]; this observation was made at almost the same time as when a large TMR was discovered in Fe/Al−O/Fe tunnel junctions. In order to distinguish these two tunnel MRs, we hereafter denote the TMR in junctions as junction TMR and that observed in granular systems as granular TMR.

Tunnel Conductance in Granular Systems

Paramagnetic granular systems exhibit a peculiar temperature dependence of the resistivity. Abeles *et al.* [205,206] have explained it in terms of a charging effect and structural characteristics. When an electron hops from one grain to a neighboring grain (see Figure 2.44), the distribution of charge changes, breaking the charge neutrality in the grains. The internal energy of these grains increases because of the Coulomb repulsion between electrons in the grains. This increase in the energy is the charging energy, which is written as E_C hereafter. A pair of grains may correspond to a condenser, and E_C depends on the size of the grains and on the distance between them. Therefore, E_C increases with decreasing size of the grains.

When the thermal energy $k_B T$ is smaller than E_C, an electron is prevented from hopping to neighboring grains due to an increase in the energy E_C; this is the so-called Coulomb blockade. At high temperatures, an electron is thermally activated to overcome the charging energy and is able to hop to neighboring grains. Therefore, the conductance of an electron between two grains is proportional to

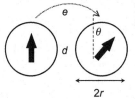

Figure 2.44 Tunneling of an electron between two grains, the magnetizations of which are canted by an angle θ.

$\exp(-E_C/k_BT)$. The transmission probability for an electron to tunnel through the barrier with a width d between two grains is proportional to $\exp(-2\kappa d)$, where $\kappa = \sqrt{2m^*(U - E_F)}/\hbar$, and where m^* is the effective mass of the electron, U is the barrier height, and E_F is the Fermi energy. The tunnel conductance between two grains is thus given by:

$$\Gamma \propto \exp(-2\kappa d - E_C/k_BT) \tag{2.49}$$

In granular films, the grain size (r) distributes, and therefore both d and r distribute. Nevertheless, Abeles *et al.* have explained the relationship between r and d in the following manner. Let us consider a relatively large region, say several 10^3 nm^3. Over such a large region, the ratio of the volumes of metallic grains and oxides will be equal to that of the entire sample with a fixed composition. In this case, large-sized grains should be surrounded by thicker oxide regions, that is, $r \propto d$. Since the charging energy is produced by the Coulomb interaction, it may be assumed that $E_C \approx e^2/r \propto 1/d$. This relation also holds when we consider the grain as a condenser. In this case, $E_C = e^2/2C \approx e^2d/2r^2 \propto 1/d$. In any case, one may write $E_C = C_0/d$, where C_0 is a constant. The conductance is then given by:

$$\Gamma \propto \exp(-2\kappa d - C_0/k_BTd) \tag{2.50}$$

We further assume that the distribution of d is sufficiently small compared with the d dependence of Γ. Under this assumption, the total conductance of the sample may be given by the conductance of a parallel circuit of these small regions. In this case, the current flows through regions that have the smallest resistance. That is, the total conductance is given by the conductance that maximizes $\exp(-2\kappa d - C_0/k_BTd)$ for a given temperature. The resultant conductance is given by:

$$\Gamma \propto \exp[-2\sqrt{2\kappa C_0/k_BT}] \tag{2.51}$$

This result shows that the resistance $R = 1/\Gamma$ increases as $\exp(1/\sqrt{T})$ with decreasing temperature, in good agreement with the experimental observations.

The temperature dependence shown above may be interpreted intuitively as follows. The Boltzmann factor $\exp(-E_C/k_BT)$ indicates competition between the charging energy and the thermal energy. At high temperatures, electrons may hop to smaller grains even when E_C is large. Therefore, electrons prefer to hop to smaller grains, since they have smaller tunneling rates $\exp(-2\kappa d)$. At low temperatures, on the other hand, electrons tend to hop to larger grains, which have smaller charging energies E_C. A schematic diagram depicting this situation is shown in Figure 2.45.

Granular TMR
The MR ratio of the granular TMR is given by [207]:

$$MR = \frac{\Gamma^{-1}(0) - \Gamma^{-1}(H)}{\Gamma^{-1}(0)} = \frac{P^2}{1 + P^2} \tag{2.52}$$

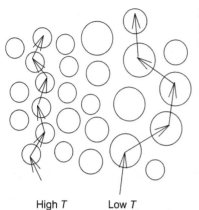

Figure 2.45 A schematic representation showing the hopping of electrons at high and low temperatures.

High *T*   Low *T*

which is just half the MR ratio of junction TMR. This is due to the random distribution of the magnetization direction of the grains when no external magnetic field is applied.

The above result is obtained as follows. Consider the two ferromagnetic grains L and R shown in Figure 2.44. The magnetizations of the R grain is canted by an angle θ with respect to that of the L grain. In this case, the transition probability for an electron on the L grain to hop to the same spin state on the R grain is given by $\cos^2(\theta/2)$, and that to the opposite spin state on the R grain is $\sin^2(\theta/2)$. Thus, the probability of an electron tunneling between L and R grains is given by:

$$T_{\uparrow\uparrow} = D_{L+}D_{R+}\cos^2(\theta/2) + D_{L+}D_{R-}\sin^2(\theta/2) \tag{2.53}$$

$$T_{\downarrow\downarrow} = D_{L-}D_{R-}\cos^2(\theta/2) + D_{L-}D_{R+}\sin^2(\theta/2) \tag{2.54}$$

where $D_{L+(-)}$ and $D_{R+(-)}$ are majority (minority) spin DOS of the L and R grains, respectively. The total tunneling probability is given by:

$$T = T_{\uparrow\uparrow} + T_{\downarrow\downarrow} Pp \propto 1 + P^2 \cos\theta \tag{2.55}$$

where

$$P^2 = \frac{(D_{L+} - D_{L-})(D_{R+} - D_{R-})}{(D_{L+} + D_{L-})(D_{R+} + D_{R-})} \tag{2.56}$$

The spin of the tunneling electron is assumed to be conserved in the tunneling process.

The tunnel conductance in the absence of a magnetic field is given by:

$$\Gamma(0) \propto \langle 1 + P^2 \cos\theta \rangle_{av} = 1 \tag{2.57}$$

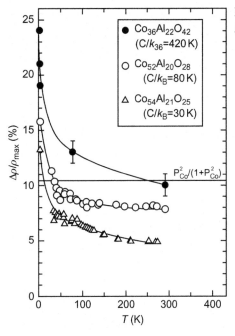

Figure 2.46 Experimental results for granular TMR [117].

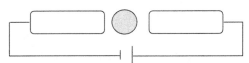

Figure 2.47 A schematic representation of a double tunnel junction for which the Coulomb blockade appears.

when an external magnetic field H is applied it is

$$\Gamma(H) = 1 + P^2 \tag{2.58}$$

We then obtain Eq. (2.52).

Figure 2.46 shows the experimental results for granular TMR [117]. They are consistent with theoretical results at high temperatures. In Figure 2.46, the MR ratio increases at low temperatures, which may be explained in terms of the Coulomb blockade.

2.5.7.2   Coulomb-Blockade TMR

Let us consider a well-controlled structure such as the one shown in Figure 2.47, in which a nanoscale grain is separated from two ferromagnetic electrodes by two insulating barriers. The grain can be either ferromagnetic or paramagnetic. When the grain is sufficiently small, the effect of charging energy, that is, the Coulomb blockade becomes important [208,209]. The terms "charging energy" and "Coulomb blockade" refer to essentially the same concept; here we adopt the latter terminology.

We assume that the grain has a charge of Ne. In this case, the electrostatic energy of the grain is given by:

$$U(N) = \frac{(Ne)^2}{2C} - Ne\varphi = \frac{1}{\sqrt{2C}}(Ne - C\varphi)^2 - \frac{C\varphi^2}{2} \tag{2.59}$$

where C is the effective capacitance of the grain and φ is the electrostatic potential. When no current flows, charge neutrality is satisfied, giving $Ne = C\varphi$. As the current begins to flow, and the number of electrons on the grain increases by one, the energy increases by $e^2/2C$. When the temperature is sufficiently low and the bias voltage is lower than the charging energy $e^2/2C$, no electron can hop onto the grain. This is the Coulomb blockade. With increasing bias voltage, the number of electron hopping onto the grain increases discretely. As a result, the current increases stepwise and the conductance changes oscillatory with increasing bias voltage. The oscillation of the conductance is referred to as the Coulomb oscillation. The current can be controlled by the gate voltage, and it is therefore possible to allow one electron to tunnel through the grain; such a device is called a single-electron transistor.

When the electrodes are ferromagnetic, TMR also appears and the MR ratio oscillates according to the Coulomb oscillation [210,211]. It is noteworthy that TMR occurs even when the grain is paramagnetic. This is because the tunneling rates of the two insulating barriers depend on the magnetization alignment of the electrodes. An example of the theoretical results of the oscillation of the TMR is shown in Figure 2.48, in which the grain is assumed to be paramagnetic.

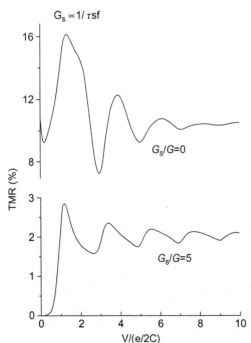

Figure 2.48 Theoretical results for the oscillation of TMR ratios in the Coulomb-blockade regime. Here the grain is assumed to be paramagnetic [211].

Now it may be questioned that a current never flows when the Coulomb blockade is effective. In reality, there exists a leakage current, which is caused by the following process. Since energy conservation need not apply over short time periods due to the uncertainty principle, a process appears to occur in which an electron moves from the grain to the R electrode almost simultaneously as another electron hops onto the grain from the L electrode. This tunneling process is called cotunneling, since two tunneling processes occur almost simultaneously. The tunnel rate of cotunneling is proportional to a product of the tunnel rates for R and L tunnel barriers. In this case, the MR ratio is twice as large as that for the usual TMR, since the spin dependence of the tunnel rates of both barriers contribute. The increase in the TMR ratio at low temperatures shown in Figure 2.46 may be explainable in terms of cotunneling [212].

2.6 Ballistic Magnetoresistance

When a current flows in a region that has a length scale shorter than the mean free path, the conductance becomes quantized. The quantization of the conductance was discovered by van Wees *et al.* [213] in a two-dimensional electron gas (2DEG). They controlled the width of a channel in which electrons flowed by applying a gate voltage to 2DEG, and observed that the conductance changed in a stepwise manner, in steps of $2e^2/h$.

This phenomenon of quantized conductance may be understood as follows. Constricted regions, in which a current flows, may be considered to be pseudo-one-dimensional. The electronic state of such constricted regions is continuous along the direction of current flow, but it is quantized perpendicular to the current flow. The quantization of the electronic state is characterized by the Fermi wavelength λ_F. When the width of the constricted region is close to λ_F, only one state is available for electrons, and therefore two electrons (having up and down spins) may contribute to a current, giving rise to a conductance of $\Gamma_0 = 2e^2/h$. If the width of the sample is increased, the conductance increases stepwise in steps of $\Gamma = 2e^2/h \times n$. A simple derivation of quantized conductance has been given in Section 2.3. The length scale of the constricted region is usually smaller than the mean free path, and the effect of scattering may be neglected. Such transport is referred to as ballistic transport.

Conductance quantization was first observed in semiconductors. Conductance quantization can be easily observed in semiconductors since λ_F is several 10 nm. By contrast, λ_F in metals is 1 nm, which makes the observation of conductance quantization difficult in conventional metals. In this section, we will briefly review (1) conductance quantization in paramagnetic metals, (2) conductance quantization in ferromagnetic metals, and (3) MR effects in ballistic transport, called BMR.

2.6.1 Conductance Quantization in Metals

2.6.1.1 Paramagnetic Metals

To realize ballistic transport in metals, the constricted region should be less than 1 nm. Potential methods for achieving this include the break-junction method in which a narrow wire is slowly bent to produce a small link just before breakdown of the wire, and the point-contact method in which the sharp tip of metallic wire is contacted onto the metal's surface. Fabrication of such a state in a controllable manner is quite difficult usually, making it necessary to make many repeated observations and then to take a statistical average in order to deduce meaningful conclusions.

In pseudo-one-dimensional systems such as 2DEG with gate control, the electronic state that is responsible for electrical conduction is well defined (it is called a channel). The quantized conductance $\Gamma_0 = 2e^2/h$ is specific to each channel. On the other hand, in the break-junction or point-contact methods in metals, each electronic state does not necessarily produce $\Gamma_0 = 2e^2/h$. This is because the constricted region in metals is too small to well define the wave vector in the current direction, that is, conservation of momentum is broken and scattering of electrons may occur in the constricted region. In other words, the constricted region itself is a scatterer, making the transmission coefficient for electrons passing through the region less than 1. Therefore, the conductance may be given by:

$$\Gamma = \sum_i \Gamma_0 \tau_i \tag{2.60}$$

where τ_i is the transmission coefficient of the ith channel.

When $\tau_i = 0$ or 1, the conductance Γ is an integer multiple of Γ_0. In metallic constrictions, since $0 < \tau_i < 1$, it is rather difficult to conclude that Γ is an integer multiple of Γ_0 from measurements. Scheer et al. [214,215] have performed realistic calculations of the conductance for point contacts and many experiments with break junctions and point contacts. They found that the characteristics of the channel in a single-atom contact is determined by the number and type of orbitals in the valence band. For example, s- and p-orbitals are responsible for conduction in Al, but d-orbitals are important in Nb. In every channel $0 < \tau_i < 1$, but the total conductance is approximately an integer multiple of Γ_0 after summing the contributions from many channels. Figure 2.49 shows the calculated results of the transmission coefficient for a single-atom contact. It shows that the transmission coefficient varies continuously with a change in the position of the Fermi energy.

2.6.1.2 Ferromagnetic Metals

The factor of 2 that appears in the quantized conductance Γ_0 indicates spin degeneracy. Since spin degeneracy is lifted in ferromagnets, conductance quantization might be expected to be given by $\Gamma_0/2 = e^2/h$. In order to confirm this expectation, many experiments have been performed with break junctions and point contacts made of ferromagnets. Figure 2.50 shows the average conductance observed for many break junctions made from Ni wires [216]. The results show that the quantized conductance is Γ_0 when an external magnetic field H of less than 50 Oe is applied, but it is $\Gamma_0/2$ when $H > 50$ Oe. The results are closely related to BMR described in the next section, and they are analyzed in detail there. Finally, it has

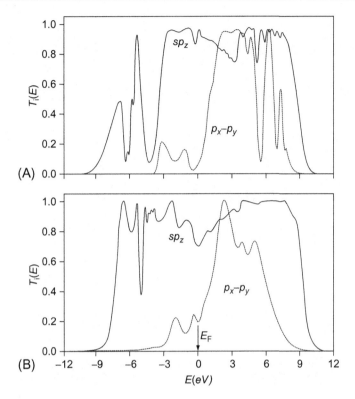

Figure 2.49 Calculated results for the transmission coefficient for Al one-atom contact in the two extreme cases of (A) short and (B) long necks [215].

been reported that a conductance quantization of $\Gamma_0/2$ was also observed in nonconstricted Au by using chemical potential control.

2.6.2   Experiment and Theory of BMR

2.6.2.1   Experiments

Garcia *et al.* [217] have measured the MR effect for Ni point contacts, and reported that the MR ratio becomes large when the conductance approaches Γ_0. The maximum MR ratio was 280% (for the pessimistic definition). The results are shown in Figure 2.51. The MR effect thus observed was termed BMR.

Later, Chopra and Hua [218,219] have reported that the MR ratio is 3000−10,000% for usual point contacts of Ni or those produced by electrodeposition. However, the resistance in their measurements is 10−100 Ω, which is much smaller than the inverse of the quantization conductance $\Gamma_0^{-1} = 12.9$ kΩ. Thus, the conductive properties of their samples may differ from those observed by Garcia *et al.* Furthermore, the MR ratio observed by electrodeposition can be either positive or negative.

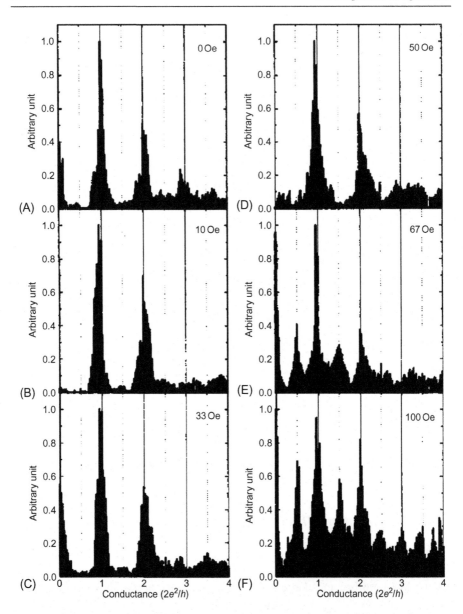

Figure 2.50 Experimental results of conductance quantization performed for many break junctions of Ni wires with and without an external magnetic field [216].

2.6.2.2 Interpretation of BMR

So far there have been several mechanisms proposed for BMR; however, complete understanding of this phenomenon has yet to be achieved. One of the theoretical interpretations of BMR is to attribute the large MR to the vanishing of the domain

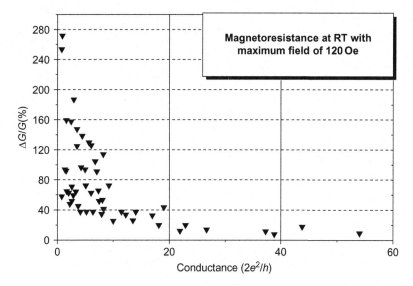

Figure 2.51 Experimental results of conductance change as a function of quantized conductance in point contacts [217].

walls (DWs) at the ferromagnetic constriction [220]. An AP alignment between the magnetizations of the ferromagnetic wire and the surface of a ferromagnetic film may produce a DW in the region of the contact. Bruno [221] has pointed out the width of a DW can be sufficiently narrow to produce a large MR effect. Tatara *et al.* have calculated the BMR ratio for narrow DWs and showed that the MR ratio increases with decreasing DW width. Both conductance quantization and shrinkage of the DW region are responsible for the large BMR. On the other hand, it should be noted that a realistic calculation of BMR gives 70% at most for a planar DW; that is for a DW in which the magnetization changes direction within a single atomic layer [222].

Imamura *et al.* [223] have studied the conductance quantization for nanocontacts of two ferromagnets, and showed that the parameter region at which a conductance quantization of $\Gamma_0/2 = e^2/h$ appears depends on the alignment of the magnetization of the two ferromagnets. A difference in the conductance quantization between the P and AP alignments of the magnetization may give rise to a large MR ratio.

These theories, however, have neglected realistic electronic states, such as multiorbitals. We have already mentioned that the quantization of conduction is strongly influenced by the number and type of orbitals. First-principles band calculations are useful to confirm this point. Such calculations have been done by the Mertig group for an atomic contact composed of transition atoms. Their results indicate that the electronic states responsible for conduction are s-like and that the spin dependence of the conduction is caused by a d-like DOS. The transmission coefficient for these states is less than 1 and changes continuously with a change in the

Fermi level. In addition, they confirmed that the MR ratio was not large and was at most 50% [224,225].

Other proposed mechanisms include a magnetoelastic mechanism and diazotization of the nanoconstricted region. While the former mechanism is easy to understand, it does not explain the experimental observations that the MR effect is independent of the direction of magnetization. The role of diazotization on the MR effect has been studied only for a uniform surface; no investigation of nanoconstriction has been performed yet [226].

2.7   Other MR Effects—Normal MR, AMR, and CMR

The MR so far described occurs in nanoscale magnets. There are other interesting MRs in bulk systems; these include the well-known phenomena of normal MR and anisotropic MR, and colossal MR, which was reinvestigated recently in perovskite manganites. In this section, we briefly describe these MR phenomena.

2.7.1   Normal MR

When both a magnetic field H and an electric field E are applied to electron systems, the electrons drift in the direction of $H \times E$. However, the external magnetic field does not produce any additional current, instead it produces only a Hall voltage perpendicular to the electric field, which is called the normal Hall effect. When there are two or more types of carriers, the MR that is produced is called normal MR [24].

In the case when two types of carrier exist say $i = 1, 2$, the total conductivity, when both H and E are applied, is given by:

$$\sigma = \sigma_1 + \sigma_2 - (\bar{J}_1 + \bar{J}_2) \cdot \frac{E \times H}{E^2} \tag{2.61}$$

where σ_i is the conductivity for the ith carrier, and $\bar{J}_i$ is determined by:

$$\sigma_i E = \bar{J}_i / n_i e c + H \times \bar{J}_i \tag{2.62}$$

where n_i is the carrier density, e is the charge of an electron, and c is the velocity of light. The second term in the expression for σ indicates the MR effect.

2.7.2   Anisotropic Magnetoresistance

Anisotropic MR is a phenomenon that occurs in ferromagnets in which the resistivity depends on the angle between the current and magnetization directions. Figure 2.52 shows the experimental results obtained for the change in resistivity of Ni when a magnetic field is applied parallel and perpendicular to the current

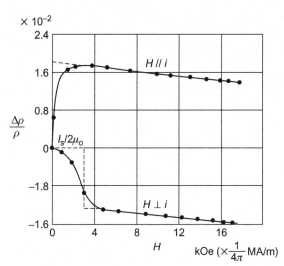

Figure 2.52 Experimental results for AMR in Ni [18].

direction. The rapid change in the resistivity at low magnetic fields is due to magnetization rotation, and the linear change at high magnetic fields is due to the so-called forced effect. The magnitude of AMR is obtained by extrapolating the resistivity change to $H = 0$.

A phenomenological theory of AMR follows. Let the magnetization M be parallel to the z-direction, and the relative angle between M and J be θ. The relationship between the electric field E and the current J is given by:

$$\begin{pmatrix} E_x \\ E_y \\ E_z \end{pmatrix} = \begin{pmatrix} \rho_\perp & -\rho_H & 0 \\ \rho_H & \rho_\perp & 0 \\ 0 & 0 & \rho_\parallel \end{pmatrix} \begin{pmatrix} J_x \\ J_y \\ J_z \end{pmatrix} \tag{2.63}$$

in general, where ρ_H is the Hall resistivity. Substituting the relation between the electric field and the current into the expression for the total resistivity, $\rho = E \cdot J / E^2$, we obtain:

$$\rho = \rho_\perp + (\rho_\parallel - \rho_\perp)\cos^2 \theta \tag{2.64}$$

When the samples are polycrystalline, it should be averaged over the angle to give:

$$\rho = \rho_\perp + (\rho_\parallel - \rho_\perp)/3 \equiv \overline{\rho} \tag{2.65}$$

When $H \parallel M$, $\rho = \rho_\parallel$ and when $H \perp M$, $\rho = \rho_\perp$, one may define the MR ratio as:

$$\frac{\Delta\rho}{\overline{\rho}} = \frac{\rho_\parallel - \rho_\perp}{\overline{\rho}} \tag{2.66}$$

Since the current direction in AMR depends on the magnetization (i.e., the spins of electrons), the SOI is a possible origin of AMR. The role of SOI is twofold; one is mixing the up- and down-spin states of the d-states, and the other is the generation of a scattering probability from the up- (down-) to down- (up-) spin states. In TM alloys, such as Co alloys, the conductivity is dominated by free electrons, such as up-spin electrons. When mixing between the up- and down-spin states occurs, scattering from the up-spin state to the down-spin state increases $\rho_\uparrow$. On the other hand, d-states at the Fermi level are reduced by the mixing, and therefore $\rho_\downarrow$ decreases. Thus, we may write the effect of SOI as:

$$\rho_{\uparrow(\downarrow)} = \rho_{0\uparrow(\downarrow)} + (-)\gamma\rho_{0\uparrow(\downarrow)} \tag{2.67}$$

where γ is a constant. The magnitude of γ depends on the angle between the current and magnetization, resulting in a MR called AMR. This expression can be approximated by:

$$\rho_{\parallel\uparrow(\downarrow)} = \rho_{0\uparrow(\downarrow)} + (-)\gamma_\perp\rho_{0\uparrow(\downarrow)} + (-)(\gamma_\parallel - \gamma_\perp)\rho_{0\uparrow(\downarrow)}$$

$$\sim \rho_{\perp\uparrow(\downarrow)} + (-)\gamma\rho_{\perp\uparrow(\downarrow)} \tag{2.68}$$

and by redefining $\gamma_\parallel - \gamma_\perp$ as γ we obtain:

$$\frac{\Delta\rho}{\bar{\rho}} \sim \frac{\rho_\parallel - \rho_\perp}{\rho_\parallel + \rho_\perp} \sim \gamma(\alpha - 1) \tag{2.69}$$

where the parameter α is given by ($\alpha = \rho_{\perp\downarrow}/\rho_{\perp\uparrow}$). Thus, the AMR ratio is related to the spin-dependent resistivity via the parameter α. The results agree well with experimental ones.

In order to estimate the value of γ, the transition probability should be calculated using, for example, the golden rule. A detailed description has been presented in a paper by Smit [227].

2.7.3 Colossal Magnetoresistance

Mn ions in $LaMnO_3$ (LMO) are trivalent and each Mn ion has one electron in the e_g state, and three electrons in the t_{2g} state. The electronic states of LMO are related with the important factors given below that govern the physical properties of manganites.

- Since each Mn ion has one electron in an e_g orbital, LMO must be metallic unless there is a strong Coulomb interaction between electrons. Since LMO is an antiferromagnetic insulator, the Coulomb interaction should play an important role in the magnetic and transport properties.
- The e_g state consists of degenerate $d_{x^2-y^2}$ and $d_{3z^2-r^2}$ orbitals. Therefore, the e_g electron can be in either the $d_{x^2-y^2}$ or $d_{3z^2-r^2}$ orbital, that is, the electron is free to choose either of

these orbitals; that is, it has an orbital degree of freedom, in addition to charge and spin degrees of freedom.

- The orbital degeneracy also gives rise to lattice distortion, for example, Jahn–Teller distortion lifts the orbital degeneracy. Therefore, the electron–lattice interaction is expected to be strong in manganites, especially for those in the insulating phase.
- Because of the octahedron configuration of the oxygen ions, t_{2g} orbitals hybridize less with oxygen p-orbitals than e_g orbitals, and electrons in the t_{2g} state behave as if they had a localized spin with $S = 3/2$. The spin of e_g electron couples ferromagnetically with t_{2g} spins due to strong Hund's rule coupling.

When La ions are replaced with Sr or Ca ions, Mn^{+4} ions are introduced, and the e_g electrons can become mobile under strong Hund's rule coupling, electron–lattice interaction, and Coulomb interaction, with the addition of an orbital degree of freedom. They produce a rich variety of phenomena, both in magnetism and transport. As for magnetism, manganites exhibit various magnetic orderings that depend on the dopant. LMO is a layered antiferromagnet (AF), which can be converted to a metallic (M) ferromagnet (F) by replacing La with Sr or Ca at approximately 30% doping. At approximately 50% doping of Sr or Ca, manganites exhibit a complicated AF (CE-type AF). Upon further doping, they show chain-type AF (C-type AF) and finally they show usual AF (G-type AF). The AF state is usually insulating, however, A-type AF often shows metallicity near 50% doping. Charge-ordered (CO) states also appear depending on the magnetic state.

The so-called CMR is also related to magnetism and charge ordering [228–233]. This phenomenon may be classified into three types.

1. The high-temperature phase is a paramagnetic insulator (PI), and the low-temperature phase is a MF. In this case, the resistivity shows a peak at the Curie temperature, which decreases dramatically with applied magnetic field, and the peak shifts toward higher temperatures. The higher the resistivity in PI phase, the sharper the peak becomes, and the larger the MR effect becomes. When the high-temperature phase is MF, the peak structure in the resistivity diminishes.
2. The high-temperature phase is a PI, which changes to MF phase on reducing the temperature. Upon a further decrease in temperature, the phase makes a first-order transition into a CO-AF phase. When an external magnetic field is applied, the first-order transition temperature shifts to lower temperatures, resulting in a spread of the metallic region. Therefore, applying a magnetic field produces a huge change in the resistivity. This may be an insulator to metal transition. In this case, however, no peak appears near the Curie temperature in the high-temperature region.
3. The high-temperature PI phase changes into a CO phase with decreasing temperature, and its resistivity becomes high. With a further decrease in the temperature, a CO-AF phase appears, and the resistivity becomes higher. When the magnetic field is applied, a first-order transition from a low-temperature CO-AF to an MF occurs. This is also an insulator–metal transition produced by the magnetic field.

Thus, CMR is related to insulator–metal transitions. Although several mechanisms of CMR of the first-type have been proposed, few theories have been proposed for the second and third ones. Typical models for explaining CMR include the so-called double exchange model, the critical scattering mechanism, double exchange with Jahn−Teller distortion, the percolation model, and the localization model. We will not describe these models in detail, rather detail descriptions can be found in various review articles [234−237].

2.8 SOI and Hall Effects

We have shown that the SOI gives rise to the MR known as AMR. Since SOI is a coupling of spin and orbital motion of electrons, it also gives rise to other interesting transport properties called as anomalous Hall effect (AHE) and spin Hall effect (SHE) as well as electric field induced spin accumulation in 2DEG. Although the theoretical studies are performed for bulk systems, the experimental measurements are usually done for mesoscopic or nanosize systems. We therefore give a brief explanation on these properties.

2.8.1 Spin−Orbit Interaction

SOI is a relativistic effect and the corresponding Hamiltonian is given by:

$$H_{SO} = \frac{e\hbar}{4m^2c^2}(\sigma \cdot [E \times p]) \tag{2.70}$$

$$\simeq -\frac{e}{2m^2c^2}\left(\frac{dV}{r\,dr}\right)(s \cdot l) \tag{2.71}$$

where c is the light velocity, E is an electric field induced by a potential gradient dV/dr, and l is the orbital angular momentum. There are several origins of the internal electric field:

- Inversion asymmetry of the lattice such as the zincblende structure. The SOI is called Dresselhaus-type SOI [238].
- Inversion asymmetry of the structure such as the 2DEG. The SOI is called Rashba-type SOI [239,240].
- $\ell - s$ coupling on atoms.
- Potential gradient caused by impurity potential and $\ell - s$ coupling of atoms.

Since the first three types of SOI are called intrinsic SOI since they reside uniformly in the system, while the last one is called an extrinsic SOI as it is caused by impurity potentials. The SOI in semiconductors is usually caused by the Dresselhaus- and/or the Rashba-type SOI, while that in metal and alloys may be related to the $\ell - s$ coupling.

The SOI shown above may be recast in different ways to clarify the role on the transport properties. Since Eq. (2.70) may be written as $H_{SO} \propto \sigma \cdot h_{eff}$, one may consider that the SOI subsists a momentum-dependent effective magnetic field. It is also rewritten as $H_{SO} \propto [\sigma \times E] \cdot p$, from which one can derive a velocity $v = dH_{SO}/dp \propto [\sigma \times E]$ for conduction electrons. The velocity is called anomalous velocity. It is also noted that the SOI results in complex wave functions of electrons, from which a phase called Berry phase appears. As shown below, Berry phase may be interpreted to be an effective magnetic field in the momentum space.

2.8.2 Anomalous Hall Effect

The Hall effect was discovered in 1879 by Edwin Hall [241] when he was studying the force acting on the charged particles under an external electric field E_{ext} and a magnetic field H_{ext}. This is the so-called "ordinary" Hall effect in which the electrons are deflected into the direction of $E_{ext} \times H_{ext}$ due to the Lorentz force. In ferromagnets, the Hall effect consists of two contributions, the ordinary Hall effect and the "anomalous" Hall effect [242] being proportional not to H_{ext} but to the magnetization M of the ferromagnet. The Hall resistivity ρ_H is thus given as:

$$\rho_H = R_0 H + 4\pi R_s M \tag{2.72}$$

The first and second terms are the ordinary and AHE, respectively. Coefficients R_0 and R_s are called as ordinary and anomalous Hall coefficient, respectively.

In experiments, with increasing H_{ext}, ρ_H changes rapidly at first, and then tends to increase in proportion to H. The initial rapid change in ρ_H is caused by the alignment of the domain magnetization. After the alignment of the domain magnetization, ρ_H changes in proportion to H_{ext}. The constant increment gives the value of R_0, and the extrapolated value of ρ_H to $H_{ext} = 0$ gives the value of $4\pi R_s M$.

Historically, an intrinsic mechanism, which results from the Berry phase, was first proposed for AHE [243]. The essential point in the theory is that there is a contribution to the velocity from the Berry phase $\Omega(k)$ as shown by,

$$\dot{x} = \frac{1}{\hbar} \frac{\partial \varepsilon(k)}{\partial k} - \dot{k} \times \Omega(k) \tag{2.73}$$

$$\hbar \dot{k} = -eE_{ext} - e\dot{x} \times B_{ext} \tag{2.74}$$

We see that the velocity $\dot{x}$ has a component perpendicular to E_{ext} even when $B_{ext} = 0$. A term $eE_{ext} \times \Omega(k)$ reminds us that $\Omega(k)$ plays a role of an effective magnetic field. Because $\Omega(k)$ is a quantity in the momentum space and is dependent on the wave vector, the effective field may be interpreted as a magnetic field in the momentum space. The AHE caused by intrinsic mechanisms is called "intrinsic" AHE.

Most of the experiments, however, have been interpreted by extrinsic mechanisms, skew scattering (SS) [244] and side jump (SJ) [245]. In SS mechanism, the up- and down-spin electrons are scattered into opposite directions as shown in Figure 2.53A, on the other hand, in SJ mechanism, there occurs a displacement of the electron path as shown in Figure 2.53B. In ferromagnets, the intrinsic spin imbalance makes the spin-up and spin-down charge Hall currents asymmetric and

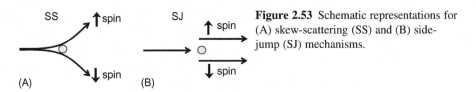

Figure 2.53 Schematic representations for (A) skew-scattering (SS) and (B) side-jump (SJ) mechanisms.

produces a Hall voltage proportional to the spin polarization, i.e., magnetization. The AHE caused by extrinsic mechanisms is called "extrinsic" AHE.

The SS mechanism is originated from electron scattering by impurity potentials and SOI. On the other hand, the SJ is caused by the anomalous velocity due to SOI. Therefore, the Hall conductivity due to SS depends on the lifetime τ, while that due to SJ is independent of τ. As for explicit expressions of SS and SJ, readers may refer to a work by Crepieux and Bruno [246] for example. Since

$$R_s = \rho_{xy} = (\sigma^{-1})_{xy} = -\frac{\sigma_{xy}}{\sigma_{xx}^2 + \sigma_{xy}^2} \sim -\frac{\sigma_{xy}}{\sigma_{xx}^2}$$

and $\sigma_{xx} \propto \tau$, SS mechanism gives $\rho_{xy} \propto \rho$, while SJ gives $\rho_{xy} \propto \rho^2$. Therefore, a relation is given as:

$$\rho_{xy} = a\rho + b\rho^2 \tag{2.75}$$

where a and b are constants, generally holds and has been used for experimental analysis. It is noted that the intrinsic spin Hall conductivity (SHC) may also be brought about in the SJ mechanism, since the SJ mechanism is caused by the anomalous velocity.

Most of the experiments have so far been analyzed by using the expression above, but the intrinsic AHE has recently been revisited to give quantitative explanations of AHE in ferromagnetic semiconductors as well as Fe [247–249].

2.8.3 Spin Hall Effect

As mentioned, the extrinsic AHE appears owing to spin-dependent scattering. The up- and down-spin electrons are scattered in opposite directions, resulting in spin-up and spin-down charge Hall currents along the perpendicular direction of $\boldsymbol{E}_{ext}$. The intrinsic spin imbalance makes the two charge Hall currents asymmetric and produces a Hall voltage. For nonmagnets, although the two charge Hall currents cancel and no Hall voltage develops, spin-dependent scattering still produces the up- and down-"spin" currents (flow of spins) that flow in the opposite directions, as long as the SOI is nonvanishing. We thus expect a Hall effect solely with spin, which is called spin Hall effect (SHE). A schematic view of SHE is shown in Figure 2.54. When the spin Hall current is originated from scattering of electrons by impurity potentials with SOI, the effect may be called "extrinsic" SHE [250,251].

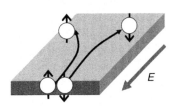

Figure 2.54 Schematic view of the SHE.

Prior to the prediction of the SHE by Hirsh using the same mechanism as that for AHE [251], the SHE was actually predicted by Dyakonov and Perel [250]. The SHE was formulated first for semiconductors and then confirmed experimentally. Subsequently, it was observed in metals, and the value of the SHC was found to be much greater for these materials than for semiconductors. Theoretical studies have also been performed.

2.8.3.1 SHE in Semiconductors

As in the case of AHE, one can conceive of an intrinsic SHE in nonmagnets on which no external magnetic field is applied. Murakami *et al.* [252] have predicted for p-type semiconductors that the effective magnetic field originated from the Berry phase makes up- and down-spin electrons drift toward opposite directions and leads to SHE. The SOI, which exists universally in any materials, may also produce the intrinsic SHE even for n-type semiconductors. Sinova *et al.* [253] have predicted a constant SHC $e/8\pi$ for 2DEG with a Rashba-type SOI produced by the asymmetry of the potential. The intrinsic SHE is a result of the inherent property, that is, a uniform SOI, of the material, as opposed to the extrinsic SHE caused by scattering.

Although the SHE does not accompany the Hall voltage, one may expect that a spin polarization of opposite signs appears at the edges even in the absence of applied magnetic fields, irrespective of extrinsic or intrinsic SHE. In the nonequilibrium state, the flows of up- and down-spin electrons into opposite directions are compensated by spin-flip scattering caused by the SO interaction itself or by other spin-dependent scattering, and give rise to a spin accumulation at the sample edges.

Elucidating the nature of the pure SHE is now an emergent issue for experimentalists. In spite of the difficulties associated with the absence of the Hall voltage in the pure SHE, two groups have succeeded in measuring the spin accumulation in nonmagnetic semiconductors by optically detecting the spin accumulation at the sample edge. Kato *et al.* [254,255] spatially resolved the Kerr rotation of the reflected light from n-type bulk GaAs and InGaAs samples and found accumulation of opposite sign at the two edges of the sample. Subsequently, Wunderlich *et al.* [256] measured the polarization of light emitted from a p−n junction placed at the edge of a structure. Kato *et al.* [254] suggested that the observed effect may be the extrinsic SHE, as the SHC is low and independent of the crystal orientation, whereas Wunderlich *et al.* concluded that the effect is consistent with the intrinsic SHE, since the magnitude of the polarization is consistent with the theoretical prediction. The interpretation of the experimental results appears not straightforward, because the current theories predict that the intrinsic SHE can strongly be suppressed by disorder effects [257], orbital Hall current [258]. Furthermore, the effect of the disorder on SHE depends on the type of SOI [259], for example, small Dresselhaus-type SOI, which is intrinsic but orientation dependent, is calculated to be sufficient to explain the observed Kerr rotation in strained InGaAs samples used by Kato *et al.* [254]. On the other hand, a recent theory on the extrinsic effect predicts the observed SHE within experimental error with no adjustable parameters [260].

The SHE has a practical relevance to the field of the spintronics, where spin polarization, manipulation, and detection are essential. Theoretical studies to link SHE with measurable quantities such as spin accumulation and optical signature are highly desired, because even if spin Hall current itself is intrinsic, the stationary spin accumulation is a result of a balance between spin Hall current and intrinsic/extrinsic effects of the spin relaxation at the edges of the sample [261]. Another interesting aspect of the SHE is the quantized SHE. Recently, quantized SHE is suggested theoretically for undoped graphene in the absence of the external magnetic field because of a gap produced by SOI [262,263], and observed in a different system [264].

2.8.3.2 SHE in Metals

The SHE is observed in not only semiconductors but also normal metals and TMs as long as the SOI exists. Valenzuela and Tinkham [265] and Kimura *et al.* [266] reported measurement of the SHE in Al and Pt, respectively. They injected a spin-polarized current into a narrow nonmagnetic metal placed in a device shown schematically in Figure 2.55. The injected current induces a spin-dependent shift of the chemical potential, which decays with increasing distance from the point of the spin injection. Because the spatial dependencies of the spin-up and spin-down chemical potentials have opposite signs, up- and down-spin currents flow in opposite directions. As a result, a spin current is induced by the spin injection into non-magnetic metals.

Because the up- and down-spin currents flow in opposite directions, they produce a charge current flowing in the same direction. The charge current is measured as a voltage change in another Hall bar attached as shown in Figure 2.55. Because the Hall bar is located far from the spin injection position, this measurement is referred to as a nonlocal measurement. The phenomenon explained above is just the opposite effect to the SHE, and therefore, it is called as an "inverse SHE." The SHC observed in Pt is much larger than that observed in semiconductors [266].

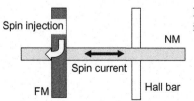

Figure 2.55 Experimental setup to measure the inverse SHE.

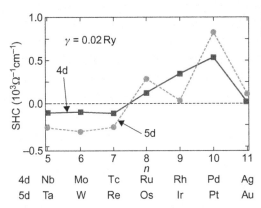

Figure 2.56 Calculated results of the SHC for TMs.

Theoretical calculations of the SHC have also been performed using an expression of Kubo formula, referred to as the Streda formula [267], with a multiorbital tight-binding model. Effects of electron scattering are included in the formalism as self-energy. Calculated results of the SHC are shown in Figure 2.56 for several TMs [268−270]. These values agree semiquantitatively with the experimental values [265,266]. The difference in sign for the SHC of the early and late TMs is attributed to Hund's third law.

Both intrinsic and extrinsic mechanisms are responsible to SHE, as in the case with AHE. The SHE and AHE originating from skew scattering are attributed to the extrinsic mechanisms and are enhanced as the relaxation time of the electron conduction increases. In fact, a large SHC has been observed for Au and Pt narrow wires [271]. Theoretical analysis using first-principles methods has also been presented [272,273]. General features of the SHE and AHE can be found in a review article [274].

2.8.4 Rashba 2DEG and Spin Accumulation

The Rashba Hamiltonian for 2DEG is given as,

$$H_0 = \begin{pmatrix} \dfrac{\hbar^2}{2m}k^2 & i\lambda\hbar k_- \\ -i\lambda\hbar k_+ & \dfrac{\hbar^2}{2m}k^2 \end{pmatrix} \tag{2.76}$$

where $k_\pm = k_x \pm ik_y$ with $\mathbf{k} = (k_x, k_y)$ the electron momentum in the 2DEG plane, λ is the spin−orbit coupling. The eigenvalues are given as:

$$E_{k\pm} = \frac{\hbar^2 k^2}{2m} \pm \lambda\hbar k \tag{2.77}$$

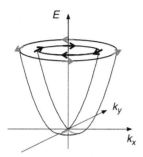

Figure 2.57 Energy dispersion relation for 2DEG with Rashba-type SOI.

Thus, the free electron states are split into two branches and the Fermi lines are two concentric circles as shown in Figure 2.57. It should be noted that the direction of spin is dependent on the momentum. The feature is a strong contrast to the Zeeman spin splitting. The momentum-dependent spin state is responsible to the SHE and spin accumulation shown below.

Former Russian scientists have predicted the spin accumulation in the Rashba split 2DEG systems with diffusive transport regime [275−277]. Inoue *et al.* [278] have confirmed the result in the Green's function method. The spin accumulation is in-plane and given as,

$$\langle s_y \rangle = 4\pi e D_0 \lambda \tau E \tag{2.78}$$

where D_0 is the DOS at the Fermi level and τ is the lifetime. The physical reason of the spin accumulation is simple. As mentioned earlier, in the equilibrium state, the spins with $k_x > 0$ point toward $+y$ direction in average, and those with $k_x < 0$ point toward $-y$ direction, and they cancel out in the equilibrium state. However, in the nonequilibrium state under external electric field along x direction, number of spins with $k_x > 0$ is more than those with $k_x < 0$, and $\langle s_y \rangle$ becomes nonzero.

Several attempts to observe the in-plane spin accumulation have been done and successful results have been reported by several groups [279−281] using optical detection.

2.9 Thermal Effects on Charge and Spin Currents

2.9.1 General Remark

2.9.1.1 Electrical and Thermal Currents

The MR effects such as GMR, TMR, CMR, and BMR that we have explained thus far are spin-dependent charge transport phenomena. That is, the electrical conduction can be regarded as a parallel circuit consisting of up- and down-spin channels, and resistivity is expressed as $1/\rho = 1/\rho_\uparrow + 1/\rho_\downarrow$ as long as the spin relaxation length is longer than the mean free path, which depends on spins in the present case. This is called Mott's two-current model.

The spin-dependent current is basically a charge current, but includes the flow of spins and may be decomposed into charge and spin currents using the following equations:

$$J_C = J_\uparrow + J_\downarrow \tag{2.79}$$

$$J_S = J_\uparrow - J_\downarrow \tag{2.80}$$

For example, in half-metals, $J_\uparrow = 0$ or $J_\downarrow = 0$. Although the spin-dependent current includes both spin and charge currents, both currents are driven by a single electric field E. The electric field is a spatial variation of a voltage V given as $E = -\nabla V$. The voltage is actually produced by a chemical potential (μ) difference between the left and right electrodes attached at both ends of a sample, that is, $\Delta\mu = \mu_L - \mu_R = eV$. The electrodes are reservoirs of electrons, and their chemical potential difference $\Delta\mu$ is the driving force of the electrical current.

It is well known that there exists another driving force for the charge current, which is the temperature difference ΔT between the two electrodes. The temperature difference is the source of thermoelectric effects including Seebeck effect and Pertier effect. The temperature difference naturally produces a flow of heat Q, or thermal conduction. Thus, ΔT produces both electrical and thermal currents. Similarly, $\Delta\mu$ produces thermal current as well as an electrical current. These effects are simply expressed in the coupled equation:

$$\begin{pmatrix} J \\ Q \end{pmatrix} = \begin{pmatrix} L_{11} & L_{12} \\ L_{21} & L_{22} \end{pmatrix} \begin{pmatrix} \Delta\mu \\ \Delta T/T \end{pmatrix} \tag{2.81}$$

Here, $L_{ij}(i,j = 1, 2)$ is a response function expressed in a tensor. Therefore, both J and Q are vectors, and for a given direction determined by the change in $\Delta\mu$ or ΔT, longitudinal and transverse components of J and Q exist. The transverse component of J is nothing but the Hall current. In addition, the relation $L_{12} = L_{21}$ holds exactly, and is well known as Onsager's reciprocity relationship [282].

2.9.1.2 Generalization

When two different metals with different chemical potentials come in contact, electrons transfer from one metal to the other so as to make the chemical potential equal, and consequently an equilibrium state is realized. In magnetic metals, up- and down-spin states are in "contact" with each other, and therefore the up-spin chemical potential $\mu_\uparrow$ equals the down-spin chemical potential $\mu_\downarrow$. For example, when an external magnetic field is applied to a paramagnetic metal, the energy states of up- and down-spin states energetically shift toward opposite directions, and thus $\mu_\uparrow \neq \mu_\downarrow$. However, such a nonequilibrium state quickly relaxes into an equilibrium state by reversing an appropriate number of electron spins. Thus, in the equilibrium state, an imbalance occurs in the number of up- and down-spin electrons, leading to the appearance of "spin polarization."

The state with $\mu_\uparrow \neq \mu_\downarrow$ survives, however, only when the nonequilibrium state is stabilized, such as by the spin injection or spin Hall current driven by an external field, as explained in the previous section. Such an imbalance in the number of up- and down-spin electrons is called "spin accumulation" rather than spin polarization. In the device shown in Figure 2.55, spin accumulation appears in NM, which is in contact with a FM, and the chemical potential difference $\Delta\mu_s \equiv \mu_\uparrow - \mu_\downarrow$ is non-zero at the contact region. Because $\Delta\mu_s(s = \uparrow, \downarrow)$ should be zero in regions that are far away from the contact region, spatial variation in $\Delta\mu_s$ occurs, resulting in a spin current. Thus, we may conclude that $\Delta\mu_s$ is a driving force of the spin current.

Now we have three "driving forces" (surely they are not real forces) $\Delta\mu$, $\Delta\mu_s$, and ΔT that produce charge current J_C, spin current J_S, and thermal current Q. These three quantities are related to each other and give the following generalized relationship [283]:

$$\begin{pmatrix} J_C \\ J_S \\ Q \end{pmatrix} = \begin{pmatrix} L_{11} & L_{12} & L_{13} \\ L_{21} & L_{22} & L_{23} \\ L_{31} & L_{32} & L_{33} \end{pmatrix} \begin{pmatrix} \Delta\mu \\ \Delta\mu_s \\ \Delta T/T \end{pmatrix} \tag{2.82}$$

Here, L_{ij} is a tensor, and the Onsager's reciprocity relationship $L_{ij} = L_{ji}$ holds. The equation above shows that the spin and heat current couple with each other. This is the basic idea behind spin caloritronics [283,284].

Below, we give a brief explanation on the basics of thermoelectric effects, followed by explanations on Seebeck and Pertier effects, which are representative thermoelectric effects [285]. When an external magnetic field is applied, more interesting effects also appears. As examples, we introduce spin-dependent Seebeck and spin Seebeck effects [286–288] that result from the coupling of spin and heat flows.

Although the charge current is conserved, the spin current is not conserved. Dissipation of spin always occurs. The dissipated spin (spin angular momentum) should be transferred to a location within the magnets according to the law of conservation of total angular momentum. This feature is characteristic of the spin current and generates the new concepts of STT and spin pumping. Because these phenomena are dealt with in other chapters of this book, we only provide a short explanation of each in the next section.

2.9.2 Thermoelectric Effects

2.9.2.1 Description in the Boltzmann Formalism

The basic equations for the Boltzmann formalism are given by two equations that determine the distribution function $f(r, k, t)$ in a nonequilibrium state,

$$\left(\frac{\partial}{\partial t} + v \cdot \nabla_r + \frac{e}{\hbar} E \nabla_k \right) f(r, k, t) = \left(\frac{\partial f}{\partial t} \right)_s \tag{2.83}$$

and the relationship between the distribution function and scattering rate:

$$\left(\frac{\partial f}{\partial t}\right)_s = -\frac{f(\boldsymbol{k}) - f_0(\boldsymbol{k})}{\tau(\boldsymbol{k})} \tag{2.84}$$

By using the temperature gradient in the sample, $\nabla_r f(\boldsymbol{r}, \boldsymbol{k}, t)$ may be rewritten as:

$$\nabla_r f(\boldsymbol{r}, \boldsymbol{k}, t) = \frac{\partial f}{\partial T} \nabla_r T = -\frac{\varepsilon - \mu}{T} \frac{\partial f}{\partial \varepsilon} \nabla_r T \tag{2.85}$$

Because the spatial dependence of the chemical potential induces an additional electric field, one may replace $\boldsymbol{E}$ with $\tilde{\boldsymbol{E}}$ defined as,

$$e\boldsymbol{E} \rightarrow e\boldsymbol{E} - \nabla\mu \equiv e\tilde{\boldsymbol{E}} \tag{2.86}$$

Substituting these relations into the first equation of the Boltzmann formalism, we obtain:

$$f(\boldsymbol{k}) \simeq f_0(\boldsymbol{k}) - \frac{e}{\hbar} \tau \tilde{\boldsymbol{E}} \nabla_k f_0(\boldsymbol{k}) + \upsilon\tau \frac{\varepsilon - \mu}{T} \frac{\partial f_0(\boldsymbol{k})}{\partial \varepsilon} \nabla_r T \tag{2.87}$$

Both the electric field $\tilde{\boldsymbol{E}}$ and temperature gradient $\nabla_r T$ produce not only electrical currents but heat currents. The change in heat δQ is related to a change in the entropy $\delta Q = T \, dS = dU - \mu \, dn$, where U and n are the internal energy and number of electrons, respectively. Let $\boldsymbol{J}_Q, \boldsymbol{J}_E$, and $\boldsymbol{J}_n$ be the heat current density, energy current density, and particle current density, respectively. Then,

$$\boldsymbol{J}_Q = \boldsymbol{J}_E \mu - \boldsymbol{J}_n \tag{2.88}$$

hold, where:

$$\boldsymbol{J}_Q = \frac{1}{8\pi^3} \int \upsilon(\boldsymbol{k})(\varepsilon - \mu) f(\boldsymbol{k}) d\boldsymbol{k} \tag{2.89}$$

$$\boldsymbol{J}_E = \frac{1}{8\pi^3} \int \upsilon(\boldsymbol{k}) \varepsilon f(\boldsymbol{k}) d\boldsymbol{k} \tag{2.90}$$

$$\boldsymbol{J}_n = \frac{1}{8\pi^3} \int \upsilon(\boldsymbol{k}) f(\boldsymbol{k}) d\boldsymbol{k} \tag{2.91}$$

The electric current is given by:

$$\boldsymbol{J} = e\boldsymbol{J}_n \tag{2.92}$$

Substituting the result of the distribution function Eq. (2.87) into J and J_Q, one may obtain,

$$J = e^2 K_0 \tilde{E} + \frac{e}{T} K_1 (-\nabla_r T) \tag{2.93}$$

$$J_Q = e K_1 \tilde{E} + \frac{1}{T} K_2 (-\nabla_r T) \tag{2.94}$$

where

$$K_n = -\frac{1}{8\pi^3} \int \upsilon(\mathbf{k}) \tau \upsilon(\mathbf{k}) (\varepsilon - \mu)^n \frac{\partial f_0(\mathbf{k})}{\partial \varepsilon} \, d\mathbf{k} \tag{2.95}$$

When the electric field is induced solely by chemical potential difference $\Delta \mu$ at both edges of the sample, Eqs. (2.93) and (2.94) agree with the Eq. (2.81). Note that K_n is a tensor in general.

Equations (2.93) and (2.94) indicate that the electrical and heat currents couple with each other. When an external magnetic field H is applied to a sample in which both electrical and heat currents exist, various effects appear. Voltage and temperature differences appear in the transverse direction to $J \times H$ as effects of the first order of H, and voltage and temperature change appears in the longitudinal direction of J as effects of the second order of H.

The effects of magnetic field on electrical currents have been called galvanomagnetic effects, and those on heat currents are called thermomagnetic effects. The coupled phenomena of electricity and magnetism are known as magnetoelectric effects.

2.9.2.2 Seebeck and Pertier Effects

Here we give a simple explanation of the Seebeck and Pertier effects [285]. Assume that two different metals A and B, each with a half-ring shape, are attached to make a ring-shaped sample. A voltage meter with high resistance is then connected to the B metal of the sample. The temperature is T_1 and T_2 at the two interfaces of A and B metals, respectively, and $T_0 (\neq T_1 \neq T_2)$ at the position where a voltage meter is connected. Because of the temperature difference $T_1 - T_2$ between the two interfaces, an electrical current should appear between them. However, no electrical current appears because of the presence of the high-resistance voltage meter. Nevertheless, an electric field E is induced as can be seen from Eq. (2.93):

$$E = \frac{1}{eT} K_0^{-1} K_1 \nabla T \equiv Q \nabla T \tag{2.96}$$

where $\nabla \mu$ in $\tilde{E}$ is neglected. The induced voltage is obtained by integrating the result shown above to give the following equation:

$$\Phi = \int_0^1 Q_B \frac{\partial T}{\partial x} + \int_1^2 Q_A \frac{\partial T}{\partial x} + \int_2^0 Q_B \frac{\partial T}{\partial x}$$

$$= \int_{T_1}^{T_2} (Q_A - Q_B) dT \tag{2.97}$$

The voltage is canceled at the high-resistance voltage meter and a stationary state (a state without current) is realized. This situation is known as the Seebeck effect.

Next, the voltage meter in the sample is replaced with a battery, and it is assumed that no temperature gradient exists in the sample. Using Eqs. (2.93) and (2.94), we obtain:

$$J_Q = eK_1 K_0^{-1} J \equiv \Pi J \tag{2.98}$$

As a result, a heat current appears when an electrical current exists. As the current J is produced by the attached battery, the heat currents in A and B metals equal $\Pi_A J$ and $\Pi_B J$, respectively, because the electrical current is conserved. Because $\Pi_A \neq \Pi_B$, the difference between these heat current $(\Pi_A - \Pi_B)J$ must come in/out at the left/right interface of A and B metals. This situation is referred to as the Pertier effect.

2.9.2.3 Spin-Dependent Seebeck and Spin Seebeck Effects

The Seebeck effect occurs when a voltage is induced by a temperature difference between the two terminals of a sample. When the sample is a ferromagnet or a magnetic multilayer, the Seebeck coefficient depends on spin. When an external magnetic field is applied to magnetic layers, for example, the magnetization rotation in magnetic layers may change the Seebeck coefficient. Furthermore, when the chemical potential depends on spin, that is, $\Delta \mu_s \neq 0$, a heat current may be induced by $\Delta \mu_s$, and spin dependence will also exist in the heat current owing to the Onsager's reciprocity relationship. These effects may be called spin-dependent Seebeck effects [286,287,289].

Slachter et al. measured the spin-dependent Seebeck effect in the following manner [287]. A crossed-bar lateral junction sample made of a ferromagnetic lead and a nonmagnetic lead was used. By passing an electrical current through the ferromagnetic lead in order to produce Joule heat at the contact, a spin accumulated state, that is, a state with $\Delta \mu_s \neq 0$, was induced at the contact. The spin-dependent chemical potential resulted in a spin-dependent heat current through the nonmagnetic lead. By attaching one more ferromagnetic Hall bar to the nonmagnetic lead, Slachter et al. [287] measured the voltage difference caused by the alternation of magnetization direction of the two ferromagnetic leads. This measurement confirmed the existence of the spin-dependent heat current. Theoretical studies have also been performed by, for example, Xiao *et al.* [290].

The samples discussed so far are made of metals. Uchida *et al.* [288] showed that the Seebeck effect exists even in insulators. When a temperature difference exists between two edges of a ferromagnetic insulator, propagation of spin waves (magnon excitation) occurs. This excitation is a type of spin current, and therefore might be observed by attaching a Hall bar. Uchida *et al.* actually observed the spin current using the inverse SHE. Appearance of spin current due to thermal effect is called spin Seebeck effect, which is distinct from the spin-dependent Seebeck effect [283].

2.9.2.4 Spin Nernst Effect

In the previous section, we briefly introduced AHE and SHE, which are transverse charge and spin currents, respectively, and are driven by an external electric field. The former appears in ferromagnetic metals, while the latter appears in nonmagnetic metals. A temperature gradient in metals also produces a Hall current, which is called the Nernst effect. In ferromagnets, a transverse charge current is produced in a manner similar to the AHE and is called anomalous Nernst effect. In nonmagnets, a transverse spin current is brought about and may be called spin Nernst effect [283,291].

2.10   Spin Transfer and Spin Pumping

2.10.1   Spin Transfer Torque

As mentioned in the previous section, the spin current and ratio of the up and down spins of the electrical current are not conserved in general. Let us consider ferromagnetic junctions made of two FMs, FM1/nonmagnetic metal/FM2, in which FM1 and FM2 have magnetization M_1 and M_2, respectively, as shown in Figure 2.58. Electrons flow from the left to the right electrodes (current flow from the right to the left), and a spin polarization is created in FM1. The direction of the spin polarization of the current is the same as that of M_1. When the two magnetizations are noncolinear, the spins of the current interact with M_2 and the direction of the spin polarization is changed. Because the total angular momentum must be conserved, the change in the direction of the spin polarization of the current, that is, the change in the spin angular momentum of the current, must be transferred to a location within FM2. The transfer may result in a rotation of M_2. The time

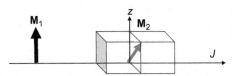

Figure 2.58 Schematic representations of a junction which gives rise to an STT on magnetization M_2 from M_1. Here J indicates the flow of electrons.

dependence of the angular momentum is nothing but a torque, and therefore, the change in the spin direction of the current may exert a torque on M_2. In other word, the spin-polarized current transfers the spin angular momentum to the magnetic moment of the ferromagnet by exerting a torque. The torque is therefore called spin transfer torque (STT) [292,293].

The magnitude of the STT T is estimated as follows [294]. Let p and m be unit vectors of M_1 and M_2, respectively. The current density is given as J. The number of electrons that flow through a unit cross section of the junction is then $J/|e|$. These electrons have a spin angular momentum of $(J/|e|)\hbar/2$. Because the direction of the spin polarization of the current changes from p to m, the STT acting on M_2 may be given as:

$$T = \frac{\hbar}{2|e|} J\eta m \times (p \times m) \tag{2.99}$$

where η is a parameter representing the spin polarization of the current. The vector T is on a plane spanned by $p \times m$. In general, there should exist a component of the torque perpendicular to the plane [295]. These two components may be written as:

$$T_\perp = g_\perp m \times p \tag{2.100}$$

$$T_\parallel = g_\parallel m \times (p \times m) \tag{2.101}$$

where $g_\perp$ and $g_\parallel$ are relevant constants.

The discussion above is based on the fact that the spin-polarized current induces the STT. It is quite interesting to note that the temperature gradient produced in the spin-polarized ferromagnet also induces STT in FTJs [296].

2.10.2 Magnetization Dynamics and Spin Pumping

The STT may have an influence on dynamics of the magnetization M_2. The magnetization dynamics is typically analyzed using the Landau−Lifsitz−Gilbert (LLG) equation:

$$\frac{dm}{dt} = -\gamma m \times H + \alpha m \times \frac{dm}{dt} \tag{2.102}$$

where $\gamma (= g\mu_B)$ is the gyromagnetic ratio and α is the Gilbert damping factor. The first term represents a precession of m around H, and the second term gives a damping of direction of m toward H. Because the first term is caused by the magnetic field and the second term represents the damping, they are called field term and damping term, respectively.

When an STT is exerted on m, T is added on the right-hand side of the LLG equation. The $T_\parallel$ term is usually called torque term because it exerts a torque on m. The $T_\perp$ term is called field term because p may be considered an effective field caused by magnetization M_1. The most important point is that the sign of $T_\parallel$ term can be either positive or negative because it is proportional to the current. When the current direction is reversed, the direction of $T_\parallel$ is also reversed. This property is the one used for magnetization reversal by a current.

Now, let us consider the physical meaning of the Gilbert damping factor α. The factor plays a role in aligning the magnetization in the direction of magnetic field. Because the change in the magnetization direction is nothing but a change in the spin angular momentum, dissipation of the angular momentum occurs in the ferromagnet via, for example, the SOI. The value of α thus depends on the ferromagnet species. In magnetic multilayers or ferromagnet/nonmagnet junctions, it also depends on the nonmagnet species. Experimental evidence for the dependence of α on paramagnetic metals is shown in Figure 2.59[297]. The value of α depends on species of the paramagnetic metals; it is small for Cu and large for Pt and Pd. The higher value for the latter may be attributed to the large SOI in heavy elements.

The fact that the value of α depends on not only the ferromagnetic metal, but also the nonmagnetic metals and indicates that the dissipation of the spin angular momentum is influenced by the nonmagnetic metal. The wave functions of the electrons in the ferromagnetic metal extend into the nonmagnetic metal that is in contact with the ferromagnet. As a result, the effects of the precession of the magnetization of the ferromagnet influence the spin state of the nonmagnetic metal. A change in the spin state of the nonmagnet in turn influences the magnetization dynamics of the ferromagnet, resulting in the dependence of α on the nonmagnetic metal.

The mechanism mentioned above has been formulated into the concept known as spin pumping. As the magnetization of a ferromagnet recesses and damps toward a certain direction, some part of the spin angular momentum is transferred to the conduction electrons via, for example, the s−d interaction. The transferred spin angular momentum may dissipate within the nonmagnet attached to the ferromagnet. The dissipation of the spin angular momentum is interpreted as a spin current

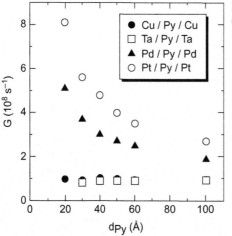

Figure 2.59 Experimental values of Gilbert damping factors.

created by the dynamics of the magnetization that flows into the nonmagnet. A part of the spin current flowing into the nonmagnet may flow back to the ferromagnet. When the outgoing and incoming spin currents are the same, there is no effect of the nonmagnet on the dissipation of the spin angular momentum. When the SOI in the nonmagnet is large, most of the spin current that flows into the nonmagnet dissipates within it, and as a result, the value of α becomes large. The creation of a spin current through the magnetization dynamics of the ferromagnet is called spin pumping [298–300]. The spin pumping is an inverse effect of the STT and is a spin-motive force.

The spin-diffusion length ℓ_{sf} may be estimated by measuring the α values. It has been reported that the value of ℓ_{sf} of Cu is 240 nm at room temperature and 2 μm at 4.5 K. The values for ℓ_{sf} for Ru, Ag, and Au are 2, 170, and 34 nm at room temperature, respectively [301–303]. An attempt to observe the spin current created by spin pumping has also been reported [304].

It is interesting that the important factor for the magnetization reversal due to a current becomes controllable by using a multilayered structure. It should also be noted that the concept of spin pumping is opposite to the concept of STT. In STT, the change in the spin current produces magnetization rotation, while in spin pumping, the magnetization rotation creates the spin current. Various aspects of the spin current have been reported in reviews, for example, by Brataas *et al.* [300].

2.10.3   Example of the Microscopic Calculation of STT

The microscopic calculation of the STT has been achieved in several ways [295,305,306]. The Kubo formalism is a representative method of the microscopic calculation of the electrical conductivity. Another commonly used method is the direct calculation of the expectation value of the current. The method is called the method of the nonequilibrium Green's function, because the expectation value is related to Green's function, which is often called the Keldysh Green's function [307]. Here, we present an example of such a microscopic calculation of the spin current.

We first describe the nonconservation of the spin current [295] using a one-dimensional lattice model for simplicity. The time derivative of a physical quantity on the site n, A_n, is given by:

$$\frac{dA_n}{dt} = \frac{\partial A_n}{\partial t} + \text{div} \, A_n \tag{2.103}$$

where the second term is the divergence of the quantity A_n at site n, that is, a difference between the incoming and outgoing values of A_n. For the spin angular momentum, the time derivative can be obtained using the Heisenberg equation for a given Hamiltonian H as:

$$\frac{dS_n}{dt} = -\frac{i}{\hbar}[S_n, H] = S_n \times h_{\text{eff}} - (J_{Sn} - J_{Sn-1}) \tag{2.104}$$

where h_{eff} is an effective magnetic field acting on the spin, and J_S indicates the spin current. The first term appears from the Zeeman term of the Hamiltonian, and the second term comes from the hopping term of the Hamiltonian. The subscript n of J_S is defined such that the spin current between sites $n + 1$ and n is J_{Sn} and the spin current between sites n and $n - 1$ is J_{Sn-1}. It should be noted that the local Coulomb interaction and the local impurity potential do not contribute to the spin current. For the charge current, the right-hand side of the equation is zero because the magnetic field has no effect on the electric charge and the current conservation holds. Or inversely, the charge conservation results in current conservation.

The equation above may be generalized for a layered structure. In this case, n is the layer index. The second term indicates the difference between the outgoing and incoming spin currents of nth layer, and $J_{Sn} - J_{Sn-1} \equiv T_n$ is the torque acting on the magnetization of the nth layer. Because the divergence of the spin current is nothing but the difference between the spin currents flowing at the left and right interfaces of the nth layer, the torque (STT) can be obtained by calculating the expectation values of the spin current at both interfaces.

There has been proposed an alternate definition of the STT for junctions made of half-infinite ferromagnets and nonmagnets. In this case, the divergence might not be well defined; however, the STT is defined by the concept of mixing conductance at the interface. Spin-polarized electrons injected from the nonmagnetic side to the interface are partially reflected at the interface, and the spin polarization of the reflected electrons is in general different from that of the incoming electrons. Therefore, nonconservation of spin angular momentum of the injected electrons occurs, and the difference gives rise to the STT acting on the magnetization of the ferromagnet [299].

The charge and spin currents may be formulated using the Keldysh formalism developed by Caroli *et al.* [40]. Formulation of the analytical expression of the STT of ferromagnetic junctions may be rather tedious. However, the analytical expression of the STT for tunnel junctions in which a ferromagnetic insulator (FI) is used as the tunnel barrier (see Figure 2.60B) is less tedious [308,309]. A ferromagnetic insulator is an insulator in which the energy gap is spin dependent as shown schematically in Figure 2.60A. Typical examples of the ferromagnetic insulator are EuS (rock salt), $BiMnO_3$ (perovskite), La_2NiMnO_6 (double perovskite),

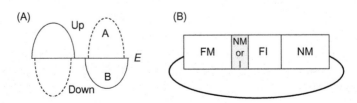

Figure 2.60 (A) Schematic DOS of a ferromagnetic insulator and (B) schematic representation of a tunnel junction with ferromagnetic insulator.

and some spinels and garnets. Because the Curie temperature of EuS, $BiMnO_3$, and La_2NiMnO_6 is 16 K, 105 K, and 280 K, respectively, they are not useful for technical applications. The Curie temperature of a garnet $Y_3Fe_5O_{12}$ (YIG) is 560 K, and those of the spinel ferrites $CoFe_2O_4$ and $NiFe_2O_4$ are 793 K and 850 K, respectively. The magnetic anisotropy of $CoFe_2O_4$ is too large to rotate the magnetization by STT. Because the magnetic anisotropy of YIG is the smallest, it may be suitable for use as a ferromagnetic insulator in such tunnel junctions.

By adopting some approximations, the analytical expression of the STT is then given as,

$$g_{\parallel} = -\int d\omega C(\omega) P_m (g_{np}^{\uparrow} - g_{np}^{\downarrow})^2 \tag{2.105}$$

$$g_{\perp} = \int d\omega C(\omega) \frac{1}{2\pi D_m^0} \left\{ (g_{np}^{\uparrow})^2 - (g_{np}^{\downarrow})^2 \right\} (R_m^{\uparrow} - R_m^{\downarrow}) \tag{2.106}$$

and

$$C(\omega) = \pi (tt')^2 (f_R - f_L) D_q^0 D_m^0 \tag{2.107}$$

Here, P_m is the spin polarization of the DOS $D_m^{\uparrow(\downarrow)} = D_m^0 (1 + (-)P_m)$ and $R_m^{\sigma} = Re(g_m^{\sigma})$ with $\sigma = \uparrow$ or $\downarrow$. Because the right electrode is an NM, $D_q \equiv D_q^0$.

There are several interesting points to be noted: first, both $T_{\parallel}$ and $T_{\perp}$ are linear in the bias voltage V, which is caused by the asymmetric structure of the junctions and the sign of $T_{\perp}$ may change depending on the magnitude of R_m^{σ}. Second, owing to the simple approximations in which the multiple scattering at the interfaces has been neglected, the STT is proportional to a product of the DOS at the left and right electrodes and to the spin polarization P_m at the left electrode, which is similar to the results obtained by Slonczewski [310]. When the multiple scattering, that is, the band mixing between the left electrode and the FI is included, higher orders of the angle dependence may appear in the STT. Although the term $f_R - f_L$ in Eq. (2.107) includes an arbitrary value of V, the expressions of T may be applicable to low bias voltages, because changes in the electronic structure due to V are disregarded in the present formalism.

Finally, it should be noted that there are two types of the spin current in magnetic multilayers. One is the flow of spin angular momentum in the nonequilibrium state, and the other is a dissipation-less spin current that exists even in the equilibrium state. The latter corresponds to the exchange coupling acting between magnetization of two ferromagnetic layers. In the calculation of STT, the latter spin current should be subtracted from the quantity calculated as the expectation value of the spin current.

2.11 Perspective

In this chapter, we briefly reviewed the relationship between spin-dependent resistivity and electronic structures in metals and alloys, and described microscopic methods for investigating electrical transport. We then reviewed the essential aspects of GMR, TMR, and BMR, emphasizing the role of the electronic structures of constituent metals of these junctions and the effects of roughness on the electrical resistivity (or resistance). It was shown that the important factors that govern GMR are the spin-dependent random potential at interfaces and band matching/ mismatching between magnetic and nonmagnetic layers. For TMR, several factors were shown to be important in determining the MR ratio, including the shape of Fermi surface of the electrodes, the symmetry of the wave functions, electron scattering at interfaces, and spin-slip tunneling. An interpretation of TMR in Fe/MgO/ Fe and of an oscillation of TMR were presented. TMR in granular films and in the Coulomb-blockade regime was also described. A brief review of normal MR, AMR, and CMR was given. AHE and SHE originated from the SOI are also briefly explained.

The effects of GMR and TMR have now been applied to several devices, such as sensors and memories. Despite a simple structure of the FTJs, they give rise to large effect, and therefore are quite suitable for the technological applications. It is interesting that huge room-temperature TMR appears in junctions with the familiar ferromagnet Fe. Recently observed oscillation of TMR in Fe/MgO/Fe junctions is another interesting phenomenon [148]. Although the MR effect is a simple phenomenon, but appears in various materials and junctions. It is expected that novel aspects of the MR may be found in near future.

The SHE, which was first predicted theoretically and only recently observed, is a novel transport property that is related to a pure spin current. Because a spin current is usually nonconserving, its spin angular momentum can be transferred, for example, to the magnetization, resulting in a current-induced magnetization reversal or the STT, which has also attracted much interests recently. We also have briefly introduced the coupled effects of spin, charge, and heat flows, which are interesting subjects in the developing field of spin caloritronics.

Acknowledgments

The author thanks many people, especially H. Itoh, S. Maekawa, G.E.W. Bauer, and Y. Asano for valuable discussions and collaborations. This work was supported by the Next Generation Super Computing Project, Nanoscience Program, MEXT, Japan, and a Grant-in-Aid for the twenty-first century COE "Frontiers of Computational Science."

References

[1] Mattis DC. *The theory of magnetism* I. Springer-Verlag; 1981.
[2] Thomson JJ. Phil Mag 1897;44:293.
[3] Gerlach W, Stern O. Z Phys 1922;9:349.
[4] Mott NF. Proc R Soc London Ser A 1936;153(699): 156, 368 (1936)
[5] Shinjo T, Takada T. Metallic superlattices. Amsterdam: Elsevier; 1987.
[6] Levy PM. Solid State Phys 1994;47:367 Ehrenreich H, Turnbull D, editors

[7] Miyazaki T. Electrochem Technol 1996;279:.
[8] Gijs MAM, Bauer GEW. Adv Phys 1997;46:285.
[9] Hartmann U, editor. Magnetic multilayers and giant magnetoresistance. Berlin: Springer-Verlag; 1999.
[10] Ziese M, Thornton MJ, editors. Spin electronics. Berlin: Springer-Verlag; 2001.
[11] Tsymbal EY, Pettifor DG. Solid State Phys 2001;56(113):.
[12] Maekawa S, Shinjo T, editors. Spin dependent transport in magnetic nanostructures. New York, NY: Taylor and Francis; 2002.
[13] Mills DL, Bland JAC, editors. Nanomagnetism. Amsterdam: Elsevier; 2006.
[14] Xu YB, Thompson SM, editors. Spintronic materials and technology. New York, NY: Taylor and Francis; 2006.
[15] Bass J, Pratt Jr. WP. J Phys Condens Matter 2007;19:183201.
[16] Tokura Y. Rep Prog Phys 2006;69:797.
[17] Harrison W. Electronic structure and the properties of solids. W. H. Freeman and Company; 1980.
[18] Chikazumi S. Physics of ferromagnetism. 2nd ed. Oxford: Clarendon Press; 1997.
[19] Slater JC. J Appl Phys 1937;8:385.
[20] Pauling L. Phys Rev 1938;54:899.
[21] Bozorth RM. Ferromagnetism. New York, NY: van Nostrand; 1951.
[22] Crangle J, Hallam GC. Proc Royal Soc London Ser A 1963;272:119.
[23] Fert A, Cambell IA. J Phys F Metal Phys 1976;6:849.
[24] Campbell IA, Fert A. In: Wohlfarth EP, editor. Ferromagnetic materials, vol. 3. North-Holland; 1982.
[25] Inoue J, Oguri A, Maekawa S. J Phys Soc Jpn 1991;60:376.
[26] Mertig I. Rep Prog Phys 1999;62:1.
[27] Anderson PW. Phys Rev 1961;124:41.
[28] Podloucky R, Zeller R, Dederichs PH. Phys Rev B 1980;22:5777.
[29] Kittel C. Solid State Phys 1968;22:1.
[30] Inoue J, Maekawa S. J Magn Magn Mater 1993;127:L249.
[31] Soven P. Phys Rev 1967;156:809.
[32] Taylor DW. Phys Rev 1967;156:1017.
[33] Velicky B, Kirkpatrick S, Ehrenreich H. Phys Rev 1968;175:747.
[34] Brouers F, Vedyayev AV. Phys Rev B 1972;5:346.
[35] Economou EN. Green's functions in quantum physics. Springer-Verlag; 1983.
[36] Datta S. Electronic transport in mesoscopic systems. Cambridge: Cambridge University Press; 1995.
[37] Kubo R. J Phys Soc Jpn 1957;17:975.
[38] Lee PA, Fisher DS. Phys Rev Lett 1981;47:882.
[39] Landauer R. IBM J Res Dev 1957;1:223.
[40] Caroli C, Combescot R, Nozieres P, Saint-James D. J Phys C Solid St Phys 1971;4:916.
[41] Grüberg P, Schreiber R, Pang Y, Brodsky MB, Sowers H. Phys Rev Lett 1986;57:2442.
[42] Baibich MN, Broto JM, Fert A, Nguyen Van Dau F, Petroff F, Etienna P, et al. Phys Rev Lett 1988;61:2472.
[43] Binasch G, Grüberg P, Saurenbach F, Zinn W. Phys Rev B 1989;39:4828.
[44] Barnas J, Fuss A, Camley RE, Grüberg P, Zinn W. Phys Rev B 1990;42:8110.
[45] Barthélémy A, Fert A, Baibich MN, Hadjoudj S, Petroff F. J Appl Phys 1990;67:5908.
[46] Mosca DH, Petroff F, Fert A, Schroeder PA, Pratt WP, Laloee R. J Magn Magn Mater 1991;94:L1−5.

[47] Petroff F, Barthélémy A, Hamzić A, Fert A, Etienne P, Lequien S, et al. J Magn Magn Mater 1991;93:95–100.
[48] Schad R, Potter CD, Beliën P, Verbanck G, Moshchalkov VV, Bruynseraede Y. Appl Phys Lett 1994;64:3500.
[49] Parkin SSP, More N, Roche KP. Phys Rev Lett 1990;64:2304.
[50] Parkin SSP. Phys Rev Lett 1991;67:3598.
[51] Parkin SSP, Mauri D. Phys Rev B 1991;44:7131.
[52] Petroff F, Barthélémy A, Mosca DH, Lottis DK, Fert A, Schroeder PA, et al. Phys Rev B 1991;44:5355.
[53] Purcell ST, Folkerts W, Jonson MT, Mcgee NWE, Jager K, aan de Stegge J, et al. Phys Rev Lett 1991;67:903.
[54] Qiu ZQ, Pearson J, Berger A, Bader SD. Phys Rev Lett 1992;68:1398.
[55] Fullerton EE, Bader SD, Robertson JL. Phys Rev Lett 1996;77:1382.
[56] Unguris J, Celotta RJ, Pierce DT. Phys Rev Lett 1991;67:140.
[57] Unguris J, Celotta RJ, Pierce DT. Phys Rev Lett 1992;69:1125.
[58] Wang Y, Levy PM, Fry JL. Phys Rev Lett 1990;65:2732.
[59] Bruno P, Chappert C. Phys Rev Lett 1991;67:1602.
[60] Edwards DM, Mathon J, Muniz RB, Phan MS. Phys Rev Lett 1991;67:493.
[61] van Schilfgaarde M, Harrison WA. Phys Rev Lett 1993;71:3870.
[62] Mathon J, Villeret M, Muniz RB, d'Albuquerque e Castro J, Edwards DM. Phys Rev Lett 1995;74:3696.
[63] Bruno P. Phys Rev B 1995;52:411.
[64] Bruno P, Kudrnovský J, Drchal V, Turek I. Phys Rev Lett 1996;76:4254.
[65] Shinjo T, Yamamoto H. J Phys Soc Jpn 1990;59:3061.
[66] Yamamoto H, Okuyama T, Dohnomae H, Shinjo T. J Magn Magn Mater 1991;99:243–52.
[67] Dieny B, Sperious VS, Parkin SSP, Gurney BA, Wilhoit DR, Mauri D. Phys Rev B 1991;43:1297.
[68] Dieny B, Speriosu VS, Metin S, Parkin SSP, Gurney BA, Baumgart P, et al. J Appl Phys 1991;69:4774.
[69] Dieny B. Europhys Lett 1992;17:261.
[70] Pratt Jr. WP, Lee S-F, Slaughter JM, Loloee R, Schroeder PA, Bass J. Phys Rev Lett 1991;66:.
[71] Yang Q, Holody P, Lee S-F, Henry LL, Loloee R, Schroeder PA, et al. Phys Rev Lett 1994;72:3274.
[72] Bass J, Pratt Jr. WP. J Magn Magn Mater 1999;200:274.
[73] Gijs MAM, Lenczowski SKJ, Giesbers JB. Phys Rev Lett 1993;70:3343.
[74] Gijs MAM, Lenczowski SKJ, van de Veerdonk RJM, Giesbers JB, Johnson MT, Faan de Stegge JB. Phys Rev B 1994;50:16733.
[75] Piraux L, George JM, Despres JF, Leroy C, Ferain E, Legras R, et al. Appl Phys Lett 1994;65:2484.
[76] Levy PM, Zhang S, Ono T, Shinjo T. Phys Rev B 1995;52:16049.
[77] Ono T, Shinjo T. J Phys Soc Jpn 1995;64:363–6.
[78] Berkowitz AE, Mitchell JR, Carey MJ, Yong AP, Zhang S, Spada FE, et al. Phys Rev Lett 1992;68:3745.
[79] Xiao JQ, Jiang JS, Chien CL. Phys Rev Lett 1992;68:3749.
[80] Xiao JQ, Jiang JS, Chien CL. Phys Rev B 1992;46:9266.
[81] Barnard JA, Waknis A, Tan M, Haftek E, Parker MR, Watson ML. J Magn Magn Mater 1992;114:L230–4.

[82] Rubinstein M. Phys Rev B 1994;50:3830.
[83] Wang J-Q, Xiao G. Phys Rev B 1994;50:3423.
[84] Hickey BJ, Howson MA, Musa SO, Wiser N. Phys Rev B 1995;51:667.
[85] Fullerton EE, Kelly DM, Guimpel J, Schuller IK. Phys Rev Lett 1992;68:859.
[86] Parkin SSP. Phys Rev Lett 1993;71:1641.
[87] Gurney BA, Speriosu VS, Nozieres JP, Lefakis H, Wilhoit DR, Need OU. Phys Rev Lett 1993;71:4023.
[88] Speriosu VS, Nozieres JP, Gurney BA, Dieny B, Huang TC, Lefakis H. Phys Rev B 1993;47:11579.
[89] George JM, Pereira LG, Barthélémy A, Petroff F, Steren L, Duvail JL, et al. Phys Rev Lett 1994;72:408.
[90] Inoue J, Itoh H, Maekawa S. J Phys Soc Jpn 1992;61:1149.
[91] Itoh H, Inoue J, Maekawa S. Phys Rev B 1993;47:5809.
[92] Bauer GEW. Phys Rev Lett 1992;69:1676.
[93] Bauer GEW, Brataas A, Schep KM, Kelly PJ. J Appl Phys 1994;75:6704.
[94] Schep KM, Kelly PJ, Bauer GEW. Phys Rev Lett 1995;74:586.
[95] Camley RE, Barnaś J. Phys Rev Lett 1989;63:664.
[96] Zahn P, Mertig I, Richter M, Eschring H. Phys Rev Lett 1995;75:2996.
[97] Kubota H, Miyazaki T. J Magn Soc Jpn 1994;18:335 [in Japanese]
[98] Saito Y, Inomata K. Jpn J Appl Phys 1991;30:L1733.
[99] Kataoka N, Saito K, Fujimori H. Mat Trans JIM 1992;33:151.
[100] Kataoka N, Saito K, Fujimori H. J Magn Magn Mater 1993;121:383.
[101] Sakakima H, Satomi M. J Magn Magn Mater 1993;121:374.
[102] Kubota H, Sato M, Miyazaki T. Phys Rev B 1995;52:343.
[103] Itoh H, Hori T, Inoue J, Maekawa S. J Magn Magn Mater 1993;121:344.
[104] Miyazaki T, Kondo J, Kubota H, Inoue J. J Appl Phys 1997;81:5187.
[105] Itoh H, Inoue J, Maekawa S. Phys Rev B 1995;51:342.
[106] Asano Y, Oguri A, Maekawa S. Phys Rev B 1993;48:6192.
[107] Asano Y, Oguri A, Inoue,and J, Maekawa S. Phys Rev B 1994;49:12831.
[108] Schep KM. Ph.D. (1997), Thesis, Technical University of Delft, The Netherland.
[109] Schep KM, Kelly PJ, Bauer GEW. J Magn Magn Mater 1996;156:385.
[110] Schep KM, van Hoof JBAN, Kelly PJ, Bauer GEW, Inglesfield JE. Phys Rev B 1997;56:10805.
[111] Valet T, Fert A. Phys Rev B 1993;48:7099.
[112] Julliere M. Phys Lett 1975;54A:225.
[113] Maekawa S, Gäfvert U. IEEE Trans Magn 1982;18:707.
[114] Miyazaki T, Tezuka N. J Magn Magn Mater 1995;151:403−10.
[115] Moodera JS, Kinder LR, Wong TM, Meservey R. Phys Rev Lett 1995;74:3273.
[116] Fujimori H, Mitani S, Ohnuma S. Mater Sci Eng B 1995;31:219.
[117] Mitani S, et al. Phys Rev Lett 1998;81:2799.
[118] Bratkovsky AM. Phys Rev B 1997;56:2344.
[119] Guinea F. Phys Rev B 1998;58:9212.
[120] Yu. E, Tsymbal, Pettifor DG. Phys Rev B 1998;58:432.
[121] Mazin II. Phys Rev Lett 1999;83:1427.
[122] Moodera JS, Mathon G. J Magn Magn Mater 1999;200:248−73.
[123] Bardeen J. Phys Rev Lett 1961;6:57.
[124] Tedrow PM, Meservey R. Phys Rev Lett 1971;26:192.
[125] Soulen Jr. RJ, Byers JM, Osofsky MS, Nadgorny B, Ambrose T, Cheng SF, et al. Science 1998;282:85.

[126] Upadhyay SK, Palanisami A, Louie RN, Buhrman RA. Phys Rev Lett 1998;81:3247.
[127] Upadhyay SK, Louie RN, Buhrman RA. Appl Phys Lett 1999;74:3881.
[128] Worledge DC, Geballe TH. Phys Rev Lett 2000;85:5182.
[129] Kaiser C, van Dijken S, Yang S-H, Yang H, Parkin SSP. Phys Rev Lett 2005;94:247203.
[130] Kaiser C, Panchula AF, Parkin SSP. Phys Rev Lett 2005;95:047202.
[131] Stearns MB. J Magn Magn Mater 1977;5:167−71.
[132] Slonczewski JC. Phys Rev B 1989;39:6995.
[133] MacLaren JM, Zhang X-G, Butler WH, Wang X. Phys Rev B 1999;59:5470.
[134] Mathon J, Umerski A. Phys Rev B 2001;63:220403.
[135] Butler WH, Zhang X-G, Schulthess TC, MacLaren JM. Phys Rev B 2001;63:54416.
[136] Itoh H, Kumazaki T, Inoue J, Maekawa S. Jpn J Appl Phys 1998;37:5554.
[137] Itoh H, Shibata A, Kumazaki T, Inoue J, Maekawa S. J Phys Soc Jpn 1999;69:1632.
[138] Itoh H, Inoue J. Trans Mag Soc Jpn 2006;30:1.
[139] Itoh H. J Phys D Appl Phys 2006;40:1228.
[140] Honda S, Itoh H, Inoue J. Phys Rev B 2006;74:155329.
[141] Inoue J, Maekawa S. J Magn Magn Mater 1999;**198−199**:167.
[142] Takada I, Itoh H, Inoue J. J Magn Soc Jpn 2006; 32: 338.
[143] Zhang S, Levy PM, Marley AC, Parkin SSP. Phys Rev Lett 1997;79:3744.
[144] Yuasa S, Sato T, Tamura E, Suzuki Y, Yamamori H, Ando K, et al. Europhys Lett 2000;52:344−50.
[145] Nagahama T, Yuasa S, Suzuki Y, Tamura E. Appl Phys Lett 2001;79:4381.
[146] Yuasa S, Fukushima A, Nagahama T, Ando K, Suzuki Y. Jpn J Appl Phys 2004;43:.
[147] Parkin SSP, Kaiser C, Panchula A, Rice PM, Hughes B, Samant M, et al. Nat Mater 2004;3:862.
[148] Yuasa S, Nagahama T, Fukushima A, Suzuki Y, Ando K. Nat Mater 2004;3:868−71.
[149] Djayaprawira DD, Tsunekawa K, Nagai M, Maehara H, Yamagata S, Watanabe N, et al. Appl Phys Lett 2005;86:092502.
[150] LeClair P, Swagten HJM, Kohlhepp JT, van de Veerdonk RJM, de Jonge WJM. Phys Rev Lett 2000;84:2933.
[151] LeClair P, Kohlhepp JT, Swagten HJM, de Jonge WJM. Phys Rev Lett 2001;86:1066.
[152] LeClair P, Kohlhepp JT, van de Vin CH, Wieldraaijer H, Swagten HJM, de Jonge WJM, et al. Phys Rev Lett 2002;88:107201.
[153] Inoue J, Itoh H. J Phys D Appl Phys 2002;35:1.
[154] Yuasa S, Katayama T, Suzuki Y. Science 2002;297:234.
[155] Nagahama T, Yuasa S, Tamura E, Suzuki Y. Phys Rev Lett 2005;95:086602.
[156] Mathon J, Villeret M, Itoh H. Phys Rev B 1995;52:R6983.
[157] Mathon J, Umerski A. Phys Rev B 1999;60:1117.
[158] Itoh H, Inoue J, Umerski A, Mathon J. Phys Rev B 2003;68:174421.
[159] Itoh H, Inoue J. J Magn Soc Jpn 2003;27:1013 [in Japanese]
[160] de Groot RA, Buschow KHJ. J Magn Magn Mater 1986;**54−57**:1377.
[161] Schwarz K. J Phys F Metal Phys 1986;16:L211.
[162] Pickett WE, Singh DJ. Phys Rev B 1996;53:1146.
[163] Wei JYT, Yeh N-C, Vasques RP. Phys Rev Lett 1997;79:5150.
[164] Hu G, Suzuki Y. Phys Rev Lett 2002;89:276601.
[165] Dowben PA, Skomski R. J Appl Phys 2004;95:7453.
[166] Sakuraba Y, Nakata J, Oogane M, Kubota H, Ando Y, Sakuma A, et al. Jpn J Appl Phys 2005;44:L1100.

[167] Sakuraba Y, Hattori M, Oogane M, Kubota H, Ando Y, Sakuma A, et al. J Phys D Appl Phys 2006;40:1221.
[168] Wollan EO, Koehler WC. Phys Rev 1955;100:545.
[169] Urushibara A, Moritomo Y, Arima T, Asamitsu A, Kido G, Tokura Y. Phys Rev B 1995;51:14103.
[170] Inoue J, Maekawa S. Phys Rev Lett 1995;74:3407.
[171] de Gennes P-G. Phys Rev 1960;118:141.
[172] Bowen M, Bibes M, Bathélémy A, Contour J-P, Anane A, Lemaitre Y, et al. Appl Phys Lett 2003;82:233.
[173] Bowen M, Barthélémy A, Bibes M, Jacquet E, Contour J-P, Fert A, et al. Phys Rev Lett 2005;95:137203.
[174] Garcia V, Bibes M, Barthélémy A, Bowen M, Jacquet E, Contour J-P, et al. Phys Rev B 2004;69:52403.
[175] Lu Y, Li XW, Gong GQ, Xiao G, Gupta A, Lecoeur P, et al. Phys Rev B 1996;54: R8357.
[176] Kwon C, Kim. K-C, Robson MC, Gu JY, Rajeswari M, Venkatesan T, et al. J Appl Phys 1997;81:4950.
[177] Obata T, Mnako T, Shimakawa Y, Kubo Y. Appl Phys Lett 1999;74:290.
[178] Itoh H, Ohsawa T, Inoue J. Phys Rev Lett 2000;84:2501.
[179] De Teresa JM, Barthrémy A, Fert A, Contour JP, Lyonnet R, Montaigne F, et al. Phys Rev Lett 1999;82:4288.
[180] Velev JP, Belashchenko KD, Stewart DA, van Schilfgaarde M, Jaswal SS, Tsymbal EY. Phys Rev Lett 2005;95:216601.
[181] Sakuraba Y, Hattori M, Oogane M, Ando Y, Kato H, Sakuma A, et al. Appl Phys Lett 2006;88:192508.
[182] Galanakis I. J Phys Condens Matter 2002;14:6329.
[183] Lezaic M, Movropoulos Ph, Enkovaara J, Bihlmayer G, Brugel S. Phys Rev Lett 2006;97:26404.
[184] Hayakawa J, Ikeda S, Lee YM, Matsukura F, Ohno H. Appl Phys Lett 2006;89:232510.
[185] Lee YM, Hayakawa J, Ikeda S, Matsukura F, Ohno H. Appl Phys Lett 2006;89:042506.
[186] Lee YM, Hayakawa J, Ikeda S, Matsukura F, Ohno H. Appl Phys Lett 2007;90:212507.
[187] Marukame T, Ishikawa T, Hatamata S, Matsuda K, Uemura T, Yamamoto M. Appl Phys Lett 2007;90:012508.
[188] Tezuka N, Ikeda N, Sugimoto S, Inomata K. Jpn J Appl Phys 2007;46:L454.
[189] Ishikawa T, Hakamata S, Matsuda K, Uemura T, Yamamoto M. J Appl Phys 2008;103:07A919.
[190] Matsumoto R, Fukushima A, Nagahama T, Suzuki Y, Ando K, Yuasa S. Appl Phys Lett 2007;90:252506.
[191] Wolf SA, Awschalom DD, Buhrman RA, Daughton JM, von Molnar S, Roukes ML, et al. Science 2001;294:1488.
[192] Kossut J, Dobrowoski W. In: Buschow KH, editor. Handbook of magnetic materials, vol. 7. Amsterdam: North-Holland; 1993.
[193] Munekata H, Ohno H, von Molnar S, Segmuller A, Chang LL, Esaki L. Phys Rev Lett 1989;63:1849.
[194] Ohno H, Munekata H, von Molnar S, Chang LL. J Appl Phys 1991;69:6103.
[195] Ohno H. Science 1998;281:951.

[196] Matsukura F, Ohno H, Shen A, Sugawara Y. Phys Rev B 1998;57:R2037.
[197] Nazmul AM, Amemiya T, Shuto Y, Sugawara S, Tanaka M. Phys Rev Lett 2005;95:17201.
[198] Shuto Y, Tanaka M, Sugahara S. J Appl Phys 2006;99:08D516.
[199] Tanaka M, Higo Y. Phys Rev Lett 2001;87:026602.
[200] Chiba D, Matsukura F, Ohno H. Phys E 2004;21:966.
[201] Gould C, Rüster C, Jungwirth T, Girgis E, Schott GM, Giraud R, et al. Phys Rev Lett 2004;93:117203.
[202] Rüster C, Gould C, Jungwirth T, Sinova J, Schott GM, Giraud R, et al. Phys Rev Lett 2005;94:27203.
[203] Giddings AD, Khalid MN, Jungwirth T, Wunderlich J, Yasin. S, Campion RP, et al. Phys Rev Lett 2005;94:127202.
[204] Saito H, Yuasa S, Ando K. Phys Rev Lett 2005;95:86604.
[205] Sheng P, Abeles B, Arie Y. Phys Rev Lett 1973;31:44.
[206] Herman JS, Abeles B. Phys Rev Lett 1976;37:1429.
[207] Inoue J, Maekawa S. Phys Rev B 1996;53:R11927.
[208] Beenakker CW. Phys Rev B 1991;44:1646.
[209] Ono K, Shimada H, Ootuka Y. J Phys Soc Jpn 1997;66:1261.
[210] Barnas J, Fert A. Phys Rev Lett 1998;80:1058.
[211] Brataas A, Nazarov YuV, Inoue J, Bauer GEW. Phys Rev B 1999;59:93.
[212] Takahashi S, Maekawa S. Phys Rev Lett 1998;80:1758.
[213] van Wees BJ, van Houten H, Beenakker CW, van der Marel D, Foxton CT. Phys Rev Lett 1988;60:848.
[214] Scheer E, Agrat N, Cuevas JC, Yeyati AL, Ludoph B, M-Rodero A, et al. Nature 1998;394:154.
[215] Cuevas JC, Yeyati AL, Rodero AM. Phys Rev Lett 1998;80:1066.
[216] Ono T, Ooka Y, Miyajima H, Otani Y. Appl Phys Lett 1999;75:1622.
[217] Garca N, Muñoz M, Zhao Y-W. Phys Rev Lett 1999;82:2923.
[218] Chopra HD, Hua SZ. Phys Rev B 2002;66:020403.
[219] Hua SZ, Chopra HD. Phys Rev B 2003;67:060401.
[220] Tatara G, Zhao Y-W, Muñoz M, Garca N. Phys Rev Lett 1999;83:2030.
[221] Bruno P. Phys Rev Lett 1999;83:2425.
[222] van Hoof JBAN, Schep KM, Brataas A, Bauer GEW, Kelly PJ. Phys Rev B 1999;59:138.
[223] Imamura H, Kobayashi N, Takahashi S, Maekawa S. Phys Rev Lett 2000;84:1003.
[224] Bagrets A, Papanikolaou N, Mertig I. Phys Rev B 2004;70:064410.
[225] Stepanyuk VS, Klavsyuk AL, Hergert W, Saletsky AM, Bruno P, Mertig I. Phys Rev Lett 2004;70:195420.
[226] Papanikolaou N. J Phys Condens Matter 2003;15:5049.
[227] Smit J. Physica 1951;16:612.
[228] Millis AJ, Shraiman BI, Mueller R. Phys Rev Lett 1996;77:175.
[229] Varma CM. Phys Rev B 1996;54:7328.
[230] Sheng L, Xing DY, Sheng DN, Ting CS. Phys Rev Lett 1997;79:1710.
[231] Moreo A, Yunoki S, Dagotto E. Science 1999;283:2034.
[232] Vergés JA, Mayor VM, Brey L. Phys Rev Lett 2002;88:136401.
[233] Ramakrisman TV, Krishnamurthy HR, Hassan SR, Venketeswara Pai G. Phys Rev Lett 2004;92:157203.
[234] Tokura Y, editor. Colossal magnetoresistive oxides. London: Gordon and Breach; 2000.

[235] Dagotto E. Nanoscale phase separation and colossal magnetoresistance. New York, NY: *Springer*; 2002.

[236] Chatterji T, editor. Colossal magnetoresistive manganites. London: Kluwer Academic Publishers; 2002.

[237] Nagaev EL. Colossal magnetoresistance and phase separation in magnetic semiconductors. London: Imperial College Press; 2002.

[238] Dresselhaus G. Phys Rev 1955;100:580.

[239] Rashba EI. Sov Phys Solid State 1960;2:1109.

[240] Bychkov YA, Rashba EI. J Phys C 1984;17:6039.

[241] Hall EH. Am J Math 1879;2:287.

[242] Hall EH. Philos Mag 1880;19:301.Kundt A. Annalen der Phys und Chemie 1893;49:257.

[243] Karplus R, Luttinger JM. Phys Rev 1954;95:1154.

[244] Smit J. Physica 1958;24:39.

[245] Berger L. Phys Rev 1970;B2:4559.

[246] Crepieux A, Bruno P. Phys Rev B 2001;64:14416.

[247] Jungwirth T, Niu Q, MacDonald AH. Phys Rev Lett 2002;88:207208.

[248] Ohno H. J Magn Magn Mater 1999;200:110.

[249] Yao Y, et al. Phys Rev Lett 2004;92:37204.

[250] D'yakonov MI, Perel VI. ZhETF Pis Red 1971;13:657.

[251] Hirsh JE. Phys Re Lett 1999;83:1834.

[252] Murakami S, Nagaosa N, Zhang SC. Science 2003;301:1348.

[253] Sinova J, Culcer D, Niu Q, Sinitsyn NA, Jungwirth T, MacDonald AH. Phys Rev Lett 2004;92:126603.

[254] Kato YK, Myers RC, Gossard AC, Awschalom DD. Science 2004;306:1910.

[255] Sih V, Myers RC, Kato YK, Lau WH, Gossard AC, Awschalom DD. Nat Phys 2005;1:31.

[256] Wunderlich J, Kaestner B, Sinova J, Jungwirth T. Phys Rev Lett 2005;94:047204.

[257] Inoue J, Bauer GEW, Molenkamp LW. Phys Rev 2004;B70:041303.

[258] Zhang S, Yang Z. Phys Rev Lett 2005;94:066602.

[259] Bernevig BA, Zhang S-C. Phys Rev Lett 2005;95:16801.

[260] Engel H-A, Halperin BI, Rashba EI. Phys Rev Lett 2005;95:166605.

[261] Shchelushkin RV, Brataas A. Phys Rev B 2005;71:045123.

[262] Kane CL, Mele EJ. Phys Rev Lett 2005;95:146802.

[263] Kane CL, Mele EJ. Phys Rev Lett 2006;95:226801.

[264] Konig M, Wiedmann S, Brune C, Roth A, Buhmann H, Molenkamp LM, et al. Science 2007;318:766.

[265] Valenzuela SO, Tinkham M. Nature 2006;442:176.

[266] Kimura T, Ohtani Y, Sato T, Takahashi S, Maekawa S. Phys Rev Lett 2007;98:156601.

[267] Streda P. J Phys C Solid State Phys 1982;15:L717.

[268] Kontani H, Naito M, Hirashima DS, Yamada K, Inoue J. J Phys Soc Jpn 2007;76:103702.

[269] Kontani H, Tanaka T, Hirashima DS, Yamada K, Inoue J. Phys Rev Lett 2008;100:09661.

[270] Tanaka T, Kontani H, Naito M, Naito T, Hirashima DS, Yamada K, et al. Phys Rev B 2007;77:165117.

[271] Seki T, Hasegawa Y, Mitani S, Takahashi S, Imamura H, Maekawa S, et al. Nat Mater 2008;7:125.

[272] Guo G-Y, Maekawa S, Nagaosa N. Phys Rev Lett 2009;102:036401.
[273] Gradhand M, Fedorov. DV, Zahn P, Mertig I. Phys Rev Lett 2010;104:186403.
[274] Nagaosa N, Sinova J, Onoda S, MacDonald AH, Ong NP. Rev Mod Phys 2010;82:1539.
[275] Vas'ko FT, Prime NA. Sov Phys Solid State 1979;21:994.
[276] Levitov LS, et al. Zh Eksp Teor Fiz 1985;88:229.
[277] Edelstein VME. Sol State Commun 1990;73:233.
[278] Inoue J, Bauer GEW, Molenkamp LW. Phys Rev B 2003;67:33104.
[279] Kato YK, Myers RC, Gossard AC, Awschalom DD. Phys Rev Lett 2004;93:176601.
[280] Ganichev SD , Bel'kov VV , Tarasenko SA, Danilov SN, Giglberger S, Hoffmann C,
       Ivchenko EL, Weiss D, Wegscheider W, Gerl C, Schuh D, Stahl J, de Boeck J,
       Borghs G, Prettl W. nature phys. 2006;2:609.
[281] Yang CL, He HT, Lu Ding LJ, Cui YP, Zeng JN, Wang, et al. Phys Rev Lett
       2006;96:186605.
[282] Onsager L. Phys Rev 1931;37:405.
[283] Bauer GEW, Saitoh E, van Wees BJ. Nat Mater 2012;11:391.
[284] Special issue of Sol Stat Commun 2010;150.
[285] Ziman JM. Principles of the theory of solids. 2nd ed. Cambridge: Cambridge
       University Press; 1972.
[286] Uchida K, Takahashi S, Harii K, Ieda J, Koshibae W, Ando K, et al. Nature
       2008;455:778.
[287] Slachter A, Bakker FL, Adams JP, van Wees BJ. Nat Phys 2010;6:879.
[288] Uchida K, Ota T, Adachi H, Xiao J, Nonaka T, Kajiwara Y, et al. J Phys Soc Jpn
       2012;111:103903.
[289] Jaworski CM, Yang J, Mack S, Awschalom DD, Heremans JP, Myers RC. Nat Mater
       2010;9:898.
[290] Xiao J, Bauer GEW, Uchida K, Saitoh E, Maekawa S. Phys Rev B 2010;81:214418.
[291] Di Xiao Y, Yao Z, Fang, Niu Q. Phys Rev Lett 2006;97:026603.
[292] Berger L. J Appl Phys 1992;71:2721.
[293] Slonczewski JC. J Magn Magn Mater 1996;159:L1.
[294] Sun JZ. J Magn Magn Mater 1999;202:157.
[295] Edwards DM, Federici F, Mathon J, Umerski A. Phys Rev B 2005;71:054407.
[296] Hatami M, Bauer GEW, Zhang Q, Kelly P. Phys Rev Lett 2007;99:066603.
[297] Mizukami S, Ando Y, Miyazaki T. Phys Rev B 2002;66:104413.
[298] Tserkovnyak Y, Brataas A, Bauer GEW, Halperin BI. Rev Mod Phys 2005;77:1375.
[299] Brataas A, Bauer GEW, Kelly PJ. Phys Rep 2006;427:157.
[300] Brataas A, Kent AD, Ohno H. Nat Mater 2012;11:372.
[301] Yakata S, Ando Y, Miyazaki T, Mizumaki S. Jpn J Appl Phys 2006;45:3892.
[302] Gerrits Th, Schneider ML, Silva TJ. J Appl Phys 2006;99:023901.
[303] Kardasz B, Mosendz O, Heinrich B, Liu Z, Freeman M. J Appl Phys
       2008;103:07C509.
[304] Saitoh E, Ueda M, Miyajima H, Tatara G. Appl Phys Lett 2006;88:182509.
[305] Theodonis I, Kioussis N, Kalitov A, Chshiev M, Butler WH. Phys Rev Lett
       2006;97:237205.
[306] Xiao J, Bauer GEW, Brataas A. Phys Rev B 2008;77:224419.
[307] Keldysh LV. Sov Phys JETP 1965;20:1018.
[308] Inoue J. Phys Rev B 2011;84:180402R.
[309] Inoue J. J Appl Phys 2012;111:07C902.
[310] Slonczewski JC. Phys Rev B 2005;71:024411.
[311] de Groot RA, Mueller FM, van Engen PG, Buschow KHJ. Phys Rev Lett
       1983;50:2024.

3 Spin Injection and Voltage Effects in Magnetic Nanopillars and Its Applications

Yoshishige Suzuki[1], Ashwin A. Tulapurkar[2], Youichi Shiota[1] and Claude Chappert[3]

[1]Department of Materials Engineering Science, Graduate School of Engineering Science, Toyonaka, Osaka, Japan, [2]Department of Electrical Engineering, Indian Institute of Technology, Powai, Mumbai, India, [3]Institut d'Electronique Fondamentale, Université Paris-Sud, Orsay cedex, France

Contents

Nanomagnetism and Spintronics. DOI: http://dx.doi.org/10.1016/B978-0-444-63279-1.00003-4

3.1 Spin Injection, Voltage Application, and Torque

3.1.1 Spin Injection

Electrons possess both charge $(-e)$ and spin-angular momentum$(\hbar/2)$. Therefore, when an electron moves in a material, it carries not only a charge but also an angular momentum. In Figure 3.1A, a schematic of electron transfer from a ferromagnetic material (FM) to a nonmagnetic material (NM)[1] through an interface is shown. In the ferromagnetic material, since electron spins are polarized, an electric current accompanies the net flow of spins, that is, "spin current." The spins traveling as the spin current in the ferromagnet are then injected ("spin injection") into the nonmagnetic material through the interface. The injected spins are subjected to spin relaxation because of spin-orbit interaction, and they lose their spin orientations as they move away from the interface. First attempts to observe the spin injection were done in metallic systems in 1980s by detecting precession of injected spins (Hanle effect) and by detecting spin-dependent chemical potentials using a transistor-like structure [1,2]. In the beginning of 2000s, Jedema et al. [3] observed a clear evidence of the spin injection into metals employing so-called nonlocal magnetoresistance geometry together with an observation of the Hanle effect. Spin injections into semiconductors have been observed by detecting an emission of circularly polarized light in compound semiconductors such as GaAs [4—7], by detecting Hanle effect and nonlocal magnetoresistance in GaAs [8] and Si [9—11], and by using a magnetoresistive/

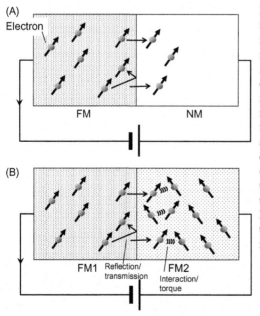

(A)
Electron

FM   NM

(B)

FM1   Reflection/   FM2
      transmission   Interaction/
                     torque

Figure 3.1 Spin injection across an interface. (A) Spin injection from a ferromagnetic material (FM) to a nonmagnetic material (NM). Application of an electric charge current through the interface results in injection of a spin-polarized current into NM. (B) Spin injection from a ferromagnetic material (FM1) to another ferromagnetic material with different magnetization direction (FM2). Spin-dependent reflection and transmission at the interface give rise to the magnetoresistance effect. Injected spins interact with spins of the host material and exert torque on it.

---

[1] In this chapter, we call both paramagnetic and diamagnetic materials as a nonmagnetic material.

filtering effect in Si [12]. The spin injection was also observed in exotic materials like carbon nanotubes [13] and graphenes [14−16].

What happens if we pass a current through two ferromagnetic materials with different magnetization orientations as shown in Figure 3.1B? It is well known that the magnetoresistance effect occurs because of spin-dependent scattering (reflection) at the interface, as described in the previous chapter. In addition, the electron spins injected from the ferromagnetic material on the left-hand side (FM1) into the ferromagnetic material on the right-hand side (FM2) interact with the electron spins in FM2 through exchange interaction. As a result, a precession can be excited [17−19] and the magnetization can be switched [20−28] in FM2.

Before providing a detailed explanation of these interesting effects, let us discuss the concepts of "spin injection" and "spin current."

In a system in which many electrons are interacting, if the interaction between the electron spins is only exchange interaction, the total spin-angular momentum of the electron system is conserved. Therefore, similar to the charge current density, which is derived from the conserved quantity, charge, one can define a flow of the spin-angular momentum density, that is, spin current density:

$$\hat{j}^S(\vec{x},t) = \sum_i \vec{v}_i(t)\vec{s}_i(\vec{x},t) + \text{(exchange mediated term)}. \tag{3.1}$$

Here, $\vec{s}_i(\vec{x},t)$ and $\vec{v}_i(t)$ are the electron spin density and velocity of the ith electron, respectively, and Σ is the sum over all the electrons concerned. The first term in Eq. (3.1) indicates that the flow of spin-polarized electrons carries spin current, that is, angular momentum. The second term indicates that the spin-angular momentum can be transferred by the exchange interaction even if the positions of the electrons are fixed. The latter term corresponds to a spin-angular momentum transfer through spin waves. This term can be expressed locally and is roughly proportional to $\sum_{i,j} \vec{s}_i(\vec{x},t) \times \vec{\nabla} \vec{s}_j(\vec{x},t)$ if spins rotate continuously in space. Here, it should be noted that the spin current density, $\hat{j}^S$, is a direct product of two vectors expressing the direction of the current flow and the direction of spin. The rule for conserving the spin-angular momentum and the charge is expressed as follows:

$$\begin{cases} \dfrac{\partial \vec{s}}{\partial t} + \text{div}\,\hat{j}^S = 0, \\[2mm] \dfrac{\partial \rho}{\partial t} + \text{div}\,\vec{j}^Q = 0, \end{cases} \tag{3.2}$$

where ρ is charge density, $\vec{j}^Q$, electric current density and $\text{div}\,\vec{A} = \left(\frac{\partial A_x}{\partial x} + \frac{\partial A_y}{\partial y} + \frac{\partial A_z}{\partial z}\right)$ expresses the divergence of flow $\vec{A}$.

Now, we consider a case where an electric field is applied to a large ferromagnetic material that is magnetized along $-\vec{e}_{spin}(|\vec{e}_{spin}| = 1)$. In the ferromagnetic materials, the spin-angular momentum density, $\vec{s}$, is nonzero since the density of the majority spin electrons, n_+, is larger than that of the minority spin electrons, n_-:

$$\begin{cases} \vec{s} = \vec{s}_+ + \vec{s}_- = \dfrac{\hbar}{2}\vec{e}_{spin}(n_+ - n_-), \\ \rho = \rho_+ + \rho_- = (-e)(n_+ + n_-). \end{cases} \tag{3.3}$$

Here, $\vec{s}_+(\vec{s}_-)$ and $\rho_+(\rho_-)$ are the spin density and charge density of the majority (minority) spin electrons. The direction of angular momentum is opposite to that of magnetization for the electrons. Hereafter, the treatment follows a derivation given by Valet and Fert [29], and suffixes $+$ and $-$ correspond to the majority and the minority spins, respectively.

As the system is large enough compared to the electron mean free path, the electrons flow diffusively by repeated scattering and acceleration. The scattering events, however, seldom change the spin orientation. Therefore, a physical picture can be constructed in which the majority and minority spin electrons flow diffusively independent of each other in each spin subchannel (Figure 3.2). Since the charge densities and drift velocities $(\vec{v}_+, \vec{v}_-)$ are different in different spin subchannels, the electric conductivities (σ_+, σ_-), which are equal to (charge density) $\times$ (mobility), are also different in different spin subchannels. As a result, the application of an

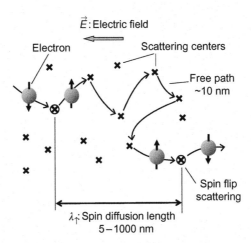

Figure 3.2 Drift flow of a free electron in a conductive material. An electron with negative charge $(-e)$ is subject to an electric field $\vec{E}$ and accelerated toward the right. After a short travel, the electron is scattered by a scattering center and its velocity changes. The electron is accelerated again until its next collision. Through the repetition of this process, the electron moves opposite to $\vec{E}$ on average. Then average distance between two scattering events is the mean free path. For metals at room temperature, the typical values of the mean free path are about 10 nm. During this drift motion, the electron seldom changes its spin direction. The average distance between two spin flip events under zero electric field corresponds to the spin diffusion length, $\lambda_\pm$.

electric field results not only in a drift charge current density, $\vec{j}^{Q,\text{Drift}}$, but also in a drift spin current density, $\hat{j}^{S,\text{Drift}}$.

$$\begin{cases} \hat{j}^{S,\text{Drift}} = \vec{v}_+\vec{s}_+ + \vec{v}_-\vec{s}_- = \frac{\hbar}{2}\vec{e}_{\text{spin}}\frac{1}{-e}(\sigma_+ - \sigma_-)\vec{E}, \\ \vec{j}^{Q,\text{Drift}} = \vec{v}_+\rho_+ + \vec{v}_-\rho_- = (\sigma_+ + \sigma_-)\vec{E}. \end{cases} \tag{3.4}$$

Here, the second term in Eq. (3.1) is neglected for simplicity.

Next, let us consider the effect of current passing through the interface between a ferromagnetic material (FM) and a nonmagnetic material (NM) as shown in Figure 3.3A. In the junction, we take two sections P and Q in the ferromagnetic material layer and the nonmagnetic material layer, respectively. We assume that both P and Q are far enough from the interface to be affected by it. The spin current should be finite at P in the ferromagnetic material but zero at Q in the

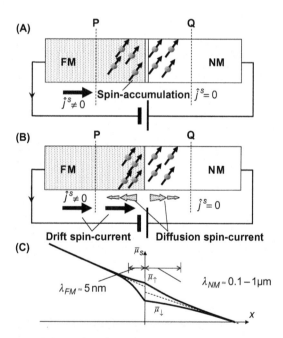

Figure 3.3 Spin accumulation, relaxation, and diffusion during spin injection from a ferromagnetic material (FM) to a nonmagnetic material (NM). (A) The spin current at section P is finite whereas that at Q is zero, that is, spins are injected into the interface region between P and Q but never leave it. As a result the spins accumulate in the interface region. (B) The number of accumulated spins is decreased because of spin flip scattering inside the interface region. Spin accumulation produces a gradient of the spin density and causes the spin diffusion current that moves spins away from the interface. (C) Associated electrochemical potential distribution. λ_{FM} and λ_{NM} are the average spin diffusion length of FM and NM. Detailed explanation is provided in text.

nonmagnetic material as consequences of Eq. (3.4). Then, from the conservation law of spin momentum (Eq. (3.2)), it is expected that the spins will accumulate around the interface as time goes by. In reality, spin accumulation does not grow indefinitely but the accumulated spins are lost through spin-orbit coupling. To take this effect into account, we add spin relaxation terms to Eq. (3.2).

$$\begin{cases} \dfrac{\partial \vec{s}}{\partial t} + \text{div}\,\hat{j}^{S} = -\dfrac{\vec{s}_{+} - \vec{s}_{+}^{\,eq}}{\tau_{+}^{sf}} - \dfrac{\vec{s}_{-} - \vec{s}_{-}^{\,eq}}{\tau_{-}^{sf}}, \\[3mm] \dfrac{\partial \rho}{\partial t} + \text{div}\,\vec{j}^{Q} = 0, \end{cases} \qquad (3.2')$$

where $\tau_{\pm}^{sf}$ and $\vec{s}_{\pm}^{\,eq}$ are the spin relaxation time and spin density in the thermal equilibrium condition in each spin subchannel. $\tau_{\pm}^{sf}$ defined in this chapter is half that defined in Ref. [29]. Since the density of states at the Fermi energy in metals, $N_{\pm}$, and the carrier density in semiconductors, $n_{\pm}$, are different for different spin subbands, the spin relaxation times are also different and are related as follows:

$$\begin{aligned} \tau_{+}^{sf}N_{-} &= \tau_{-}^{sf}N_{+} : \text{Metals} \\ \tau_{+}^{sf}n_{-} &= \tau_{-}^{sf}n_{+} : \text{Nondegenerate semiconductors} \end{aligned} \qquad (3.5)$$

The accumulated spins also flow from the interface to the bulk by diffusion resulting in a diffusion spin current (Figure 3.3B):

$$\begin{cases} \hat{j}^{S,Diffusion} = -D_{+}\vec{\nabla}\vec{s}_{+} - D_{-}\vec{\nabla}\vec{s}_{-}, \\[2mm] \vec{j}^{Q,Diffusion} = -D_{+}\vec{\nabla}\rho_{+} - D_{-}\vec{\nabla}\rho_{-}, \end{cases} \qquad (3.6)$$

where $D_{\pm}$ is the diffusion constant of the electron in each spin subchannel. The Einstein relation holds between the diffusion constants and the conductivities:

$$\begin{aligned} \sigma_{\pm} &= N_{\pm}e^{2}D_{\pm} : \text{Metals}, \\ \sigma_{\pm} &= n_{\pm}e^{2}D_{\pm}/kT : \text{Nondegenerate smiconductors}. \end{aligned} \qquad (3.7)$$

The gradient of the electron density in each spin subchannels can be expressed by using spin-dependent chemical potentials, $\mu_{\pm}$, as follows:

$$\begin{aligned} \vec{\nabla}n_{\pm} &= N_{\pm}\vec{\nabla}\mu_{\pm} : \text{Metals}, \\ \vec{\nabla}n_{\pm} &\approx \pm n_{\pm}\vec{\nabla}\mu_{\pm}/kT : \text{Nondegenerate smiconductors}. \end{aligned} \qquad (3.8)$$

A positive (negative) sign in the second line in Eq. (3.8) corresponds to electrons (holes) in the semiconductor. The above equations are valid only for a small change in $\mu_\pm$. In addition, we introduce an electrochemical potential, $\overline{\mu}_\pm = \mu_\pm - e\phi$, to treat drift and diffusion at the same time. Here, ϕ is the electric scalar potential. By using $\overline{\mu}_\pm$, the charge current density in each spin subchannel can be expressed simply as follows:

$$\overset{\to Q}{j_\pm} = \overset{\to Q,Drift}{j_\pm} + \overset{\to Q,Diffusion}{j_\pm} = \sigma_\pm \vec{\nabla} \overline{\mu}_\pm / e. \tag{3.9}$$

The total spin and charge current densities can be expressed in terms of the current densities appearing in Eq. (3.9) as

$$\begin{cases} \hat{j}^S = \dfrac{\hbar}{2} \vec{e}_{spin} \dfrac{\vec{j}_+^Q - \vec{j}_-^Q}{-e}, \\ \\ \vec{j}^Q = \vec{j}_+^Q + \vec{j}_-^Q. \end{cases} \tag{3.10}$$

The first equation in Eq. (3.10) is identical to Eqs. 2.80 and 2.79 in Chapter 2.

For a given constant current, the growth of the spin accumulation continues until it balanced by spin diffusion and relaxation, and, then the system falls into a steady state. In the steady state, we obtain the following set of equations from Eqs. (3.2'), (3.5), (3.9), and (3.10):

$$\begin{cases} \Delta(\overline{\mu}_+ - \overline{\mu}_-) = \dfrac{\overline{\mu}_+ - \overline{\mu}_-}{\lambda_{spin}^2}, \\ \\ \Delta(\sigma_+ \overline{\mu}_+ + \sigma_- \overline{\mu}_-) = 0, \end{cases} \tag{3.11}$$

where

$$\lambda_{spin}^{-2} = \frac{\lambda_+^{-2} + \lambda_-^{-2}}{2},$$

$$\lambda_\pm = \sqrt{D_\pm \tau_\pm^{sf}}.$$

Here, $\Delta = (\partial^2/\partial x^2) + (\partial^2/\partial y^2) + (\partial^2/\partial z^2)$ is the Laplacian. $\lambda_\pm$ is the spin diffusion length in each spin-subchannel. λ_{spin} is the average spin diffusion length. For nondegenerate semiconductors, it is often required to solve the Poisson equation in parallel with Eq. (3.11) to take the band bending and charge redistributions into account [30]. For semiconductors, if charge recombination and creation take place, the sources of the charges and spins must be taken into account by adding appropriate terms to Eq. (3.11) or to Eq. (3.2')[30].

Equation (3.11) can be solved easily, and are widely used to evaluate spin-dependent transport in the diffusive regime. For these calculations, the boundary conditions at the interfaces must be considered. In the simplest model, we assume the continuity of the electric current density, spin current density, and electrochemical potential at the interface. If there is a particular spin relaxation at the interface, the discontinuity of the spin current density must be considered. On the other hand, for tunneling junctions, the discontinuity of the electrochemical potential caused by spin-dependent interface resistances must be taken into account.

In Figure 3.3C, the behavior of the electrochemical potential is shown for the simplest case in which the electrical current, spin current, and electrochemical potential are continuous at the interface. It is clear that the slopes of the electrochemical potentials are discontinuous at the interface to make the spin current density in each spin subchannel continuous. Under this condition, spin polarization of the injected current is calculated as follows:

$$(\text{spin polarization of the injected current}) = \frac{j_\uparrow^Q(+0) - j_\downarrow^Q(+0)}{j_\uparrow^Q(+0) + j_\downarrow^Q(+0)} = \frac{r^{FM}}{r^{FM} + r^{NM}} \beta^{FM},$$

$$(3.12)$$

where

$$\beta^{FM} \equiv \frac{\sigma_+^{FM} - \sigma_-^{FM}}{\sigma_+^{FM} + \sigma_-^{FM}}, r^{FM} \equiv \left(\frac{1}{\sigma_+^{FM}} + \frac{1}{\sigma_-^{FM}}\right)\lambda_{spin}^{FM}, r^{NM} \equiv \left(\frac{1}{\sigma^{NM}/2} + \frac{1}{\sigma^{NM}/2}\right)\lambda_{spin}^{NM}.$$

Here, $+0$ indicates the position at the right-hand side of the interface. $\uparrow$ and $\downarrow$ are the up and down spins in the nonmagnetic material (NM), respectively. The up (down) spin in the NM is parallel to the majority (minority) spin in the ferromagnetic material (FM). $\sigma_+^{FM}(\sigma_-^{FM})$ is the electric conductivity at the majority(minority)-spin subchannel in the FM. σ^{NM} is the total electric conductivity of the NM. β^{FM} is the spin asymmetry of the electric conductivity in the FM. A large β^{FM} results in large spin polarization of the injected current. $\lambda_{spin}^{FM}(\lambda_{spin}^{NM})$ is the spin diffusion length in the FM (NM). $r^{FM}(r^{NM})$ expresses the difficulty in spin injection into the FM (NM) and is often regarded as the "spin (interface) resistance." As is clear from Eq. (3.12), spin injection from a FM to a NM is difficult if $r^{NM} \gg r^{FM}$. For example, if NM is a semiconductor and FM is a ferromagnetic metal, larger resistivity and spin diffusion length in the semiconductor provides a much larger spin resistance than that in the ferromagnetic metal. Therefore, spin injection from the FM to the semiconducting NM will be difficult. This problem is known as "the conductance mismatch problem" [31].

Spin accumulation at the interface is obtained as follows:

$$(\text{spin accumulation}) = \frac{\hbar}{2}\frac{N^{NM}}{2}(\mu_\uparrow(+0) - \mu_\downarrow(+0)) = -\frac{\hbar}{2}\frac{N^{NM}}{2}\frac{r^{FM}r^{NM}}{r^{FM} + r^{NM}}\beta^{FM}\frac{ej^Q}{2}.$$

$$(3.13)$$

Here, N^{NM} is the total density of states at Fermi energy in the NM. If NM is a semiconductor, N^{NM} should be replaced by $\pm n/kT$. Here, n is the total carrier density and $+(-)$ is for electrons (holes). Spin accumulation can be large if N^{NM} and β^{FM} are large, and both r^{FM} and r^{NM} are large.

A way to overcome the conductance mismatch problem of spin injection into a semiconductor is to use a ferromagnetic semiconductor as a spin source [4,5]. Ohno et al. [4] used ferromagnetic GaMnAs as a spin source to inject spin-polarized holes into an InGaAs layer through a GaAs layer. In ZnS type direct gap semiconductors like InGaAs, light emitted by the recombination of the spin-polarized carriers, is up to 50% circularly polarized because of the selection rule at the Γ point ($S_{1/2}$ to $P_{3/2}$ transition). Therefore, by observing the degree of circular polarization one can estimate the spin polarization of the injected carriers. Ohno et al. [4] observed about 10% circular polarization of the emitted light at 6 K and could conclude a significant spin polarization (more than about 20%) of the injected carriers. Since GaMaAs has a relatively large spin resistance, they could achieve a reasonable conductance matching to a semiconductor.

Another way to inject spins into semiconductors is to insert a spin-dependent interface resistance [32]. A tunnel barrier at the interface, for example, may work as such a spin-dependent interface resistance. If the inserted interface resistance is sufficiently larger than the spin resistances, the polarization of the injected current will be determined by the spin asymmetry of the interface resistance, γ^I.

$$(\text{spin polarization of the injected current}) = -\gamma^I, \tag{3.14}$$

where $\gamma^I \equiv (r_+^I - r_-^I)/(r_+^I + r_-^I)$. In addition, spin accumulation is governed by the interface resistance as follows:

$$(\text{spin accumulation}) = \frac{\hbar}{2} \frac{N^{NM}}{2} r^{NM} \gamma^I \frac{ej^Q}{2}. \tag{3.15}$$

Using a tunnel barrier, large spin accumulation in the semiconductors can be expected. Motsnyi et al. [33] successfully injected spin-polarized current from the CoFe to the GaAs layer through an AlO_x barrier layer. In addition, they proved it undoubtedly by showing precession of the injected spins under an applied magnetic field with an oblique angle by again observing the polarization of the emitted light.

As shown above, the spins injected into III−V semiconductors have been detected by polarization analysis of the emitted light [4−6]. On the other hand, to detect the spins injected into nonmagnetic metals, the detection of the spin-dependent chemical potential using a second ferromagnetic electrode has been often used. This method is called "nonlocal magnetoresistance (MR) measurement," and it offers an explicit proof of spin injection.

In Figure 3.4A and B, submicron-scale wire junctions made by Jedema et al. [3] are shown. To measure the nonlocal MR effect, current was applied between Co1 and Al electrodes and the voltage generated between the Co2 and Al electrodes was measured. In this structure, there is no charge current between the Co1 and Co2 electrodes. By current injection, however, spin accumulation takes place at the interface between Co1 and Al electrodes and generates a spin diffusion current

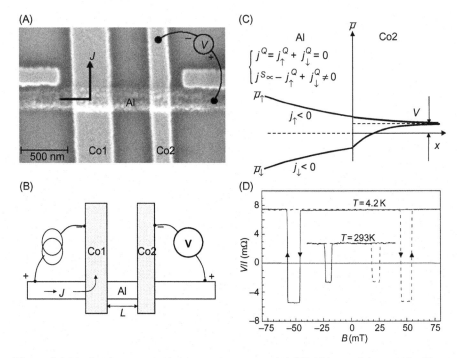

Figure 3.4 Nonlocal magnetoresistance measurement. (A) SEM image of the sample. The sample consists of two parallel Co wires (Co1 and Co2) and an Al wire vertical to the Co wires. (B) Current, J, was applied between Co1 and the left side of the Al wire and voltage, V, that appeared between Co2 and the right side edge of the Al wire was measured. Nonlocal resistance, R, was defined as $R = V/J$. (C) Electrochemical potential distribution from the Al wire to the Co2 interface. On the Al side, the electrochemical potential splits symmetrically. The slope of the electrochemical potential drives the diffusion spin current. The spin current is absorbed by the Co2 electrode. Asymmetric splitting in the chemical potential in Co2 causes a voltage shift of the electrode. (D) Nonlocal resistance as a function of the applied magnetic field. Nonlocal resistance is always positive except in a particular range of the applied field where magnetizations in Co1 and Co2 are antiparallel. The results are well explained by employing, $D_{Al} = 0.43 \times 10^{-2}$ m^2/s (4.2 K), 0.27×10^{-2} m^2/s (293 K), $\lambda_{sf} = 0.65$ μm (4.2 K), 0.35 μm (293 K).
Source: From Ref. [3].

toward the Co2 electrode. In Figure 3.4C, the electrochemical potentials associated with the above spin accumulation and spin current are shown together with the detection part (Co2). $\overline{\mu}_{\uparrow}$ and $\overline{\mu}_{\downarrow}$ split symmetrically at the interface between Co1 and Al electrodes (left-hand side in the Figure 3.4C) because of the spin accumulation. Here, the up spin is parallel to the majority spin in Co1. The gradients of the electrochemical potentials in the Al wire produce diffusion charge currents in both spin subchannels. While these charge currents have the same magnitude, they are opposite in sign. Therefore, while there is no net charge current between the Co1 and Co2 electrodes, a spin current flows.

In the Co2 electrode, where the injected pure spin current is completely absorbed, different diffusion constants for different spins and the continuity of the current in each spin subchannel require different slopes of the electrochemical potentials (right-hand side in Figure 3.4C). As a consequence, a finite electromotive force, V, appears between the Co2 wire and the Al wire. This "nonlocal voltage" is expressed as follows assuming a large contact resistance between Co2 to Al contact,

$$V = \pm \frac{1}{2} e^{jQ} \beta_{Co}^2 \frac{r^{Co} r^{Al}}{r^{Co} + r^{Al}} \exp\left(-\frac{L}{\lambda_{spin}^{Al}}\right). \tag{3.16}$$

Here, a positive (negative) sign corresponds to the parallel (antiparallel) alignment of magnetization. In Figure 3.4D, the signal detected by Jedema et al. is shown. Antiparallel alignment is achieved in a particular range of the applied field and it shows a negative voltage output, as predicted by Eq. (3.16). The observed voltage was less than 1 μV for a 100 μA input current.

If we apply an external field perpendicular to the sample plane, a precession of the injected spins occurs. If the precession angle for travel from Co1 to Co2 equals 90°, the nonlocal signal disappears. Moreover, if the angle is greater than 90°, the signal changes in sign. As a result, the nonlocal signal shows damped oscillation as a function of the external field strength. By using this Hanle effect [1], the spin diffusion length can be estimated correctly. The effect has been observed not only in metals [2,3] and semiconductors [8,33] but also in new materials like carbon nanotubes [13] and graphene [14,16], thus providing clear evidence of spin injection. In Figure 3.5, the MR effect and the Hanle effect observed for a multilayer graphene are illustrated [16].

Kimura et al. [34] developed a structure similar to that shown in the Figure 3.4A but they replaced the Co2 electrode with a small permalloy (FeNi alloy) pad and injected a pure spin current into the pad. In this way, they were able to transfer sufficient angular momentum to the pad to reverse its magnetization. This is the first demonstration of magnetization reversal using a pure spin current.

By using spin accumulation or spin pumping [35,36], we can inject a pure spin current that does not accompany a charge current, into nonmagnetic materials. If there is spin-orbit interaction in the nonmagnetic material into which a pure spin current is injected, the flow of the spin is scattered asymmetrically by impurities and it produces a voltage along a direction that is perpendicular to both the current flow direction and the spin direction [36−39] (Figure 3.6A and B). This phenomenon is called the "inverse spin Hall effect." In addition, a charge current inside a nonmagnetic material may produce a spin current and spin accumulation along a direction perpendicular to the current direction. This "spin Hall effect" has been observed in semiconductors [40,41] and metals [38,42] (Figure 3.6C and D). Moreover, Murakami et al. [43] predicted the existence of an "intrinsic" spin Hall effect, in which an electric field produces a dissipation-less spin current. Wunderlich et al. claimed that their observations agree with latter idea [41]. In Ta,

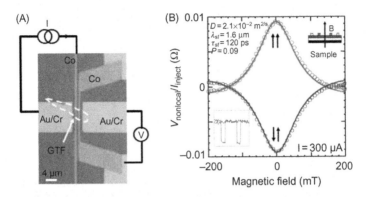

Figure 3.5 Spin-dependent transport in a multilayer grapheme. (A) Sample structure. The multilayer of graphene (a stack of several graphene sheets) is under four electrodes, two of which are ferromagnetic electrodes (Co1 and Co2). (B) Nonlocal resistance signal as a function of the external field that is applied perpendicular to the film plane. At zero applied field, the magnetizations of Co1 and Co2 electrodes are either parallel (upper curve) or antiparallel (lower curve). Precession of the injected spins results in damped oscillation of the nonlocal resistance signals as a function of the external field (Hanle effect). Solid curves are fitted using the diffusive transport model with $D_{graphene} = 2.1 \times 10^{-2}$ m²/s (RT), $\lambda_{sf} = 1.6$ μm (RT). Insets show the nonlocal signal as a function of the in-plane external field and the sample structure.
Source: From Ref. [15,16].

since the spin Hall effect is very large, it is possible to reverse a magnetization using a spin current produced by the effect [42].

3.1.2 Spin-Transfer Torque

Next, we discuss what happens when spins are injected from one ferromagnetic layer (FM1) into another ferromagnetic layer (FM2). This spin injection gives rise to a torque on the FM layers, and as a result, one of the FM layers may switch its direction (called as spin-injection magnetization switching or SIMS) or undergo continuous oscillations (called as spin-transfer oscillation or STO). SIMS and STO are usually performed by employing magnetic nanopillars made of magnetic multilayers, as shown in Figure 3.7. A typical device consists of two ferromagnetic layers (FM1 and FM2: Co, for example) and a nonmagnetic layer (NM1: Cu or MgO, for example) inserted between them. When current is passed through this device, the electrons are first polarized by the FM1 and then injected into the FM2 through the NM1. The spins of the injected electrons interact with the spins in the host material by exchange interaction and exert torque. If the exerted torque is large enough, magnetization in FM2 is reversed or continuous precession is excited.

To simplify the problem, let us imagine an electron system in which the conduction electrons (*s*-electrons) and the electrons that hold local magnetic moments

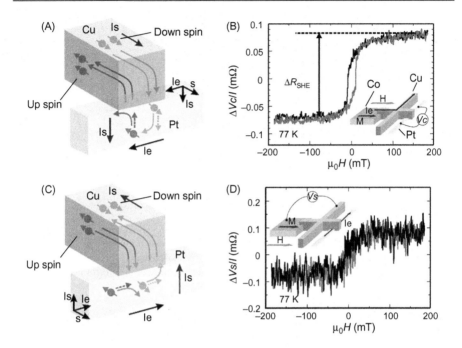

Figure 3.6 Inverse spin Hall effect and spin Hall effect in Pt wires. (A) Graphical illustration of the inverse spin Hall effect. Pure spin current (I_s) is injected from a Cu electrode into a Pt wire. The spin current consists of up spin and down spin electrons moving in opposite directions. These electrons are scattered around the same side (left-hand side) of the Pt wire because of large spin-orbit coupling in the Pt, and produces voltage V_C. (B) The change in Hall resistance (V_C/I_e) due to the inverse spin Hall effect at 77 K. The inset shows the sample structure and the measurement configuration. Spin-polarized charge current (I_s) is injected from the Co electrode to the Cu electrode and it produces spin accumulation in the latter. The accumulated spins produce diffusion spin current and are injected into the Pt wire. The voltage along to the Pt wire (V_C) was measured. (C) Graphical illustration of the spin Hall effect. In the Pt wire, because of a charge current (I_e) the up spin and down spin electrons move along the same direction. These electrons, however, are subjected to the scattering toward the opposite direction and they produce a spin current (I_s). (D) The spin current injected into the Cu electrode produces a potential change in the Co electrode (V_s). The graph shows the spin accumulation signal (V_S/I_e) generated by the spin Hall effect at 77 K. The intensity of the signal is similar to that in (b) as a consequence of Onsager's reciprocal theorem.
Source: From Ref. [34].

(d-electrons) interact with each other through exchange interactions (Figure 3.8A). The exchange interaction (s-d exchange interaction) conserves the total spin-angular momentum. Therefore, a decrease in the subtotal angular momentum of the conduction electrons equals the increase in the subtotal angular momentum of the d electron system. In the magnetic pillar, if the spin-angular momentum of a

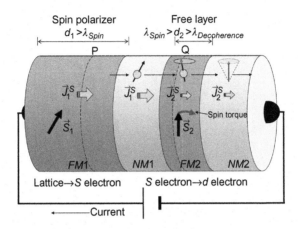

Figure 3.7 Schematic illustration of a magnetic nanopillar consisting of two ferromagnetic layers and two nonmagnetic layers. The electrons are spin polarized by the thick ferromagnetic layer (FM1: spin polarizer) and then injected into the thin ferromagnetic layer (FM2: free layer) through the nonmagnetic layer (NM1). The injected spins undergo precession in FM2 and lose their transverse components. The lost spin components are transferred to the local spin moment in FM2, $\vec{S}_2$. Here, current arrows indicate directions of the positive charge (spin) current. Details are provided in the text.

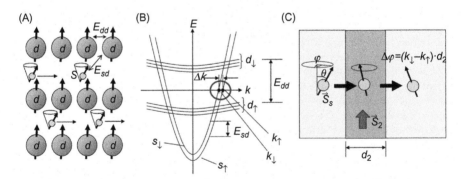

Figure 3.8 (A) A simple s-d model to describe the spin-transfer effect. s-electrons flow among the localized d-electrons and contribute to a charge and spin current, while d-electrons create a single large local magnetic moment because of strong d-d exchange interaction. s-d exchange interaction causes a precession of s- and d-electrons. Since d-electrons create a single large spin moment, the precession angle of the d-electron system is considerably smaller than that of s-electrons. (B) Schematic band structure of a ferromagnetic 3-d transition metal. s-Electrons are in the free electron like s-bands with small spin splitting whereas d-electrons are in the relatively narrow d-bands with large spin splitting. Because of the spin splitting, Fermi k-vectors for the majority and minority spin bands are different. (C) The injected spin shows a precession in a ferromagnetic layer as a consequence of different k-vectors for the majority and minority spin bands.

conduction electron changes because of the *s-d* interaction during transport through the FM2 layer, this amount of angular momentum should be transferred to the *d*-electrons in the FM2 layer. From Eq. (3.2),

$$\frac{d\vec{S}_2}{dt} = \vec{J}_1^S - \vec{J}_2^S,$$ (3.17)

where $\vec{S}_2$ is the total angular momentum in the FM2. The spin currents $\vec{J}_1^S$ and $\vec{J}_2^S$ are obtained by integrating the spin current density flowing in NM1 and NM2, respectively, over the cross-sectional area of the pillar. Since FM2 is very thin, we neglected the spin-orbit interaction in FM2. Equation (3.17) indicates that a torque can be exerted on the local angular momentum as a result of spin transfer from the conduction electrons. This type of the torque, which appears in Eq. (3.17), is called the "spin-transfer torque."

Next, we consider how $\vec{J}_1^S$ and $\vec{J}_2^S$ are determined for metallic junctions [20]. We assume that the FM1 layer is thicker than its spin diffusion length. Therefore, the conduction electrons after passing through the FM1 layer becomes spin-polarized along the angular momentum direction of the FM1, $\vec{S}_1$. They are then injected into the NM1 layer. If $\vec{S}_1$ is oriented along (θ, φ) in a polar coordinate system, the spin function of the injected spin can be expressed as follows:

$$|(\theta, \varphi)\rangle = \cos\frac{\theta}{2}|\uparrow\rangle + e^{i\varphi}\sin\frac{\theta}{2}|\downarrow\rangle \quad \text{or} \quad \begin{pmatrix} \cos\dfrac{\theta}{2} \\ e^{i\phi}\sin\dfrac{\theta}{2} \end{pmatrix}.$$ (3.18)

Here, $|\uparrow\rangle$ and $|\downarrow\rangle$ are the spin eigenstates along the $+z$ and $-z$ directions, respectively. It is easy to understand that Eq. (3.18) expresses the spin state pointing in (θ, φ) direction if one examines that the expectation values of the Pauli matrices, $(\sigma_x, \sigma_y, \sigma_z)$, for the state are $(\cos\varphi\sin\theta, \sin\varphi\sin\theta, \cos\theta)$. Since FM2 is magnetized along the $-z$ direction, the *s*-electron bands are split into $s_\uparrow$ and $s_\downarrow$ bands (Figure 3.8B). Therefore, the wave function of the injected *s*-electron is split into $s_\uparrow$ and $s_\downarrow$ partial waves (Bloch states) that belong to different wave vectors, that is, $k_\uparrow$ and $k_\downarrow$. As a result, the phase acquired during the travel in the FM2 layer with thickness d_2 will be $k_\uparrow d_2$ and $k_\downarrow d_2$ for each partial wave. Therefore, after a ballistic travel through a very thin FM2 layer the spin functions of the electron will be as follows:

$$\begin{pmatrix} e^{ik_\uparrow d_2} & 0 \\ 0 & e^{ik_\downarrow d_2} \end{pmatrix} \begin{pmatrix} \cos\dfrac{\theta}{2} \\ e^{i\varphi}\sin\dfrac{\theta}{2} \end{pmatrix} = e^{ik_\uparrow d_2} \begin{pmatrix} \cos\dfrac{\theta}{2} \\ e^{i(\varphi+(k_\downarrow-k_\uparrow)d_2)}\sin\dfrac{\theta}{2} \end{pmatrix}.$$

In the above expression, we can see that φ is altered by $\varphi + (k_\downarrow - k_\uparrow)d_2$. This means that the spins of the conduction electrons were subject to a precession around $\vec{S}_2$ by $(k_\downarrow - k_\uparrow)d_2$ (rad) (Figure 3.8C). For realistic cases, since most films are polycrystalline and each conduction electron travels along different crystal orientations, the phases and therefore the precession angles should be different for different electrons. As a result, the transverse components (x and y components) of the injected spins cancel each other and disappear on average. For this case, the change in the spin current that produces the spin-transfer torque, will be as follows:

$$\frac{d\vec{S}_2}{dt} = g(\theta)\frac{J^Q}{-e}\left(\frac{\hbar}{2}\begin{pmatrix} \cos\varphi\sin\theta \\ \sin\varphi\sin\theta \\ \cos\theta \end{pmatrix} - \frac{\hbar}{2}\begin{pmatrix} 0 \\ 0 \\ \cos\theta \end{pmatrix}\right) = g(\theta)\frac{J^Q}{-e}\frac{\hbar}{2}\vec{e}_2 \times (\vec{e}_1 \times \vec{e}_2),$$

(3.19)

where $\vec{e}_1 = (\cos\varphi\sin\theta, \sin\varphi\sin\theta, \cos\theta)$ and $\vec{e}_2 = (0, 0, 1)$ are unit vectors along the angular momenta in FM1 and FM2, respectively. $J^Q/(-e)$ is the number of electrons flowing per unit time (J^Q is a charge current). $g(\theta)$ expresses the efficiency of spin transfer and is dependent on the spin polarization[2], P, of the conduction electron in the ferromagnetic layers and the relative angle between $\vec{S}_1$ and $\vec{S}_2$, that is, θ. Slonczewski [44] proposed a formula for the spin-transfer efficiency, $g(\theta)$, that is suitable for the CPP-GMR junctions:

$$g(\theta) = \frac{1}{2}\frac{\Lambda P}{\Lambda\cos^2(\theta/2) + \Lambda^{-1}\sin^2(\theta/2)}.$$

(3.20)

Here, P is a spin polarization of the spin-channel resistances. Λ is a parameter that is determined by a ratio between the Sharvin ballistic conductance of the spacer metal and the spin-channel conductance. In the above formula, the effects of electron reflection at the NM1/FM interface, which have not been discussed in this section, are also taken into account.

Next, we consider the spin-transfer torque exerted magnetic tunnel junctions (MTJs) according to the derivation given by Slonczewski [45,46]. In this case, NM1 is a barrier layer made up of MgO or Al-O. Again, we assume that FM1 is thick enough; therefore, at point P in FM1 (Figure 3.7), the conduction spins are relaxed and aligned parallel to $\vec{S}_1$. In addition, we assume that at point Q inside FM2, the spins of the conduction electrons have already lost their transverse spin component because of the above-mentioned decoherence mechanism and that the spins have aligned parallel to $\vec{S}_2$. Since the spins of the conduction electrons at P and Q are either the majority or minority spin of the host materials, the total

---

[2] The definition of efficiency in this text is twice of that of Slonczewski's.

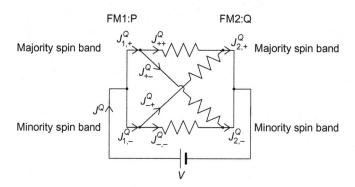

Figure 3.9 Circuit model of an MTJ.
Source: From Ref. [46].

charge current in the MTJ can be expressed as a sum of the following four components (Figure 3.9):

$$J^Q = J^Q_{++} + J^Q_{+-} + J^Q_{-+} + J^Q_{--}. \tag{3.21}$$

Here, suffixes $+$ and $-$ indicate the majority and minority spin channels, respectively. For example, J^Q_{+-} represents a flow of electrons from the FM2 minority spin band into the FM1 majority spin band. These charge currents are expressed using the conductance for each spin subchannel, $G_{\pm\pm}$.

$$\begin{cases} J^Q_{\pm\pm} = V\, G_{\pm\pm} \cos^2 \dfrac{\theta}{2}, \\[2mm] J^Q_{\mp\pm} = V\, G_{\mp\pm} \sin^2 \dfrac{\theta}{2}. \end{cases} \tag{3.22}$$

Here, V is the applied voltage. The angle dependence of the conductions can be derived from the fact that the spin functions in the FM1 are $|\text{maj.}\rangle = \cos(\theta/2)|\uparrow\rangle + \sin(\theta/2)|\downarrow\rangle$ for the majority spins and $|\text{min.}\rangle = \sin(\theta/2)|\uparrow\rangle - \cos(\theta/2)|\downarrow\rangle$ for the minority spins, and those in the FM2 are $|\uparrow\rangle$ and $|\downarrow\rangle$, respectively. Since the spin quantization axes at P and Q are parallel to $\vec{S}_1$ and $\vec{S}_2$, respectively, the spin currents at P and Q are obtained easily as follows:

$$\begin{cases} \vec{J}'^S_1 = \dfrac{\hbar}{2} \dfrac{1}{-e} (J^Q_{++} + J^Q_{+-} - J^Q_{-+} - J^Q_{--})\vec{e}_1, \\[3mm] \vec{J}'^S_2 = \dfrac{\hbar}{2} \dfrac{1}{-e} (J^Q_{++} - J^Q_{+-} + J^Q_{-+} - J^Q_{--})\vec{e}_2. \end{cases} \tag{3.23}$$

Therefore, after a straightforward calculation, the total current and spin-transfer torque are obtained as follows:

$$
\begin{cases}
J^Q = \dfrac{1}{2}V[(G_{++} + G_{--} + G_{+-} + G_{-+}) + (G_{++} + G_{--} - G_{+-} - G_{-+})\vec{e}_2 \cdot \vec{e}_1] \\[2mm]
\quad = \dfrac{1}{2}V[(G_P + G_{AP}) + (G_P - G_{AP})\cos\theta], \\[4mm]
\dfrac{d\vec{S}_2}{dt} = \dfrac{\hbar}{2}\dfrac{1}{-e}\dfrac{1}{2}V[(G_{++} - G_{--}) + (G_{+-} - G_{-+})](\vec{e}_2 \times (\vec{e}_1 \times \vec{e}_2)) \equiv T_{ST}V(\vec{e}_2 \times (\vec{e}_1 \times \vec{e}_2)),
\end{cases}
$$

$$(3.24)$$

where G_P and G_{AP} are the conductance in the parallel and antiparallel configurations, respectively. In the first equation in Eq. (3.24), $\cos\theta$ dependence of the tunnel conductance can be seen. In the second equation, $\sin\theta$ dependence of the spin torque appears, similar to the case in Eq. (3.19). The important difference, however, is that in Eq. (3.19), the efficiency of spin transfer, $g(\theta)$, also has an angular dependence, while in the MTJs, the $\sin\theta$ dependence is exact since G_{++} does not have angular dependence. From this property, Slonczewski termed the coefficient in the second line of Eq. (3.24), $T_{ST}(\vec{e}_2 \times (\vec{e}_1 \times \vec{e}_2))$, as a "torquance," which is an analogue of the "conductance." This simple expression was obtained because the torque was expressed as a function of voltage in Eq. (3.24). If we rewrite the second line of Eq. (3.24) as function of the current, the spin-transfer efficiency is obtained as follows from comparison with Eq. (3.19):

$$
g(\theta) = \frac{1}{2}\frac{\Lambda P}{\Lambda\cos^2(\theta/2) + \Lambda^{-1}\sin^2(\theta/2)},
$$

$$
\text{where}\begin{cases}
\Lambda \equiv \sqrt{\dfrac{G_{++} + G_{--}}{G_{+-} + G_{-+}}}, \\[4mm]
P \equiv \dfrac{G_{++} - G_{--} + G_{+-} - G_{-+}}{G_{++} + G_{--}}.
\end{cases}
$$

$$(3.25)$$

The structure of the equation is the same as that in Eq. (3.20), but the parameters are expressed by using spin-channel-dependent conductances. As shown above, $g(\theta)$ for the MTJs also depends on the angle between $\vec{S}_1$ and $\vec{S}_2$.

Another important feature in the MTJs is that the torque has a bias voltage dependence because G_{++} has bias voltage dependence. In Figure 3.10A, the theoretically predicted spin-transfer torque is plotted as a function of the bias voltage by fine lines [47]. As shown in the figure, the bias dependence of the spin-transfer torque is neither monotonic nor symmetric. The torque will be much higher at a large negative bias even if the magnetoresistance is smaller at such a high bias. This slightly complicated behavior can be explained as follows from the second

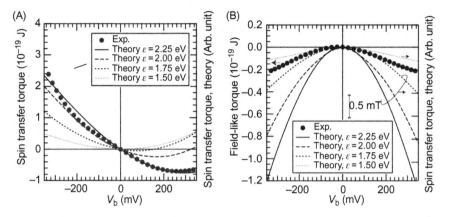

Figure 3.10 Bias dependence of the spin torques. (A) Bias dependence of the spin-transfer torque. (B) Bias dependence of field-like torque. Fine curves were obtained by theoretical model calculations with different spin splitting, ε. Points show the experimental result obtained for a CoFeB/MgO/CoFeB MTJ using the spin-torque diode effect. *Source*: From Refs. [47] and [48].

line in Eq. (3.24). Assume that the FM1 and FM2 are made of the same material. Now if we apply voltage, due to the symmetric conditions for tunneling, the conductances G_{++} and G_{--} do not depend on the sign of the voltage. Thus the contribution to the torque from $(G_{++} - G_{--}) \times V$ term is odd with respect to the voltage. The conductances, G_{+-} and G_{-+} are equal at zero bias. For positive voltage, G_{+-} decreases, whereas G_{-+} increases, thus giving a positive contribution to the spin torque. For negative voltage, due to the symmetry, we have opposite situation and $(G_{+-} - G_{-+})$ changes sign. But as the voltage is also negative, the net contribution is again positive. Therefore, $(G_{+-} - G_{+-}) \times V$ term is an even function of the voltage. This combination of odd and even terms, gives an asymmetry in the spin torque as a function of voltage.

Kubota et al. [48] and Sankey et al. [49] experimentally observed the bias dependence of the torques. In Figure 3.10A, the experimentally obtained bias voltage dependence of the spin-transfer torque in a CoFeB/MgO/CoFeB MTJ is plotted by large circles [48]. The torque was measured by using the "spin-torque diode effect," which will be explained later in this Chapter (Section 3.2). The experimental observations [48−50] essentially agree with the model calculation.

3.1.3 Field-Like Torque and Rashba Torque

Even if it is assumed that $\vec{S}_2$ does not change in size, it may change in direction in two different ways. One is along the direction parallel to the spin-transfer torque, $(\vec{e}_2 \times (\vec{e}_1 \times \vec{e}_2))$. Another is the direction parallel to $(\vec{e}_1 \times \vec{e}_2)$. If the torque is parallel to $(\vec{e}_1 \times \vec{e}_2)$, it has the same symmetry as a torque exerted by an external field.

Therefore, the latter torque is called "field-like torque." It can also be called "perpendicular torque" based on its direction with respect to the plane that is spanned by $\vec{e}_1$ and $\vec{e}_2$.

It has been pointed that one of the important origins of the field-like torque in the MTJs is the change in the interlayer exchange coupling through the barrier layer at a finite biasing voltage [47,51]. In Figure 3.10B, the theoretically obtained strength of the field-like torque is plotted as a function of the bias voltage [47]. As theoretically predicted for the symmetrical MTJs, the field-like torque is an even function of the bias voltage. Its strength itself is less than 1/5th that of the spin-transfer torque. The experimental results obtained so far [48−50] seems to agree with this prediction.

The field-like torque could have originated from other mechanisms that are similar to those responsible for the "β-term" in magnetic nanowires (see Chapters 5 and 6). Several mechanisms, such as a spin relaxation [52,53], Gilbert damping itself [54], momentum transfer [55], and current-induced Ampere field have been proposed for the origin of the "β-term."

Recently, a torque because of spin-orbit interaction is also proposed as an origin of the field-like torque [56]. The torque is called Rashba torque. Miron et al. [57] claimed an observation of the Rashba torque in AlO/Co/Pt system with perpendicular magnetization and in-plane current. In the system, vertical asymmetry produces a vertical built-in electric field and causes a spin-orbit interaction onto flowing electrons. The interaction produces nonequilibrium spin accumulation, which causes an in-plane Rashba field.

3.1.4 Electric Field-Induced Anisotropy Change and Torque

The electric current-based manipulation of the magnetization direction has been investigated as a writing process in the research field of magnetic memories and logic applications. However, the current consumes a much larger energy compared with the thermal stabilization energy for magnetic bit information, which is often estimated as $(50 - 60) \times k_B T$, due to the Joule heat loss, for example. Since the voltage-based manipulation does not consume high energy as demonstrated in complementary metal-oxide-semiconductor field-effect transistor (CMOS FET), voltage control of the magnetization will be a promising candidate as a future writing technique. Most studies so far have been carried out using the magnetic semiconductors [58−60], single-phase multiferroic materials [61], two-phase multiferroic system including artificial ferroelectric/ferromagnetic heterostructures [62,63], or ferromagnetic/piezoelectric hybrids [64,65]. However, many of these studies are limited in low temperature or low writing endurance. In contrast, $3d$ transition metals such as Fe, Co, and their alloys have very high Curie temperatures and show extremely long rewrite endurance; these properties are necessary in magnetic random access memory (MRAM) applications. However, the electric field cannot penetrate into the metal due to the electron screening. The accumulated electric charges at the surface screen the bulk of the metal from the applied field over the length of a

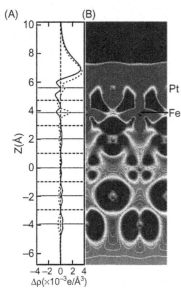

(A) (B)

Z(Å)

$\Delta\rho(\times10^{-3}e/\text{Å}^3)$

Figure 3.11 (A) Electric field-induced majority (solid curves) and minority (dashed curves) spin charge densities and (B) contour maps of induced spin densities along the thickness direction for Pt/Fe/Pt(001).
Source: From Ref. [69].

several Å. Therefore, the electric field effect in the ferromagnetic layer is localized at the interface of only a few atomic layers.

Voltage-induced magnetic anisotropy change in metallic system is originally expected by the fist principle calculations [66−69]. In Figure 3.11, the change of spin charge density by an applied electric field in Pt/Fe/Pt(001) calculated by Tsujikawa et al. is shown [69]. At the position of Fe atom, the minority spin density is induced, and the increase of $d_{3z^2-r^2}$, $d_{x^2-y^2}$ component and decrease of d_{xy}, d_{yz} are found in Figure 3.11A and B. These electric field modulations in electronic structure can be related with the magnetic anisotropy through the spin-orbit interaction. Although the magnitude of voltage-induced magnetic anisotropy change is roughly in agreement between experiments and calculations, there is no identical match including the sign of voltage effect. For a definitive comparison between experiment and calculation, it is important to control the interfacial conditions, such as hybridization, oxidation, interdiffusion, roughness, etc [70−72].

An important pioneering work of voltage-induced magnetic anisotropy change was reported by Weisheit et al. [73]. They used the liquid electrolyte to enhance the surface charging because of the large dielectric constant of the electrolytic double-layer, and observed up to 4.5% coercivity change in FePt(FePd) films with the application of electric field. After that Maruyama et al. [74] demonstrated the voltage-induced magnetic anisotropy change in all-solid-state materials with Au/ultrathin-Fe/MgO junctions. Then larger voltage effect on perpendicular magnetic anisotropy was observed in the ultrathin FeCo alloy layer [75].

As schematically shown in Figure 3.12A, the multilayers of MgO(10 nm)/Cr (10 nm)/Au(50 nm)/$Fe_{80}Co_{20}$ (t_{FeCo})/MgO(10 nm) were grown on a single crystal MgO(001) substrate by molecular beam epitaxy. After depositing the top MgO layer, the sample was removed from the deposition chamber and the surface was

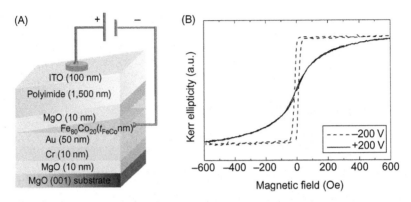

Figure 3.12 (A) Schematic of sample structure used for voltage-induced magnetic anisotropy change. (B) Magnetic hysteresis curves for $t_{FeCo} = 0.58$ nm measured under bias voltage. Solid line, $+200$ V; dashed line, -200 V.
Source: From Ref. [75].

coated with a polyimide layer (1500 nm in thickness) by using a spin coater to ensure a pinhole-free barrier over an extended area. An indium thin oxide (ITO) electrode of 1 mm diameter was fabricated using a metal mask. The voltage was applied between the top ITO and the bottom Au electrodes. The bias voltage direction was defined with respect to the top ITO electrodes. In Figure 3.12B, the polar-Kerr hysteresis curves with a 0.58-nm-thick $Fe_{80}Co_{20}$ layer under the two bias voltage applications of ± 200 V is shown. When positive voltage is applied, perpendicular magnetic anisotropy is suppressed. On the other hand, negative bias voltage application induces perpendicular magnetic anisotropy, and clear square hysteresis was observed, which means that the film was magnetized in the perpendicular direction. This indicates that the electrical manipulation of the magnetic easy axis between the in-plane and out-of-plane directions was realized in this experiment. In thinner region below the critical thickness, electric field-induced magnetic anisotropy change can be observed as a change in coercivity. And voltage-induced magnetization switching under assisted magnetic field has been demonstrated [75].

In the above-mentioned experiment, as high as ± 200 V has been required to observe the significant change due to the thick polyimide insulating layer. However, since the origin of magnetic anisotropy change may be an increase/decrease of the electron density at the interface, which is determined by the product of effective permittivity and electric field, low voltage drive is possible by decreasing the thickness of insulating layer using a pinhole-free insulator. The voltage-induced magnetic anisotropy change with the hundreds of mV order of an applied voltage has successfully been demonstrated in the MTJs [76]. The realization of voltage effect in MTJ structure should provide a useful approach for voltage-driven spintronic devices.

Next, we consider the voltage torque. The change in the effective field by the voltage application produces the torque on the ferromagnetic layer. The voltage torque is given by,

$$\frac{d\vec{S}_2}{dt} = \vec{S}_2 \times V \frac{\partial}{\partial V} \gamma \vec{H}_{\text{eff}} \tag{3.26}$$

where $\vec{H}_{\text{eff}}$ is the effective magnetic field obtained by the derivative of the magnetic energy E_{mag}. The difference between voltage torque and other spin torques is its direction. The spin-transfer and field-like torques depend on the direction of the spin polarizer, while the voltage torque is determined by the electric field direction. To identify the voltage-induced torque, the voltage torque-induced ferromagnetic resonance (FMR) was experimentally observed in MTJs [77,78]. Using the spin-torque diode effect, the existence of voltage torque was clarified from the shape and amplitude of the homodyne detection spectra.

3.2 Spin-Injection Magnetization Reversal

3.2.1 Amplification of the Precession

In this and subsequent sections, we mainly focus on the two types of magnetic nanopillars shown in Figure 3.13 (A) pillars with magnetization perpendicular to the film plane (perpendicularly magnetized pillars) and (B) pillars with magnetization parallel to the film plane (in-plane magnetized pillars). These structures provide all the essential features necessary to understand the physics of spin-transfer effects and they are also important for practical applications. In addition, we assume that the perpendicularly magnetized and in-plane magnetized nanopillars have a round and elliptical cross sections, respectively. The in-plane magnetized pillars have their magnetization parallel to the long axis of the ellipse.

As discussed in the previous section, an electric current passing through a magnetic nanopillar composed of magnetic multilayers (Figure 3.7) transfers spin-angular momentum to the magnetic free layer and changes the direction of the local spins in the layer (Eqs. (3.19) and (3.24)). The dynamic properties of the local spins can be expressed by the following Landau−Lifshitz−Gilbert (LLG) equation, which includes a spin-transfer torque term [20] and field-like torque term:

$$\frac{d\vec{S}_2}{dt} = \gamma \vec{S}_2 \times \vec{H}_{\text{eff}} - \alpha \vec{e}_2 \times \frac{d\vec{S}_2}{dt} + T_{ST} V \vec{e}_2 \times (\vec{e}_1 \times \vec{e}_2) + T_{FT} V (\vec{e}_2 \times \vec{e}_1). \tag{3.27}$$

The first term is the effective field torque; the second, Gilbert damping; the third, the spin-transfer torque, and fourth field-like torque. $\vec{S}_2 = S_2 \vec{e}_2$ is the total spin-angular momentum of the free layer and is opposite to its magnetic moment, $\vec{M}_2$. $\vec{e}_2(\vec{e}_1)$ is a unit vector that expresses the direction of the spin-angular momentum of

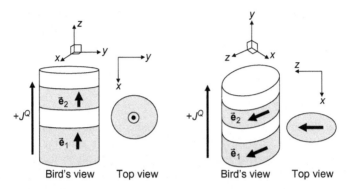

Bird's view   Top view   Bird's view   Top view

Figure 3.13 Typical structures of magnetic nanopillars designed for SIMS (spin injection magnetization switching) experiments. In the bird's-eye-view pictures, the upper magnetic layer acts as a magnetically free layer, whereas lower magnetic layers are thicker than the free layer and act as a spin polarizer. The spins in the lower layers are usually pinned by an exchange interaction at the bottom interface to an antiferromagnetic material. Therefore, the spin polarizer layer is often called the "pinned layer," "fixed layer," or "reference layer." In GMR nanopillars, the layer between two ferromagnetic layers is made of a nonmagnetic metal such as Cu. In MTJ nanopillars, this interlayer is made of insulators such as MgO. The diameter of the pillar is around 100 nm, presently. The free layer is typically a few nanometers thick. (A) A magnetic nanopillar with perpendicular magnetization. Because of the crystalline uniaxial anisotropy, the film has remnant magnetization and spin momentum perpendicular to the film plane. The pillar has a circular cross section. (B) A magnetic nanopillar with in-plane magnetization. The pillar has an elliptical cross section with the long axis along parallel to the z-axis. Because of its shape anisotropy the film has in-plane magnetization and spin momentum parallel to the z-axis.

the free layer (fixed layer). For simplicity, we neglect the distribution of the local spin-angular momentum inside the free layer and assume that the local spins within each magnetic cell are aligned in parallel and form a coherent "macrospin" [79,80]. This assumption is not strictly valid since the demagnetization field and current-induced Oersted field inside the cells are not uniform. Such nonuniformities introduce incoherent precessions of the local spins and causes domain and/or vortex formation in the cell [80–82]. Despite the predicted limitations, the macrospin model is still useful, because of both its transparency and its validity for small excitations. The orbital moment, which is very small for 3d transition metals, is neglected in this treatment. γ is the gyromagnetic ratio, where $\gamma < 0$ for electrons ($\gamma = -2.21 \times 10^5 \ m/A \cdot$ sec for free electrons). The effective field, $\vec{H}_{eff}$, is a sum of the external field, demagnetization field and anisotropy field. It should be noted that the demagnetization field and the anisotropy field depend on $\vec{e}_2$. $\vec{H}_{eff}$ is derived from the magnetic energy, E_{mag}, and the total magnetic moment, M_2, of the free layer:

$$\vec{H}_{eff} = \frac{1}{\mu_0 M_2} \frac{\partial E_{mag}}{\partial \vec{e}_2}, \tag{3.28}$$

where $\mu_0 = 4\pi \times 10^{-7}\ H/m$ is the magnetic susceptibility of vacuum.

The first term determines the precessional motion of $\vec{S}_2$. In the second term, α is the Gilbert damping factor ($\alpha > 0$), $T_{ST}V = g(\theta)\frac{J^Q}{-e}\frac{\hbar}{2}$. $g(\theta)$ is a coefficient that expresses the efficiency of the spin-transfer process as a function of the relative angle, θ, between $\vec{S}_1$ and $\vec{S}_2$. Explicit expressions for $g(\theta)$ was introduced in the previous section for giant magnetoresistance (GMR) junctions (Eq. (3.20)) and MTJs (Eq. (3.25)).

To understand the effects of the different torques, we first discuss the spin dynamics in a perpendicularly magnetized pillar with a cylindrical symmetry under an external magnetic field parallel to the axis of symmetry (Figure 3.13A). The magnetic energy in this system comprises uniaxial magnetic anisotropy energy and Zeeman energy. For the cylindrical pillar with uniaxial anisotropy,

$$E_{mag} = -\frac{1}{2}\mu_0 M_2 H_u \cos^2\theta + \mu_0 M_2 H_{ext}\cos\theta + (\text{const.}), \tag{3.29}$$

where H_u and H_{ext} are the effective uniaxial anisotropy field and external field, respectively. The magnetic uniaxial anisotropy energy is a sum of the crystalline anisotropy energy and the demagnetization energy,

$$\frac{1}{2}\mu_0 M_2 H_u = K_u v + \frac{1}{2}\mu_0 M_2 \frac{N_{demag} M_2}{v}, \tag{3.30}$$

where K_u is a (crystalline) uniaxial anisotropy constant; v the volume of the free layer, and N_{demag}, the demagnetizing factor. From Eq. (3.28), the effective field is $\vec{H}_{eff} = -H_u\cos\theta\vec{e}_1 + H_{ext}\vec{e}_1$. Substituting this in Eq. (3.27), we get following equation of motion for the cylindrical pillar with uniaxial anisotropy:

$$\begin{cases} \dfrac{d\vec{e}_2}{dt} \cong \gamma(-H_u\cos\theta + H_{ext})(\vec{e}_2 \times \vec{e}_1) - \alpha_{eff}(\theta)\gamma H_u\cos\theta\vec{e}_2 \times (\vec{e}_1 \times \vec{e}_2), \\[2mm] \alpha_{eff}(\theta) \equiv \alpha + \dfrac{1}{(-\gamma)H_u\cos\theta}\dfrac{T_{ST}V}{S_2}. \end{cases}$$

$$\tag{3.31}$$

Since α is small for 3d transition metals ($\alpha_{FeNi} = 0.007$, e.g., [83]), small terms on the order of $O(\alpha^2)$ are neglected in the above equation. $\alpha_{eff}(\theta)$ expresses the effective Gilbert damping coefficient of the free layer under spin-transfer torque. The directions of the torques are illustrated in Figure 3.14. The effective field torque promotes a precession motion of $\vec{S}_2$ around $-\vec{H}_{eff}$, while the damping torque tends to reduce the opening angle of the precession. By the effective field and damping torques, $\vec{S}_2$ exhibits a spiral trajectory and finally aligns antiparallel to the effective field if a junction current, J^Q, is absent (Figure 3.14A). It must be noted that the direction of $\vec{S}_2$ is opposite to that of its magnetic moment. Direction of the spin-transfer torque is also illustrated in Figure 3.14B for the case where both $g(\theta)$ and J^Q are positive. If the current, J^Q, is sufficiently large, the spin-transfer torque

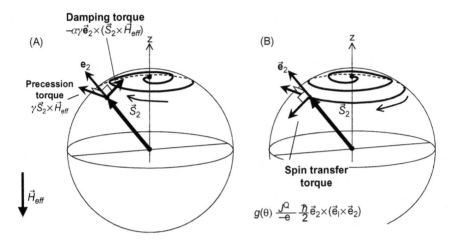

Figure 3.14 Illustration of the direction of each torque and trajectory of the free layer spin momentum for a nanopillar with perpendicular remnant magnetization. (A) In the absence of an electric current, the precession of the free layer spin is damped. (B) Under an electric current, if the spin-transfer torque overcomes the damping torque, the precession of the free layer spin is amplified.

overcomes the damping torque, resulting in negative effective damping. This negative damping results in an increase in the opening angle of the precession motion, that is, an amplification of the precession takes place. Depending on the angular dependence of the effective damping, the amplification of the precession motion leads to a limit cycle (i.e. STO) or to a total magnetization reversal (SIMS) [84].

In Figure 3.15, a trajectory for SIMS (A) is compared with trajectory for a magnetic field-induced magnetization switching (B) in a nanopillar with in-plane magnetization. The figure also illustrates the magnetic potential shapes during switchings. In the absence of a current and an external magnetic field, the potential shows a double minimum for parallel (P) and antiparallel (AP) configurations of the local spin. For a particular case of SIMS, the spin-transfer torque does not affect the shape of the magnetic potential but amplifies the precession thereby providing energy to the local spin system. Once the orbital crosses the equator, it converges rapidly to opposite direction since the spin-transfer torque extracts energy from the local spin system. In other words, the spin-transfer torque amplifies the precession in the front hemisphere, while enhancing the damping in the back hemisphere. In contrast to this process, the external magnetic field deforms the magnetic potential and the minimum on the P side disappears. Therefore, the local spin turns toward AP side. The local spin system, however, keeps excess energy in the back hemisphere. As a result, it cannot stop at once and shows precessional motion (ringing) in the back hemisphere.

The critical current at which the system becomes unstable to small deviations from equilibrium position is given by $\alpha_{\text{eff}}(\theta) = 0$. This "instability current" is

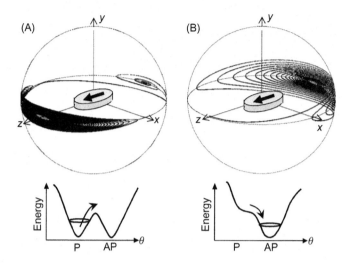

Figure 3.15 Comparison of the magnetization processes driven by (A) spin-transfer torque and (B) external magnetic field for in-plane magnetized nanopillars.

obtained from Eq. (3.31) as follows for the cylindrical pillar with uniaxial anisotropy at zero Kelvin,

$$J_{c0}^Q = \frac{e}{g(\theta)\hbar/2}\alpha(-\gamma)(H_u\cos\theta - H_{ext})S_2, (\theta = 0 \text{ or } \pi). \tag{3.32}$$

If there is no stable limit cycle in between the parallel (P) and antiparallel (AP) states, the above equation gives the threshold current of SIMS, with $\theta = 0$ for P to AP switching and $\theta = \pi$ for AP to P switching.

3.2.2 Linearized LLG Equation and Instability Current

To analyze the dynamics of the macrospin system without cylindrical symmetry, we develop Hamilton's equation of motion of the system. Using spherical coordinate, that is, (ϕ, θ), Lagrangian and Rayleigh's dissipation function of the LLG equation (without spin torques) are expressed as follows [85]:

$$\begin{cases} \mathcal{L}(\phi,\dot{\phi},\theta,\dot{\theta}) = S_2\dot{\phi}(\cos\theta - 1) - E_{mag}(\phi,\theta), \\ W(\dot{\phi},\dot{\theta}) = \frac{\alpha}{2}S_2(\dot{\theta}^2 + \dot{\phi}^2\sin^2\theta). \end{cases} \tag{3.33}$$

The kinetic energy term in the Lagrangian is known as a spin Berry phase term. In classical mechanics, this term also results equation of motion of angular momentum. From above Lagrangian, we may find following Hermitian conjugate valuables.

$$\begin{cases} x^1 \equiv \phi, \\ x^2 \equiv S_2(\cos\theta - 1). \end{cases} \tag{3.34}$$

Using this new coordinate system, Eq. (3.27) can be rewritten as,

$$\dot{x}^i = \sum_{j=1}^{2} \varepsilon^{ij}(\partial_j E_{mag} - T_{FT}V\partial_j(\vec{e}_2 \cdot \vec{e}_1) + \alpha S_2 \dot{x}_j) + S_2^{-1} T_{ST} V \partial^i(\vec{e}_2 \cdot \vec{e}_1), \quad (i = 1, 2), \tag{3.35}$$

where,

$$\left(\begin{array}{l} \partial_i \equiv \dfrac{\partial}{\partial x^i}, \partial^i \equiv \sum_{j=1}^{2} g^{ij}\partial_j \\[2mm] (\varepsilon^{ij}) \equiv (\varepsilon_{ij}) \equiv \begin{pmatrix} 0 & 1 \\ -1 & 0 \end{pmatrix} \\[2mm] (g_{ij}) = \begin{pmatrix} \sin^2\theta & 0 \\ 0 & \dfrac{1}{S_2^2 \sin^2\theta} \end{pmatrix} = (g^{ij})^{-1} \end{array} \right.$$

Here, ε is the Levi-Civita's symbol and g a metric tensor. Explicit form of Eq. (3.35) with respect to $\dot{x}^i$ can be obtained easily.

$$\frac{dx^i}{dt} = F^i, \tag{3.36}$$

where

$$\begin{cases} F^i(t) \cong \displaystyle\sum_{j=1}^{2} \varepsilon^{ij}\partial_j E_{mag+FT} - \alpha S_2^{-1}\partial^i E_{mag+ST}, \\[2mm] E_{mag+FT} \equiv E_{mag} - T_{FT}V(t)(\vec{e}_2 \cdot \vec{e}_1), \\[1mm] E_{mag+ST} \equiv E_{mag} - \alpha^{-1}T_{ST}V(t)(\vec{e}_2 \cdot \vec{e}_1). \end{cases}$$

Here, terms with α^2, αT_{ST} and αT_{FT} are neglected. We also assumed $\partial_j T_{FT} = \partial_j T_{ST} = 0$. In Eq. (3.36), we clearly see that the spin-transfer torque term that is a consequence of the spin current, take a part of the damping term. This Hamilton type equation of motion on the orthogonal curvilinear coordinate is useful to obtain analytic understanding of the dynamics under spin-transfer torque.

Equations (3.27), (3.35), and (3.36) are all equivalent and describe nonlinear response of a macrospin in the junction under an applied magnetic field and a voltage. Before a discussion about nonlinear behavior like switching, we derive a linearized equation of motion and discuss stability of the system to infinitesimal perturbations.

The equilibrium point of the macrospin, (x_0^1, x_0^2), under static external field and dc bias voltage, (H_0^{ext}, V_0), can be obtained by solving,

$$F^i(x_0^1, x_0^2; H_{ext,0}, V_0) = 0. \tag{3.37}$$

The linearized equation of motion is obtained taking deviation from the equilibrium point as new coordinates, that is, $(\delta x^1(t), \delta x^2(t)) = (x^1(t) - x_0^1, x^2(t) - x_0^2)$. Then, linear expansion of Eq. (3.36) yields,

$$\frac{d\delta x^i(t)}{dt} = \sum_{j=1}^{2} (\partial_j F^i)\delta x^j(t) + \left(\frac{\partial F^i}{\partial H_{ext}}\right)\delta H_{ext}(t) + \left(\frac{\partial F^i}{\partial V}\right)\delta V(t), \tag{3.38}$$

where $\delta H_{ext}(t)$ and $\delta V(t)$ are a time-dependent part of the external field and the bias voltage, respectively. In Eq. (3.38), the derivatives of the force are evaluated at equilibrium point.

The time evolution of the solution to the homogeneous equation is expressed by $e^{\lambda t}$. In general, λ is a complex number. The imaginary part of λ corresponds to the precession frequency of the free layer, whereas its real part represents the time evolution of the precession angle. If $\text{Re}[\lambda] < 0$, precession is damped and the equilibrium point is stable. On the other hand, if $\text{Re}[\lambda] > 0$, the precession is amplified by the spin-transfer torque, and a magnetization switching or an auto-oscillation will take place. λ is an eigenvalue of the matrix, $(\partial_j F^i)$, and is obtained as follows:

$$\lambda \equiv -\frac{\Delta\omega}{2} \pm i\omega_0, \tag{3.39}$$

where

$$\begin{cases} \omega_0^2 \cong \det[\hat{\Omega}] = \det[\partial_i \partial_j E_{mag}], \\ \Delta\omega \cong \alpha S_2^{-1} \sum_{i=1}^{2} \partial_i \partial^i E_{mag+ST} \cong \frac{\alpha}{S_2}\left(\sum_{i=1}^{2} \partial_i \partial^i E_{mag}\right) + \frac{2T_{ST}V}{S_2}(\vec{e}_2 \cdot \vec{e}_1). \end{cases}$$

Here, ω_0 is a resonance frequency of the system. Positive $\Delta\omega$ provides a line width of the magnetic resonance. If $\Delta\omega$ is negative the system is unstable. Above equation show that the spin-transfer term may linearly increase or decrease $\Delta\omega$ depending on the sign of the applied voltage.

Applying $Re[\lambda] = 0$ to above equations, the instability current in the general system is estimated as,

$$J_{c0}^{Q} = \frac{e}{\hbar/2} \frac{\alpha}{g(\theta_0)(\vec{e}_2 \cdot \vec{e}_1)} \frac{1}{2} \sum_{i=1}^{2} \partial_i \partial^i E_{mag}. \tag{3.40}$$

For an elliptical cell with in-plane magnetization, the magnetic energy is expressed as follows,

$$E_{mag} = \mu_0 M_2 H_{ext} \cos\theta + \frac{1}{2}\mu_0 M_2 (H_{//}\cos^2\varphi + H_{\perp}\sin^2\varphi)\sin^2\theta + (\text{const.}), \tag{3.41}$$

where $H_{//}$ and $H_{\perp}$ are the in-plane and out-of-plane effective anisotropy fields, respectively. The external field is assumed to be parallel to the z-axis in the Figure 3.11B. From Eqs. (3.40) and (3.41), we get an instability current at zero Kelvin for an elliptical cell with in-plane magnetization,

$$J_{c0}^{Q} = \frac{e}{\hbar/2} \frac{\alpha}{g(\theta_0)(\vec{e}_2 \cdot \vec{e}_1)} S_2(-\gamma)\left(\frac{H_{\perp} + H_{//}}{2} - H_{ext}\cos\theta_0\right). \tag{3.42}$$

Here again, we should take $\theta_0 = 0$ for P to AP switching and $\theta_0 = \pi$ for AP to P switching. The instability current for the pillars with in-plane magnetization is large because of the large out-of-plane anisotropy field, which mainly consists of the demagnetization field.

3.2.3 Spin-Injection Magnetization Switching

SIMS was first predicted theoretically [20,21] and was then experimentally demonstrated for a Co/Cu/Co nanopillar with in-plane magnetization [22,24]. Subsequently, SIMS was also observed in the case of MTJs with Al-O barriers [25] and MgO barriers [26,27]. All those magnetic pillars had in-plane magnetization. And their cross sections were ellipses or rectangles. In 2006, SIMS was also observed in magnetic nanopillars with perpendicular magnetization [28]. They employed Co/Ni multilayers to give a perpendicular crystalline anisotropy to the film.

In Figure 3.16, a typical fabrication process for nanopillars from an MTJ (for research purposes) is shown. First, (A) a magnetic multilayer including an MTJ is sputter deposited. The multilayer consists of a bottom electrode layer, an antiferromagnetic exchange bias layer (e.g., MnPt), a synthetic antiferromagnetic pinned layer (e.g., CoFeB/Ru/CoFe), an MgO barrier layer, a magnetic free layer (e.g., CoFeB), and a capping layer. The multilayer is then covered by a resist layer using a spin coater and transferred to an electron beam lithography machine. (B) After exposure and development, the sample with micro patterned resist is transferred to

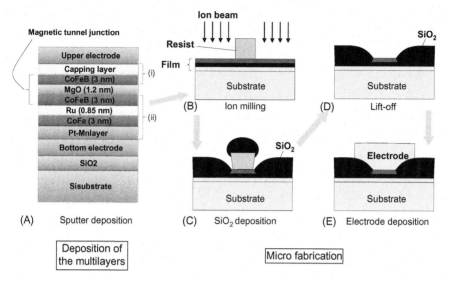

Figure 3.16 An example of the sample fabrication process for a SIMS (spin injection magnetization switching) experiment. (A) A magnetic multilayer including the tunneling barrier is first deposited under vacuum by a sputtering method. For memory applications, the film is deposited onto a CMOS and wiring complex after a chemical–mechanical planarization process. (B) After resist coating is applied using a spin coater, the resist is patterned by the electron beam lithography. Using the patterned resist, a part of the film is etched by ion beam bombardment. (C) Interlayer insulator (SiO$_2$) deposition using a self-alignment technique. (D) A lift-off process to open a contact hole. (E) Deposition of the upper electrode.

an ion beam milling machine to remove parts of the multilayers and form magnetic pillars. (C) The outer side of the pillar is filled by a SiO$_2$ insulating layer. (D) The SiO$_2$ layer on the junction is lifted off with the resist by using a chemical solvent and ultrasonic scrubbing. Finally, (E) the top electrodes are deposited onto the junction under the vacuum.

A hysteresis loop obtained for a magnetic nanopillar comprising a CoFeB/MgO/CoFeB tunneling junction is shown in Figure 3.17 [26]. The pillar has in-plane magnetization and elliptical cross section with the dimensions 100 nm × 200 nm. A current was applied as a series of 100 ms wide pulses. In between the pulses, the sample resistance was measured to check the magnetization configuration while the pulse height was swept between − 1.5 mA and + 1.5 mA. By this method, the effect of temperature increase during the application of the current on the resistance measurement could be eliminated. For the data shown in Figure 3.17, the hysteresis measurement started at a zero pulse height for the P state (285 Ω). An increase in the pulse height caused a jump from the P state to the AP state (560 Ω) at + 0.6 mA. Further increase in the pulse height followed by a reduction to zero current did not affect to the state. Subsequently, negative pulses were applied to the sample. At

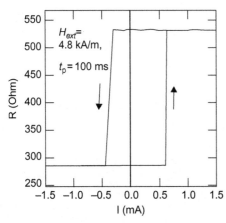

Figure 3.17 A typical SIMS (spin injection magnetization switching) hysteresis loop obtained for a CoFeB/MgO/CoFeB MTJ. The junction area, free layer thickness, resistance area product, and MR ratio are 100 nm × 200 nm, 3 nm, 3 $\Omega\mu m^2$, and about 100%, respectively. Measurements were performed at room temperature using electric current pulses of 100 ms duration. The resistance of the junction was measured after each pulse to avoid the effect of the heating on the sample resistance.
Source: From Ref. [26].

-0.35 mA, the sample switched its magnetization from the AP state to the P state. The average switching current density was about 6×10^6 A/cm^2. An intermediate resistance states between the P and the AP states were not observed during either of the two switching events: the switching events were always abrupt and complete. The slope of the hysteresis loop at the switching point is only due to discrete measurement points that were not regularly placed because of the large change in the resistance. P to AP and AP to P switching events occurred at different current levels because of the dipole and the so-called orange peel coupling field from the pinned layer. In the experiment, an external field of -4.8 kA/m was applied to cancel these coupling fields. After the cancellation of the coupling fields, the hysteresis still exhibited a certain shift because of the following intrinsic mechanisms. For the MTJ nanopillars, the asymmetrical voltage dependence of the torque, which was discussed in the previous section, causes a horizontal shift in the hysteresis curve. For the GMR nanopillars, in contrast, the angle dependence of the spin-transfer efficiency results in a significant shift in the hysteresis loop.

Many efforts have been done to reduce threshold current of the switching. The first attempt is to reduce total spin-angular momentum, S_2, in the free layer. SIMS requires effective injection of spin angular momentum that is equal to that in the free layer. Therefore, reduction in S_2 results a reduction in J_{c0}. Albert [86] showed that threshold current of the SIMS is proportional to the free layer thickness. Reduction in the free layer thickness reduces S_2 and J_{c0}. S_2 can be also reduced by reducing the magnetization of the ferromagnetic material. Especially in the nanopillar with in-plane magnetization, since magnetization also affects to the size of the anisotropy field, J_{c0} is a quadratic function of the magnetization. Yagami et al. [87] reduced J_{c0} considerably by changing a material of the free layer from CoFe (1.9×10^6 A/m) to CoFeB (0.75×10^6 A/m) and obtained 1.7×10^7 A/cm^2. Second attempt is to use double spin filter structure. This method was originally proposed by Berger [88]. By using this structure, Huai et al. [89] observed substantial reduction of the threshold current to 2.2×10^6 A/cm^2. Third attempt is to use perpendicular magnetic anisotropy, which can reduce the size of the anisotropy field. By this method, Nagase et al. [90] obtained a significant reduction of the threshold current under the required thermal stability factor for the MTJ nanopillars with the CoFeB/[Pd/Co] × 2/Pd free layer and FePt/CoFeB pined layer.

3.2.4 Switching Time and Thermal Effects

Switching time of the SIMS has been investigated to test its applicability as an information writing technique and to clarify the dynamics of the SIMS itself. Observations of changes in the magnetic configuration in magnetic nanopillars after an application of short pulses revealed the pulse width dependence of the threshold current and the probabilistic nature of the switching [91−93]. Very high-speed switching down to 200 ps [92] and precessional nature of the switchings [93,94] have been also clarified.

An example of the pulse width dependence of the switching current for a cylindrical shape magnetic nanopillar with perpendicular magnetization is shown in the Figure 3.18 [95]. The switching current increases for pulses with shorter length. This is because the switching requires a certain number of injected spins, and the required time is longer for a smaller current. This switching time, in principle, can be obtained by integrating the LLG equation with spin-transfer torque (Eq. (3.27)) [79]. For a cylindrical pillar with uniaxial magnetic anisotropy, the equation of motion for the opening angle of precession, θ, is derived from Eq. (3.31) as follows,

$$\frac{1}{\sin\theta}\frac{d\theta}{dt} \cong -\alpha(-\gamma)(H_u\cos\theta - H_{ext}) - g(\theta)\frac{J^Q}{-e}\frac{\hbar}{2S_2}, \tag{3.43}$$

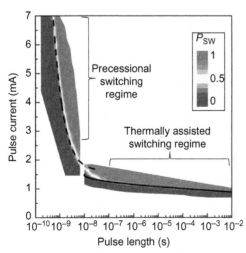

Figure 3.18 Pulse width dependence of the SIMS (spin injection magnetization switching) in a Co/Pd-based CPP-GMR nano pillar. The junction diameter, free layer thickness, and resistance change are 100 nm, 2 nm, and about 0.1 Ω, respectively. Measurements were performed at room temperature using electric current pulses with duration from 200 ps to 10 ms.
Source: From Ref. [95].

To obtain the switching time, τ_{sw}, this equation is integrated analytically by neglecting the θ-dependence of the efficiency. The precessional switching time for the cylindrical pillars with uniaxial anisotropy is[3][79].

$$\tau_{sw} \cong \frac{J_{c0}^Q}{J^Q - J_{c0}^Q} \frac{1}{\alpha(-\gamma)H_u} \log\left(\frac{2}{\delta\theta_0}\right), \tag{3.44}$$

where, J^Q and J_{c0}^Q are the applied charge current and the critical current (Eq. (3.32)), respectively. $\delta\theta_0$ is the angle between the direction of magnetization and the easy axis at the beginning of amplification. For the thermal equilibrium, the average value of $\delta\theta_0$ is estimated as,

$$\delta\theta_0 = \sqrt{\frac{k_B T}{\mu_0 M_2 H_u}} = \sqrt{\frac{1}{2\Delta}}, \tag{3.45}$$

where, k_B is Boltzmann constant and T is the absolute temperature. Δ is the thermal stabilization factor. Since Eq. (3.44) only treats thermal distribution of the initial states but neglects thermal effects during switching, this is a precessional switching under an adiabatic approximation.

For the nanopillars with in-plane magnetization, an approximate expression for switching time is obtained replacing H_u and J_{c0}^Q with $(H_\perp + H_{//})/2$ and J_{c0}^Q for the in-plane pillar (Eq. (3.42)), respectively in Eq. (3.44).

$$\tau_{sw} \cong \frac{J_{c0}^Q}{J^Q - J_{c0}^Q} \frac{2}{\alpha(-\gamma)(H_\perp + H_{//})} \log\left(\frac{2}{\delta\theta_0}\right), \tag{3.46}$$

In Figure 3.18, the theoretically expected switching time in precessional switching regime is shown by a dotted curve. The steep increase in the switching current for short switching times is well fitted by the theoretical curve taking Δ, J_{c0}^Q, and $\alpha(-\gamma)\omega_u$ as three independent fitting parameters, whereas the gradual decrease in the switching current for long pulse application is not explained by the theory. In addition, the switching current seems to be much smaller than that expected from Eq. (3.46) in the long switching time region.

Koch et al. [96] and Li et al. [97] explained the reduction in the switching current for long pulse durations by considering the thermal excitation of the macrospin fluctuation [96] induced by the stochastic field, $H_{stochastic}(t)$. Because of the stochastic field, direction of the free layer spin-angular momentum, $\vec{e}_2$, is now also a stochastic variable. Therefore, we can only discuss its probability distribution,

---

[3] An accurate expression under the adiabatic approximation is available in Ref. [97].

$p(\vec{e}_2, t)$, taking into account the statistical nature of $H_{stochastic}(t)$. By Langevin's method [98], $H_{stochastic}(t)$ is related with the Gilbert damping constant.

$$\alpha \cong \frac{(-\gamma)\mu_0 M_2}{kT} \frac{1}{2} \int_{-\infty}^{\infty} d\tau \langle H_{stochastic}(t) H_{stochastic}(t+\tau) \rangle, \tag{3.47}$$

where $\langle \rangle$ expresses an thermodynamical ensemble average. Using this fluctuation-dissipation theorem, the time evolution of $p(\vec{e}_2, t)$ is expressed as follows,

$$\begin{cases} \dfrac{\partial p(\vec{e}_2, t)}{\partial t} + \vec{\nabla}_{\vec{e}_2} \cdot \vec{J}(\vec{e}_2, t) = 0 \\[2mm] \vec{J}(\vec{e}_2, t) = (1 - \alpha \vec{e}_2 \times) \left(\vec{e}_2 \times \gamma \vec{\mathbf{H}}_{eff} + g(\theta) \dfrac{J^Q}{-e} \dfrac{\hbar/2}{S_2} \vec{e}_2 \times (\vec{e}_1 \times \vec{e}_2) \right) p(\vec{e}_2, t), \\[3mm] \qquad\qquad + \alpha \dfrac{kT}{S_2} (\vec{e}_2 \times (\vec{e}_2 \times \vec{\nabla}_{\vec{e}_2} p(\vec{e}_2, t))) \end{cases}$$

$$\tag{3.48}$$

where $\vec{J}(\vec{e}_2, t)$ is the probability density current and $\vec{\nabla}_{\vec{e}_2} = \partial/\partial \vec{e}_2$ derivative with respect to the spin-angular momentum direction. The first line in the expression of the probability density current is a drift current and is proportional to a sum of the three torques, that is, the precession torque, the damping torque, and the spin-transfer torque. The second line expresses a thermal diffusion current of the probability density and is proportional to the gradient of the probability density. This is the Fokker–Planck equation [98] adapted for the LLG equation including spin-transfer torque [99–103].

For a cylindrical pillar with perpendicular uniaxial anisotropy, Eq. (3.48) can be reduced as follows using Hamiltonian equation of motion,

$$\begin{cases} \dfrac{\partial}{\partial t} p(x^2, t) + \partial_2 \left\{ -\dfrac{1}{1+\alpha^2} \dfrac{\alpha}{S_2} [(\partial^2 E_{\text{eff}}) + k_B T \partial^2] p(x^2, t) \right\} = 0 \\[3mm] E_{\text{eff}}(x^2) = E_{\text{mag}} - \dfrac{1}{S_2} \int_0^{x^2} (T_{FT} V + \dfrac{T_{ST} V}{\alpha}) dx^2 \end{cases} \tag{3.48'}$$

where $E_{eff}(x^2)$ is the effective magnetic energy, and $x^2 = S_2(\cos\theta - 1)$ is a canonical valuable in the Hamiltonian equation of motion. $\cos\theta$ is z-component of $\vec{e}_2$ and $\vec{e}_1 = \vec{e}_z$. Without charge current, both the equations have the Boltzmann distribution, $p(\vec{e}_2) \propto e^{-(E_{mag}(\vec{e}_2)/k_B T)}$, as a solution for the thermal equilibrium. Under a current, the system is no more in equilibrium and the probability distribution deviates from the Boltzmann distribution in general. For cylindrical symmetry case, however, the system under a constant current may show Boltzmann type distribution, $p(z) \propto e^{-(E_{eff}(z)/k_B T)}$, taking the effective magnetic energy defined in Eq. (3.48') as for a steady state. If the system is in out of equilibrium because of a charge current application, the probability density currents start to flow to approach to a steady state.

The probabilistic distribution of the switching time, $p_{sw}(t)$, and its integration, $P_{sw}(t) = \int_0^t p_{sw}(t_1)dt_1$, in the thermal activation regime can be obtained by applying the Kramer's method [99] to the Eqs. (3.48) or (3.48'), for a case of the high thermal barrier, that correspond to large thermal stability factor, small current and small external field, as follows:

$$\begin{cases} p_{sw}(t) = \tau_{sw}^{-1} e^{-(t/\tau_{sw})} \\ P_{sw}(t) = 1 - e^{-(t/\tau_{sw})} \\ \tau_{sw}^{-1} \cong \tau_0^{-1} e^{-\Delta(1-h)^2} \\ h \equiv \dfrac{J^Q}{J_{c0}^Q} + \dfrac{H_{ext}}{H_c} \end{cases} \tag{3.49}$$

where t is the elapsed time, and τ_{sw} is the average switching time, $\tau_0^{-1} = \alpha(1+h)(1-h)^2(-\gamma)H_u\sqrt{(\Delta/\pi)}$ attempt frequency [102]. Δ is a thermal stability factor calculated for zero current and zero external field. Reader can find that the Néel−Brown's exponential law is valid under a current injection. The exponent of 2 is valid for a system with cylindrical symmetry [102].

In Figure 3.18, the theoretical switching time expressed by Eq. (3.49), which includes current dependence of the attempt frequency, is the plotted on the experimental data. For the calculation, the same parameter set that was obtained from a fitting of precessional switching regime were used. In both regimes, experimental data were well fitted employing a common parameter set $(\Delta, J_{c0}^Q, \alpha(-\gamma)\omega_u)$.

An approximate expression for the pillars with in-plane magnetization is obtained by replacing H_u and J_{c0}^Q with $(H_\perp + H_\parallel)/2$ and J_{c0}^Q for the in-plane pillar (Eq. (3.41)), respectively into Eq. (3.48). Li et al. solved Fokker−Plank equation for in-plane magnetization with small current and proposed a slightly different formula that includes a parameter, β [97]. In their paper, the term $(1-(J^Q/J_{c0}^Q)-(H_{ext}/H_c))^2$ in Eq. (3.49) is replaced with $(1-(H_{ext}/H_c))^\beta(1-(J^Q/J_{c0}^Q))$. Taniguchi et al. showed that the exponent on the current, δ in $(1-(J^Q/J_{c0}^Q))^\delta$, for the magnetic cell with in-plane magnetization may also vary from 1 to 3 depending on the situation using numerical simulation [103] and analytical treatment [104].

In the thermal activation regime, SIMS can occur at currents smaller than the critical current at 0 K because of thermal activations. And the switching time is distributed widely following an exponential law for a given applied current. In another way, if we apply current pulses with constant widths but different in heights (current), we will see the following switching current distribution:

$$p_{sw}(J^Q) = \frac{2\Delta}{J_{c0}^Q} \frac{\tau_{pulse}}{\tau_{sw}} e^{(-\tau_{pulse}/\tau_{sw})}, \tag{3.50}$$

where τ_{pulse} is the width of the applied pulse current, and τ_{sw} is a function of the applied current (Eq. (3.49)). Experimentally, J_{c0}^Q and Δ can be determined either from the pulse width dependence of the average switching current or from the switching current distribution measured using pulses with constant width.

In contrast to the thermal activation regime, switching occurs almost adiabatically for short and high-current pulses. Even for such adiabatic switching, the switching time is scattered because of the thermal distribution of the initial angle between the directions of the free layer magnetization and pinned layer magnetization. One may avoid this problem by applying a small external field perpendicular to the easy axis of the system to force a definite initial angle [93].

For memory applications, a small switching current, small distribution of the switching current and switching time, and high thermal stability are required. From Eq. (3.32), the critical current is related to the thermal stability factor, Δ, as follows [79] for a pillar with perpendicular magnetization at room temperature:

$$|J_{c0}^Q| = \left| e\alpha \frac{\gamma H_u}{g(\theta_0)} \frac{S_2}{\hbar/2} \right| \simeq \frac{\alpha\Delta}{|g(\theta_0)|} \frac{8\pi e k_B T}{h} \simeq \frac{\alpha\Delta}{|g(\theta_0)|} \times 24[\mu A], \tag{3.51}$$

where h is the Planck's constant. To reduce the critical current while keeping Δ large, a material with a small damping constant and/or high spin-transfer efficiency should be developed. It should be noted, however, that a small damping constant will result a long switching time. Domain formation, that is, a smaller activation volume than the magnetic cell volume, reduces thermal stability factor and may make critical current to thermal stability factor ratio large. To avoid this problem, the cell volume should be smaller than the single domain limit. To obtain the expression for the elliptical pillar with in-plane magnetization, in which $H_\perp \gg H_{//}$, we should multiply $H_\perp/(2H_{//})$ to the right side of Eq. (3.51). Therefore, the critical current to the thermal stabilization factor ratios for the in-plane magnetized pillars are normally larger than that of the pillar with perpendicular magnetization.

3.2.5 High-Speed Measurements

The interesting aspects of the spin injection magnetization switching (SIMS) phenomenon are the small energy consumption and very high precession speed. To investigate the high-speed properties of the SIMS, time domain high-speed electrical observations have been performed [94,95,105–108]. The first observation was performed by Krivorotov et al. for a $Ni_{80}Fe_{20}$ 4 nm/Cu 8 nm/$Ni_{80}Fe_{20}$ 4 nm GMR nanopillar at 40 K [105]. A free layer was microfabricated in an elliptical shape with dimensions of 130×60 nm^2. To obtain reproducible trajectories for adiabatic switching, they maintained the initial angle between the fixed layer spin and the free layer spin at about $30°$ by using an antiferromagnetic under layer to pin the spins in the fixed layer. Since the GMR nanopillars provide a very small output voltage, the authors averaged more than ten thousands of traces using a sampling oscilloscope with a 12-GHz bandwidth. After a background subtraction process, they obtained a transient signal that corresponds to the adiabatic switching of the free layer spin, as shown in Figure 3.19 [105]. The precession of the free layer spin

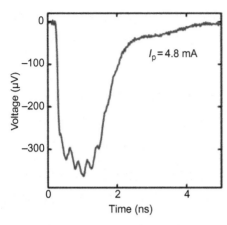

Figure 3.19 Time resolved measurement of the spin injection magnetization switching (SIMS) at 40 K. The observation was performed using a sampling oscilloscope. More than ten thousands traces were averaged for a NiFe/Cu/NiFe nanopillar with elliptical cross-sectional dimensions of 130×60 nm^2. The initial angle between the free layer spin and the fixed layer spin was about 30°. *Source*: From Ref. [94].

was clearly observed. The amplitude of the precession was amplified in the early stage of the switching and was then damped before the transition from the P state to the AP state at around 2 ns. The observed behavior was slightly different from that predicted by the simple macrospin theory according to which continuous amplification of the precession should be observed until the transition (Figure 3.20A [82]). Krivorotov et al. explained this deviation by a dephasings among the traces. If the precession contains phase noise, the averaging process carried out by the sampling oscilloscope decreases the observed precession amplitude. The authors stated that the spectrum linewidth of about 10 MHz obtained from the dephasing rate agreed with that obtained from precession noise spectrum measurement. This fact implies that in their sample the phase noise dominated the spectrum linewidth of the precession. Krivorotov et al. also clearly showed [94] that, for large applied current, the switching time becomes multiples of the precession period (200 ps, for example) as that was already pointed by Devolder et al. from their high-speed pulse measurements [93]. This also means that only one extreme point out of two in the orbital (Figure 3.15A) was responsible for the switching. They explained this fact from their asymmetrical configuration of the magnetization.

The real-time single-shot observation of the SIMS has been performed by employing GMR pillars [94] and MTJs with a large electrical signal output [95,106,107]. In Figure 3.21, a circuit to observe real-time switching signal (A) and an example of a switching signal (B) obtained by Tomita et al. [107] are shown. For the measurement, a CoFeB/MgO/CoFeB nanopillar was placed at an end of a radio frequency (rf) wave guide. A 10-ns pulse from a pulse generator was supplied to the MTJ through a power divider. The pulse was partially reflected by the MTJ depending on its resistance states and was provided to a high-speed storage oscilloscope with a 16-GHz bandwidth after being made to pass through the power divider again. In Figure 3.21B, both the direct pulse signal and the reflected pulse signal are shown. Since the pulse reflectivity is dependent on the sample resistance, a magnetization switching event can be observed as a step in the reflected signal, as shown in the inset of Figure 3.21B. To cancel parasitic signals generated by pulse reflections

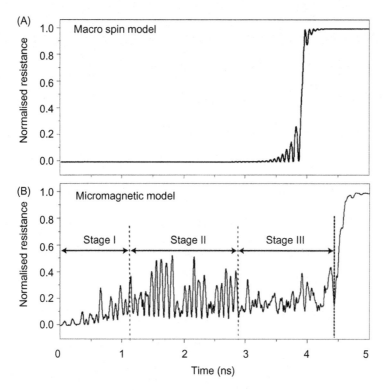

Figure 3.20 (A) Macromagnetic and (B) micromagnetic simulation of spin injection magnetization switching (SIMS) at 0 K. A magnetic pillar with elliptical cross-sectional dimensions of 130×60 nm^2 was considered.
Source: From Ref. [82].

from the pulse generator and the cable connections, the obtained signal was numerically divided by the signal obtained in the absence of switching; the resulting signals are shown in Figure 3.22 for a current slightly below J_{c0}^Q. The signal shows a long waiting time followed by a short transition time (about 500 ps) as expected for a thermally assisted SIMS. The waiting time seems to vary considerably. Tomita et al. [107] also analyzed 1000 single-shot data and obtained the distribution of the non-switched probability, $1 - P_{sw}(t)$, which should be a simple exponential function of the elapsed time according to Néel–Brown's law. The obtained distribution was not explained by simple Néel–Brown's law, but well fitted by considering a certain initial period in which the switching probability was negligibly small as shown in Figure 3.23. Tomita et al. [107] named this initial period as a nonreactive time and explained that this was the time needed to complete a transition from the initial thermal equilibrium without current to a quasi-thermal equilibrium state under a finite current. This behavior was well reproduced by a micromagnetic simulation (inset in Figure 3.23). Lee et al. also showed such a transition in his micromagnetic simulation [82] at 0 K. It is shown as "Stage I" in Figure 3.20B. This means that even at

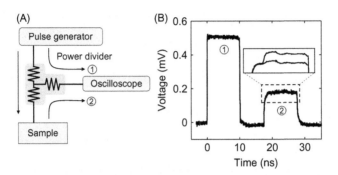

Figure 3.21 (A) Electric circuit and (B) an example of an obtained signal for a real-time single-shot observation of the SIMS process. Since an MTJ terminates an rf wave guide with certain mismatch in impedance, the supplied voltage pulse is partially reflected by the MTJ and observed by a high-speed single-shot oscilloscope with a 16-GHz band width. The amplitude reflectivity of the MTJ is $(R_T - Z_0)/(R_T + Z_0)$, where R_T is the MTJ resistance and Z_0 is the characteristic impedance of the wave guide (50 Ω). Transition during a pulse causes a jump in the reflectivity and observed as a jump in the reflected voltage wave as shown in the inset in (B).
Source: From Ref. [107].

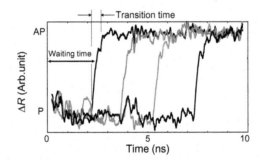

Figure 3.22 Real-time single-shot observation of the spin injection magnetization switching (SIMS) process in a CoFeB/MgO/CoFeB nanopillar. All the traces are obtained for the same value of current but show differences due to the probabilistic nature of switching. The employed current magnitude is 2.7 mA (2.4×10^7 A/cm^2), which is slightly smaller than J_{c0}^Q. The signals are normalized by the signal obtained without switching.
Source: From Ref. [107].

0 K, the system makes a transition from the quiet initial state to an excited state, where the system shows the spatial distribution of the local spins in the cell, mainly because of the nonuniform demagnetization field.

The micromagnetic characteristics of the SIMS have been directly observed by Acremann et al. [109] and Strachan et al. [110] using transmission X-ray microscopy (STXM) with high spatial (30 nm) and temporal (70 ps) resolutions. Using bunches of X-ray from the synchrotron light source, that was matched to the L$_3$ edge of Co

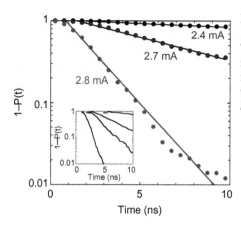

Figure 3.23 Non-switched probability, $1 - P_{sw}(t)$, as a function of elapsed time. The data deviates from Néel−Brown's law in the initial stage and are well fitted by the function $1 - P_{sw}(t) = e^{-(t-\tau_0)}$, where τ_0 is the nonreactive time.
Source: From Ref. [107].

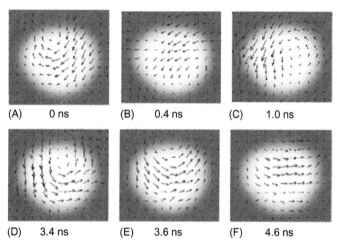

Figure 3.24 Spatial spin distribution in a GMR nanopillar with $Co_{0.86}Fe_{0.14}$ free layer $(100 \times 150 \times 4 \text{ nm}^3)$ during magnetization switching observed by scanning transmission X-ray microscope (STXM) in a pump-probe manner. The top part shows the current pulses applied to the sample, and the various time points at which the images were taken by X-ray bunches. The sample switches from AP to P state during the positive pulse, and P to AP state during the negative pulse. The switching proceeds via a vortex motion during both the reversals. The current intensity was 1×10^8 A/cm^2.
Source: From Ref. [110].

atom, they could image the spin distribution of a GMR nanopillar with in-plane magnetized $Co_{0.86}Fe_{0.14}$ free layer. They found that the magnetization reversal takes place via motion of a magnetic vortex. The position of the vortex core changes from inside the sample to outside, with decreasing sample size. Images of spin distribution during magnetization reversal of 180×110 nm^2 sample are shown in Figure 3.24. In this case, the vortex core lies inside the sample during both AP to P and P to AP switching. Such a nonuniform reversal was attributed to nonuniformities initiated by the Oersted field generated by current. A dynamic process involving vortex formation and annihilation was predicted by Lee et al. [82]. In Figure 3.20B "Stage III" corresponds to a random domain state where a vortex can be generated and annihilated if the current is sufficiently large to produce a rotational magnetic field. The vortex formation should be avoidable in the case of a nanopillar with a small J_{c0}^Q and small dimensions. Inhomogeneous precession caused by inhomogeneous demagnetization fields is serious in the nanopillars with in-plane magnetization. However, it is expected to be less important in the nanopillars with perpendicular magnetization.

3.3 High-Frequency Phenomena

3.3.1 Spin-Transfer Oscillation

From the beginning of spin-injection research using magnetic multilayers, it was thought that the electric current inside a ferromagnetic material may interact with the collective modes of spins and excite spin waves [17,20,21]. Actually, before the confirmation of the spin injection magnetization switching (SIMS) [22−24], spin dynamics in magnetic nanopillars resulting from spin injection were observed as anomalies in derivative conductance spectra [18,22,111]. The first and complete observation of microwave emission from magnetic nanopillars with in-plane magnetization was performed by Kiselev et al. in 2003 [19]. They employed Co/Cu/Co GMR nanopillar with a $130 \times 70 \times 2$ nm free layer and applied a direct current (more than J_{c0}^Q) and an external magnetic field (more than H_c) at the same time. The external field preferred parallel (P) configuration of the spins, while the direct current preferred antiparallel (AP) configuration. Under such a situation, the P state is unstable and the switching from P to AP state is prevented by the external field. As a result, the free layer spin is driven into a cyclic trajectory (limit cycle) with frequency typically in GHz range. Because of the GMR effect, the resistance of the pillar also oscillates with the continuous precession of the free layer spins. The oscillation of the resistance under a direct current bias results an rf voltage that can

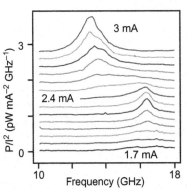

Figure 3.25 Rf power density spectra normalized by square of the injected current. The emission was observed for a GMR nanopillar with in-plane magnetization under −2 kOe of the external field. *Source*: From Ref. [19].

be detected by a spectrum analyzer or rf diode. In Figure 3.25, the observed rf spectra taken under an external field of -2 kOe are shown. For current up to 2.4 mA, the spectrum intensity normalized by the square of the current is almost unchanged. The peak frequency matches with the FMR frequency of the free layer and does not shift significantly under this magnitude of the current. A further increase in the applied current, however, results in a strong increase in the peak height and significant lowering of the peak frequency (red shift). Such behavior was understood as the spontaneous excitation of the precessional motion of the macrospin. The maximum rf power obtained was about several tens of pW. This is the spin-transfer oscillation (STO).

Clearer evidence of the onset of auto-oscillation in an MTJ is shown in Figure 3.26 [112]. When the injection current is less than J_{c0}^Q, an increase in the injection current results in a linear reduction in the peak width, as it was explained previously from Eq. (3.39) (Figure 3.26B). The threshold current (J_{c0}^Q), which is indicated by an arrow in Figure 3.26B, corresponds to the current at which the peak width reduces to zero, if a linear reduction holds until J_{c0}^Q. In practice, when the injection current is around the threshold current, there is a sudden increase in the peak width. The peak width has its maximum value slightly below J_{c0}^Q. Further increase in the injection current reduces the peak width and results a sudden increase in the output power. These observations provide clear evidence of the threshold properties, which are in good agreement with the theory developed by Kim et al. [113]. The width of the spectral lines are, however, very wide when compared to the width of the spectral lines for the CPP-GMR nanopillars and nanocontacts. Often MTJs provide much larger output power but also much larger linewidth compared with those in GMR nanopillars and point contacts [114].

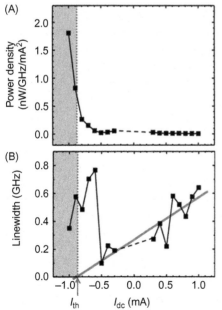

Figure 3.26 Rf auto-oscillation properties observed in Fe/MgO/Fe single crystalline MTJs (AP state). The junction size is 220×420 nm^2. (A) Peak power density as a function of the biasing current. The power is normalized by $(J^Q)^2$. (B) FWHM as a function of the biasing current.
Source: From Ref. [112].

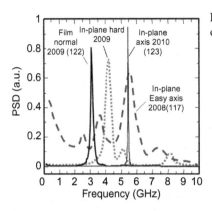

Application of an external field along hard axis in the magnetic cell, fabrication of MTJ with high current density contribute to obtain narrower lines keeping output power large [115−120] (Figure 3.27).

Condition to have a limit cycle can be understood using the LLG equation involving the spin-transfer torque (Eq. (3.27)) and magnetic energy of the macro-spin (Eq. (3.28)). Change in the magnetic energy during one cycle of iso-energy trajectory of the free layer spin is estimated as follows [80],

$$\Delta E_{mag}(E) = \mu_0 M_2 \oint_{E_{mag}=E} \left\{ -\alpha(-\gamma)|\vec{H}_{eff} \times \vec{e}_2|^2 + g(\theta)\frac{J^Q}{-eS_2}\frac{\hbar}{2}(\vec{e}_1 \times \vec{e}_2)\cdot(\vec{H}_{eff} \times \vec{e}_2) \right\} dt,$$

(3.52)

where integral should be done for one cycle of an iso-energy trajectory with energy E by taking time as a parameter. The first term in the integral is always negative and expresses energy consumption through the Gilbert damping. The second term in the integral can be positive depending on the sign of the current and expresses energy supply from the current source through the spin-transfer torque. The condition to have a stable limit cycle at energy E is as follows:

$$\begin{cases} \Delta E_{mag}(E) = 0 \\ \dfrac{d\Delta E_{mag}(E)}{dE} < 0 \end{cases}.$$

(3.53)

As it can be seen in above equations, the condition is sensitive to the angle θ dependence of E_{mag} and $g(\theta)J^Q$. Especially, in the nanopillar with perpendicular magnetization, a higher order crystalline anisotropy can also play a role. These conditions describing threshold currents of the SIMS (Eqs. (3.35), (3.44), and (3.48)) and the STO (Eq. (3.53)) separate possible dynamic phases appearing in magnetic nanopillars.

The phase diagram of a nanopillar with in-plane magnetization under external field and a current injection obtained by Kiselev et al. [19] is illustrated in

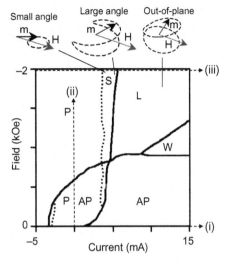

Figure 3.28 Phase diagram observed for the GMR nanopillar with in-plane magnetization. *Source*: From Ref. [19].

Figure 3.28. For a zero external field and zero current the system is in the bistable state (P/AP in the figure). The application of a positive (negative) current causes the SIMS to undergo a transition from P (AP) to AP (P) state and stabilizes AP (P) state (dotted line (i)). The system shows hysteresis along the line (i). For zero current, if we apply a negative (positive (not shown)) external field, the system switches to P (AP) state (dotted line (ii)). The system again shows a hysteresis along the line (ii). Now, we apply large negative external field, −2 kOe for example. At zero applied current, the system is in P state with small precession of the spin caused by a thermal excitation. Under such large field, even if we supply a positive current larger than the threshold current of the SIMS, switching does not occur. Alternatively, the precession starts to be enhanced significantly and spontaneous oscillation starts. Further increase in current changes the orbital form from small angle oscillation to large angle oscillation and then out-of-plane oscillation (dotted line (iii)). Corresponding to the change in the orbital form, the oscillation frequency first shows a significant red shift (see also Figure 3.25) and then a blue shift. Along the line (iii), the system does not show hysteresis.

In Figure 3.29, iso-energy contours of a rectangular shape magnetic nanopillar with in-plane magnetization are shown. Magnetic energy is lowest if magnetization is at P or P'. Those points are stable equilibrium points. The small angle oscillation trajectory corresponds to an iso-energy contour around P and P'. Larger bias current may sustain higher energy orbital and it may approach R point (Large angle orbital). R point places at a saddle point in energy landscape. This is an unstable equilibrium point. The trajectory that includes R point separates a region that includes small angle oscillation orbits and a region that includes out-of-plane orbits. Therefore, it is called as a separatrix. Since infinite time is needed to approach a saddle point, period in the separatrix is also infinite. As a result, red shift occurs when trajectory approaches to the separatrix. A large enough bias voltage may excite an out-of-plane orbit. The out-of-plane orbit shows blue shift when

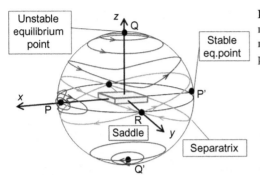

Figure 3.29 Iso-energy contours of a magnetic nanopillar with in-plane magnetization. Several equilibrium points are listed.

it estranges from the separatrix. Both north and south poles are also unstable equilibrium points.

When an intermediate negative field and a large positive current were applied, a new phase "W" shown in Figure 3.28 appeared. The very wide spectra observed in region W were attributed to a chaotic motion of the spins [19]. The overall phase diagram was well explained by the micromagnetic simulation including the region "W" [82]. It was shown that vortex generations and annihilations were the main origin of the chaotic behavior in "W" [82]. Deac extended the phase diagram to positive field case using a nanopillar with pinned layer and showed that a combination of a positive field and a negative current also produces STO [121]. Phase diagram of a nanopillar with perpendicular magnetization was obtained by Mangin et al. [28] and was quite different from the in-plane case. STO from a nanopillar consisting of a free layer with in-plane magnetization, a perpendicularly magnetized polarizer and a reference layer with in-plane magnetization was observed by Houssameddine et al. [122]. The phase diagram of this system is in the Ref. 122.

Apart from magnetic nanopillars, the STO have also been observed in the case of magnetic nanocontacts. A schematic structure of the magnetic nano contact is shown in Figure 3.28A. In 2004, Rippard et al. demonstrated that the line width of the rf emission spectrum emitted by a magnetic nanocontact can be as narrow as 1.89 MHz by applying a perpendicular magnetic field [123]. The obtained line width corresponds to a very large Q-factor of about 18,000. Here Q is defined as Q = (peak frequency)/(line width). After this report the line width of the STO was investigated both experimentally [114,116,124−127] and theoretically [113,128−130]. Kim et al. employed a general model of nonlinear oscillator and showed that the special point in STO compared to the other oscillators is a strong amplitude dependence of the oscillation frequency. This nonlinear coupling and thermal fluctuations produce a significant phase noise and dominate the line width. Therefore, the line width is proportional to the absolute temperature and depends on the size of the nonlinear coupling between amplitude and frequency. By finding a configuration with small amplitude-frequency coupling one may achieve in principle very small line width.

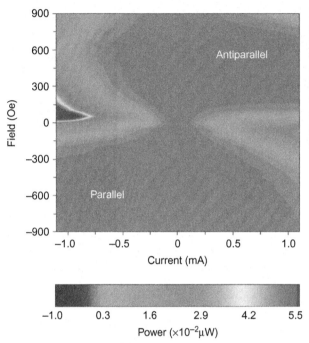

Figure 3.30 Rf power mapping observed for a nanopillar made from CoFeB/MgO/CoFeB MTJ with in-plane magnetization. Large rf power emission was observed for positive applied field (AP) and negative current. *Source*: From Ref. [114].

To obtain larger output power, it is important to use pillars with large MR ratio and with resistance that matches to the impedance of the wave guide (usually 50 Ω). Replacing a GMR pillar by the MTJ, we can expect significant increase in the output power [114,116]. Deac et al. showed large output power of 0.14 μW from a single CoFeB/MgO/CoFeB nanopillar with 70×160 nm^2 cross section [114]. The junction was specially designed to have the high MR ratio of about 100% (MR' = 67% at zero bias) and the very low resistance area product (RA = 4 $\Omega\mu$m^2) [131]. Figure 3.30 is a color mapping of the output power as a function of the current and the external field. Large output was observed for two cases in which the external field and the current preferred opposite configurations of the spins. Among them, the combination of the positive field (AP) and the negative current gave larger power. This phenomenon was understood from the asymmetrical behavior of the spin-transfer torque to the bias voltage. Because of the bias dependence of the spin current in MTJ, the spin-transfer torque is larger for the negative bias voltage in the CoFeB/MgO/CoFeB MTJs [48,49] (Figure 3.10).

Another method to obtain higher output power was demonstrated by Kaka et al. [132] and Mancoff et al. [133]. They made two magnetic point contacts in very small distance (about 100 nm) (Figure 3.31B and C) and observed mutual coupling between them. By changing current passing through one of the oscillators, the oscillation frequencies of these two oscillators were approached. When the difference of two frequencies became enough small, the frequency were suddenly unified and two oscillators started to oscillate coherently at the same frequency.

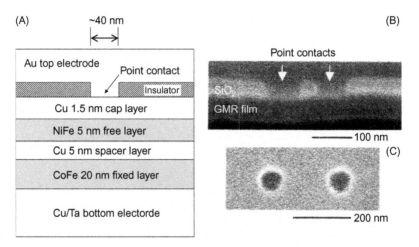

Figure 3.31 (A) Schematic cross section of the magnetic point contact. (B) Cross-sectional SEM image of the coupled point contacts that was used to observe mutual coupling of the STO. (C) SEM top image of the coupled point contacts.
Source: From Ref. [133].

In addition, the total output power of these two oscillators was doubled by this coupling. By this way, if we can achieve coherent coupling of n oscillators, we may get n^2 time larger output power. Mutual coupling was mediated by spin waves in their oscillators, therefore, it was difficult to connect many oscillators. The electrical mutual coupling, however, is more realistic and also possible if the electric output of each oscillator is large enough [134].

3.3.2 Spin-Torque Diode Effect

Recent advances have made it possible to fabricate MTJs with up to 500% magnetoresistance at room temperature [135–139]. Thus even a small change in the spin direction of MTJs, can be easily detected by measuring the change in resistance. In this section, we introduce a new concept that MTJs may possess a rectification function [140], if we combine their large TMR with spin-torque effect. We will show below that this "spin-torque diode effect" provides a direct measure of the spin torque and contributes to an elucidation of the physics of the spin-torque phenomena [48–50,140,141].

To observe the spin-torque diode effect, we may prepare a special nanopillar in which both free layer and fixed layer magnetizations are lying in-plane but perpendicular to each other as shown in Figure 3.32B. We then apply a negative current that induces a preferential parallel configuration of the spins. Thus resistance of the junction becomes smaller and we observe only a small negative voltage across the junction for a given current (Figure 3.32A). Next we apply a positive current of the same amplitude. This current induces preferential antiparallel configuration and

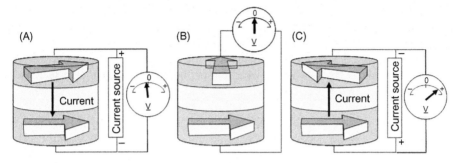

Figure 3.32 Schematic explanation of the spin-torque diode effect. (A) Negative current. (B) Null current. (C) Positive current [142].

the resistance becomes higher. We observe a larger positive voltage appearing across the junction for the given current amplitude (Figure 3.32C). This is the "spin-torque diode effect." The effect can be large if the frequency of the applied current matches with FMR frequency of the free layer. In other words, this effect provides sensitive FMR measurement technique of the nanopillar moment excited by the spin torque. Therefore, the effect is useful to investigate the high-frequency dynamics of the nanopillar and the physical mechanism of the spin torque itself.

To treat a linear response of the magnetic cell to the alternative current injection, we express linearized LLG equation (Eq. (3.38)) using Fourier transformation.

$$-\sum_{j=1}^{2}(i\omega g_j^i + \partial_j F^i \delta x^j(\omega)) = \left(\frac{\partial F^i}{\partial H_{ext}}\right)\delta H_{ext}(\omega) + \left(\frac{\partial F^i}{\partial V}\right)\delta V(\omega),$$

$$(3.54)$$

$$\text{where} \quad \begin{cases} \delta x^i(t) = \int_{-\infty}^{+\infty} d\omega \delta x^i(\omega)e^{-i\omega t}, \\ g_j^i:\text{Kronecker delta.} \end{cases}$$

Taking $\vec{e}_1 = \vec{e}_z$ and $\Delta H_{ext}\omega = 0$ for simplicity, the above equation can be solved easily.

$$\begin{pmatrix} \delta x^1(\omega) \\ \delta x^2(\omega) \end{pmatrix} \cong \frac{1}{\omega^2 - \omega_0^2 + i\omega\,\Delta\omega}\begin{pmatrix} i\omega - \Omega_{12} & -\Omega_{22} \\ \Omega_{11} & i\omega + \Omega_{21} \end{pmatrix}\begin{pmatrix} \partial_2 E'_{mag} - T'_{FT}/S_2 \\ -\partial_1 E'_{mag} + T'_{ST}\sin^2\theta \end{pmatrix}\delta V(\omega),$$

$$(3.55)$$

where

$$E'_{mag} \equiv \frac{\partial E_{mag}}{\partial V}, T'_{FT} \equiv \frac{\partial T_{FT}V}{\partial V}, T'_{ST} \equiv \frac{\partial T_{ST}V}{\partial V}.$$

From above equation, we see that the spin-transfer torque, the field-like torque and voltage-induced uniaxial anisotropy field, can excite a uniform mode (FMR mode) in the free layer. The field-like torque and the spin-transfer torque shows a

90° difference in the phase, whereas the voltage-induced anisotropy field may take both phases. The difference in the precessional phase is a consequence of the different directions of the respective torques (Figure 3.14). In addition, the width of the resonance, $\Delta\omega$, is affected only by the spin-transfer torque exerted by the direct voltage, V_0. This is the (anti)damping effect of the spin-transfer torque that was already discussed in the previous section. The field-like torque and voltage-induced anisotropy field exerted by a direct voltage V_0, changes the resonance frequency, ω_0, in a similar manner to an external field.

When the rf voltage across the MTJ is $\delta V \cos \omega t$, from Eqs. (3.24) and (3.55) the precession motion and the oscillating part of the junction resistance, $\delta R \cos(\omega t + \varphi)$, which is linear in δV, are determined. Rf current through the junction can be approximated as $\delta V \cos \omega t / R(\theta_0)$. Thus from Ohm's law, the following additional voltages appear across the junction.

$$\delta R \cos(\omega t + \varphi) \times \frac{\delta V}{R(\theta_0)} \cos \omega t = \frac{\delta R}{R(\theta_0)} \frac{\delta V}{2} (\cos \varphi + \cos(2\omega t + \varphi)).$$

Here, we see that the frequencies of the additional voltages are zero (dc) and 2ω. It means that, under spin-torque FMR excitation, the MTJs may possess a rectification function and a mixing function. Because of these new functions, A. A. Tulapulkar et al. referred to these MTJs as spin-torque diodes and these effects as "spin-torque diode effects" [140]. This is a nonlinear effect that results from two linear responses, that is, the spin-torque FMR and Ohm's law.

For the case when the MTJ is placed at the end of an emission line, the explicit expression of the rectified dc voltage under a small bias voltage is as follows:

$$V_{dc} \cong \frac{\eta}{4} \frac{R(\theta_0)(G_P - G_{AP})}{S_2}$$

$$\mathrm{Re}\left[\frac{\Omega_{11}(-\partial_2 E'_{mag} + T'_{FT}/S_2) + (i\omega + \Omega_{21})(\partial_1 E'_{mag} - T'_{ST} \sin^2 \theta_0)}{\omega^2 - \omega_0^2 + i\omega \, \Delta\omega}\right] V_\omega^2, \tag{3.56}$$

where V_ω is the rf voltage amplitude applied to the emission line. η is the coefficient used to correct impedance matching between the emission line with a characteristic impedance of Z_0.

$$\eta = \left(\frac{2R(\theta_0)}{R(\theta_0) + Z_0}\right)^2. \tag{3.57}$$

If the emission line and the MTJ include some parasitic impedances (capacitance for most of the cases), we should employ an appropriate value of η to correct the effect [48,49].

This is a type of homodyne detection and is, thus, phase-sensitive. The motion of the spin, illustrated in Figure 3.32, corresponds to that excited by the spin-transfer

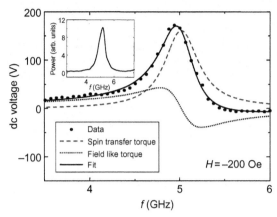

Figure 3.33 Spin-torque diode spectra for a CoFeB/MgO/CoFeB MTJ. Data (closed dots) are well fitted by a theoretical curve considering contributions from both the spin-transfer torque and the field-like torque. The inset shows the rf voltage dependence of the dc output voltage [142].

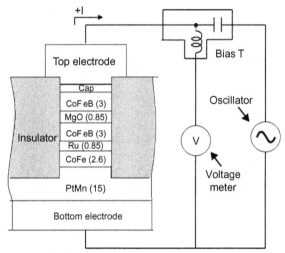

Figure 3.34 Measurement setup for the spin-torque diode effect measurements [142].

torque at the resonance frequency. However, the motion of the spin excited by the field-like torque shows a 90° difference in phase. As a consequence, only the resonance excited by the spin-transfer torque can rectify the rf current at the resonance frequency. In Figure 3.33, the dc voltage spectra predicted for the spin-transfer torque excitation and for the field-like torque are both shown. The spectrum excited by the spin-transfer torque exhibits a single, bell-shaped peak (dashed line) but that excited by the field-like torque is of a dispersion type (dotted line). This very clear difference provides us with an elegant method to distinguish spin-transfer torque from field-like torque [48,49,140,141]. For a case of voltage-induced anisotropy field, both type of spectra may appear depending on the geometry [77], Therefore, by changing geometry, we may identify all three torques from spin-torque diode spectra.

Figure 3.34 shows a schematic illustration of the measurement setup for spin-torque diode effect measurements with a cross-sectional view of the MTJ employed in

Ref. [140]. The rf voltage was applied through a bias-T from a high-frequency oscillator and the dc voltages across the MTJ was detected using a dc nanovoltmeter.

In Figure 3.33, an example of the diode spectrum (closed dots) is shown together with a fitting curve based on the theoretical expression (Eq. (3.56)) [142]. The data were taken at room temperature (RT) without applying dc bias voltage. The observed spectrum has an asymmetrical shape and was well fitted by Eq. (3.56). By this fitting, the spectrum was decomposed to a contribution from the spin-transfer term and from the field-like term. The intensity and even the sign of the field-like term contribution varied from sample to sample, while those of the spin-transfer term were reproducible. Therefore, it is thought that the contribution from the field-like term at zero bias voltage is very sensitive to small defects in the magnetic cell. By taking a sample that does not show a contribution from the field-like term at zero bias voltage, H. Kubota et al. [48] have investigated the dc bias voltage dependence of the spectra. The results, that were already shown in 3.1.2 (Figure 3.10), were well explained by band theory, in which bias-dependent spin-subchannel conductivities and bias-dependent interlayer magnetic coupling were taken into account [47]. According to Eq. (3.56), V_{dc} is proportional to V_ω^2.

An expression for the rectified dc voltage at the peak of the spectrum for $\theta_0 = \pi/2$ is shown below together with that for p-n junction semiconductor diodes:

$$
(V_{dc}(\theta_0))_{peak} = \begin{cases} \dfrac{1}{4}\dfrac{G_P - G_{AP}}{G_P + G_{AP}}\dfrac{V_\omega^2}{V_c}; & \text{(spin-torque diode)} \\[2ex] \dfrac{1}{4}\dfrac{V_\omega^2}{k_B T/e}; & (p\text{-}n \text{ junction semiconductor diode)} \end{cases}, \quad (3.58)
$$

where $k_B T/e$ is the thermal voltage (25 mV at RT). For both cases, the rectified voltage is a quadratic function of the applied rf voltage. Therefore, these detectors are referred to as quadratic detectors. Output voltage is scaled by the critical switching voltage, V_c, for the spin-torque diode and by $k_B T/e$ for the p-n junction semiconductor diode. A typical critical switching voltage for MTJs was about 300 mV and was about 10 times larger than $k_B T/e$ for the experiments in Refs. [48,140]. Therefore, the output of the spin-torque diode was smaller than that of the semiconductor diode. Increase in the MR ratio and reduction of the critical switching voltage will enhance the performance of the spin-torque diode.

Wang et al. [143] found a nonlinear FMR excitation by an rf current. Since the nonlinear FMR associates a change in average resistance of the junction, it can be detected by passing a dc current to the junction. Under such configuration, Ishibashi et al. found a very large dc voltage output for a small rf signal input [144]. At finite temperature, by controlling a magnetic potential shape, the system can be unstable and may show a telegraph noise as a consequence of spontaneous switching between two quasi-stable states. Under such condition, Krivorotov et al. [145] applied an rf signal and found big response of the system. They claimed that it is a kind of stochastic resonance. Such efforts are providing better diode sensitivity and will break a limitation in semiconductor diodes, since that of the spin torque diode is still far below the fundamental limitation.

3.3.3 Electric Field-Induced Dynamic Switching

An important and basic issue for practical application of the voltage effect is the realization of two-way switching using only a voltage. However, reversible magnetization switching possesses intrinsic difficulties, because an electric field does not break time-reversal symmetry. So far, voltage-induced magnetization switching by controlling the coercive field has been demonstrated in a ferromagnetic semiconductors [59] or ferromagnetic metals [75]. This method requires changing the direction of assisting magnetic field to achieve reversible switching. To realize the reversible switching using only voltage, two methods have been proposed in the simulation. The first method uses the ratchet-type motion by successive anisotropy change along the different axes [146]; however, this method is difficult to apply on nanoscale magnetic cells. The second method uses the precessional motion excited by an abrupt change in the magnetic anisotropy [74,147].

Shiota et al. have reported the observation of coherent magnetization switching induced by the application of pulse voltages under a constant-bias magnetic field [147]. A fully epitaxial MTJ structure comprising Au (50 nm)/$Fe_{80}Co_{20}$ (0.70 nm)/MgO (1.5 nm)/Fe (10 nm) was deposited on MgO(001) substrate using molecular beam epitaxy. The film was patterned into a pillar with 800×200 nm^2.

A macrospin model simulation based on the LLG equation was performed to evaluate the possibility of voltage-induced magnetization switching. The magnetic energy in this system comprises the magnetic anisotropy energies and the Zeeman energy,

$$E_{\text{mag}} = \mu_0 M_2 \vec{e}_2 \cdot \vec{H}_{\text{ext}} + \frac{1}{2}\mu_0 M_2 (H_c \sin^2 \phi \sin^2 \theta + H_{d,\text{eff}}(V)\cos^2 \theta) \qquad (3.59)$$

where $H_{d,\text{eff}}(V)$ is the effective demagnetization field depending on the bias voltage. The perpendicular magnetic anisotropy component has significant linear voltage dependence, therefore, $V(\partial/\partial V)\vec{H}_{\text{eff}}$ in Eq. (3.26) can be expressed as $V(\partial/\partial V)\vec{H}_{\text{eff}} = (0, 0, V(\partial/\partial V)H_{d,\text{eff}}(V) \cos \theta)$. From the magnetoresistance curve and FMR measurement under the perpendicular magnetic field, the effective demagnetization field at zero bias voltage is 1400 Oe, and the voltage dependence of it is 533 Oe/V, which corresponds to perpendicular magnetic anisotropy energy per electric field of 38 fJ/(V · m). In the calculation, the amplitude of voltage pulse is -1.5 V, and rise and fall times of 70 ps with pulse length, τ_{pulse}, is used.

Figure 3.35 shows examples of the calculated trajectories for various τ_{pulse}. Here the influence of the thermal fluctuation is omitted ($T = 0$ K). The initial and final magnetization states are denoted by I.S. and F.S., respectively, and the line indicates the trajectory of the magnetization. When $\tau_{\text{pulse}} = 0.4$ ns, the final state is stabilized to the reversed position to the initial state. On the other hand, when $\tau_{\text{pulse}} = 0.8$ ns, it is switched back to the initial state by rotation through an angle

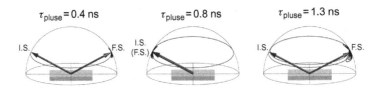

Figure 3.35 The calculated motion of magnetization under various pulse duration, τ_{pulse}. Initial state (I.S.) and final state (F.S.) represent the magnetization state before and after pulse voltage application.
Source: From Ref. [147].

of 360°. Again it can be reversed when $\tau_{\text{pulse}} = 1.3$ ns. As for the relaxation process after the pulse is turned off, the precession angle becomes larger for longer pulse duration. From these observations, one predict that the switching probability should oscillate depending on the pulse duration and become smaller for the longer pulse duration, because a large precession angle may be susceptible to the effect of thermal magnetization fluctuation at room temperature.

Figure 3.36A shows the tunneling resistance measured after 50 successive pulse voltage applications with the amplitude of −0.76 V and the length of 0.55 ns. The applied voltage to the MTJs corresponds to twice because of the reflected voltage caused by the impedance mismatch. Here the external magnetic field is 700 Oe at a tilt angle of 6° from the film's normal direction toward the in-plane easy axis. The tilted external magnetic field cancels the in-plane dipole field from the thick Fe layer on the top and tilts the magnetization in the perpendicular direction to increase the voltage torque effectively. The two-way toggle magnetization switching was observed between P and AP state in all pulse voltage application. This switching by the unipolar pulses oppose to the bipolar pulsed used in the usual spin-transfer torque switching. Figure shows the contour plots of switching probabilities for AP to P state as function of pulse duration and external magnetic field. Red and blue color indicates the high and low switching probability, respectively. From this diagram, several tendencies were observed. One is that switching probability is low in small magnetic field region. Another is that the switching probability clearly oscillates with τ_{pulse} and become feeble with increase of τ_{pulse}. This is because that small magnetic field stabilized the magnetization to AP state and AP to P switching requires large excitation energy, and increase of τ_{pulse} promotes incoherency by the thermal fluctuation of magnetization. To understand the observed behavior, switching probability was calculated using the macrospin model simulation discussed above. The influence of room temperature was introduced as the effective thermal fluctuation field with an isotropic Gaussian distribution. The values used in the calculation are identical to those in the calculation used in Figure 3.35. The calculation results for AP to P switching are shown in Figure 3.36B. The oscillation period and phase were reproduced completely.

Here we need to consider carefully the influence of the current—for example, the influence of current-induced magnetic field and the spin-transfer effect.

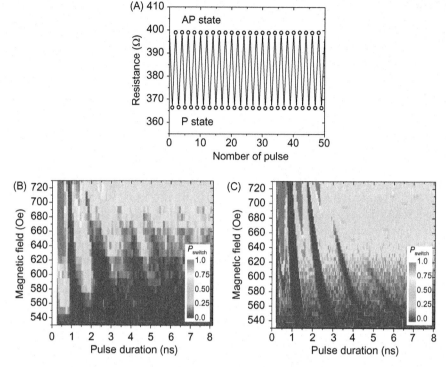

Figure 3.36 (A) Resistance measured after pulse voltage application of -0.76 V with 0.55 ns duration. The external magnetic field is 700 Oe at tilted angle of $6°$. The switching probability (P_{switch}) diagram from AP to P state as functions of pulse duration and external magnetic field for (B) experimental result and (C) calculated results. The measurements and calculations were repeated 100 times in each condition; the red and blue region represent high and low probability. (For interpretation of the references to color in this figure legend, the reader is referred to the web version of this book.)
Source: From Ref. [147].

However, this magnetization switching does not depend on the current density, as demonstrated in high-resistance MTJs. Therefore, the voltage-induced perpendicular magnetic anisotropy change is effective in magnetization switching.

3.4 From Spin-Transfer Torque RAM to Magnetic Logic

The last decades have seen a continuous race for miniaturization and cost reduction of electronic components, usually described by the "Moore's law." But this race is now approaching serious hurdles such as:

- The explosion of fabrication costs (for instance a set of lithography masks in today's CMOS logic circuits costs about 1 M$).

– The increasing variability of device properties (such as the threshold voltage of MOS transistor) on a single chip, that threatens reliability.

– Enhanced leakage currents in smaller transistors: starting from the 45 nm CMOS node, more energy is dissipated in standby mode than needed for actually operating (switching) the logic gates [148].

In parallel, the development of spin transfer in high magnetoresistance MgO tunnel junctions opens a new paradigm for fabricating very compact, nonvolatile magnetoresistive cells, written by bipolar voltage pulses and compatible with the electrical characteristics of the smallest CMOS transistors. Moreover, the integration of such cells is done "above CMOS," that is, on planarized SiO_2 layers well above the Si wafer, which in practice can save wafer area, and make the integration easier.

The first application proposed was of course solid state magnetic storage in the MRAM. Whether the spin-transfer version of MRAM, also called Spin-RAM or STT-MRAM, is able to compete with NAND Flash for mass storage of data is still debated, but this is by no way the only possible application of the Spin-RAM.

Microelectronic circuits are based on a few components: the microprocessor or central processing unit (CPU), a primary memory storing the information immediately needed by the processor, a series of input/output controllers, and peripheral units such as nonvolatile memories to store data and codes. For optimum operation, the different elements must be adapted to each other, and so different levels of memory are used as "cache" between the processor and the main memory, ranging from small capacity but high speed near the processor (typically an SRAM), to higher capacity but lower speed for the main memory (typically a DRAM). Beyond, several kinds of nonvolatile memories are used to store code (typically, Flash NOR) or data (Flash NAND or hard disk). In a microcontroller, all these elements are integrated on the same chip, including eventually the nonvolatile memory. One speaks of "embedded memory." By associating the speed of SRAM with the much higher density of DRAM, Spin-RAM has the potential to become a "universal" embedded memory, able to take on nearly all memory functions in a microcontroller. Note that, because MRAM is nonvolatile, such circuits could be powered off and on very rapidly, without loss of information, a key issue to get rid of standby power.

Finally, the spin-transfer magnetoresistive cell can also be used in programmable logic circuits, at different levels of dissemination in the circuit, to bring for instance instant on/off capability for reduced energy cost and zero standby power. At longer term the objective could even be to replace logic gates (up to a few tens of transistors) by purely magnetic devices, reducing the effective number of transistors per gate and thus contributing to extend Moore's law.

In the text below, we will review the MRAM development, and discuss the main evolutions brought by the Spin-RAM cell for potential applications to MRAM and logic circuits.

3.4.1 The Magnetic Random Access Memory

Even before the discovery of the MTJ at room temperature, MRAM had been developed using AMR and then GMR effects [149]. The development of TMR

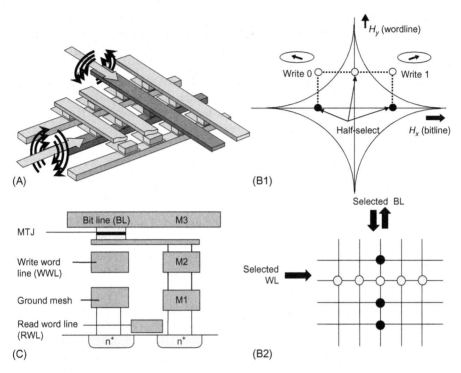

Figure 3.37 (A) Schematic diagram of MRAM. The MTJ marked by red color can be switched by combination of magnetic fields produced by passing currents through the top and bottom lines (B) The asteroid switching curve of the MTJ in response to magnetic fields along easy and hard axes. The curve is obtained assuming single domain particle with uniaxial anisotropy. The MTJ denoted by red color is switched by the magnetic field, as it crosses the asteroid boundary. The magnetic field on the half select MTJs, marked by full and hatched colors, remains within the asteroid and no switching takes place. (C) The reading scheme in one-transistor one-MTJ (1T1MTJ) architecture. (For interpretation of the references to color in this figure legend, the reader is referred to the web version of this book.)
Source: From Ref. [152].

promised much higher speed and density, and very rapidly the MRAM was considered as a major candidate for realizing a "universal memory technology," that is, a single technology that could finally replace all others in microcontrollers and "Systems on Chip" (SoC) [150]. One of the first working MRAMs, using magnetic field for writing, was developed at IBM in 2000 [151].

A schematic diagram of MRAM is shown in Figure 3.37A. As shown in the figure, MRAM consists of an array of MTJs [153]. Each MTJ is situated between two perpendicular wires labeled as word line and bit line. The magnetization direction of the free layer of each MTJ can be changed by the magnetic fields created by passing currents through the word line and bit line. In principle, the vector sum of the magnetic fields is sufficient to change the magnetization of only the selected MTJ (shown by darker gray in Figure 3.37A). The switching behavior of the MTJ can be described by an asteroid curve, which shows how the easy axis magnetic field required for switching changes with the hard axis magnetic field. For the asteroid curve shown in Figure 3.37B, the easy axis of the free layer is assumed to be along x-axis. The magnetic field generated by the current in the bit line is along x-axis, and by word line is along y-axis. Thus starting from zero field, if the applied field is increased beyond the asteroid boundary and then returns to zero while remaining in the x-positive half space, the free layer will be oriented along the positive x-axis. Whereas, if the applied field is increased beyond the asteroid boundary, and returns to zero while remaining in the x-negative half space, the free layer will be oriented along the negative x-axis. The MTJs marked by full and hatched dots are called half-selected, and their magnetization should not change during writing the selected MTJ, as the magnetic field remains within the asteroid.

The reading of MTJs is done by measuring their resistance. A diode or a transistor must be inserted in series with the MTJs to select the bit to read and suppress nondirect conduction paths. A "one-transistor one-MTJ" (1T1MTJ) cell architecture is shown in Figure 3.37C. The MTJ is connected to a FET whose drain contact is grounded. The gate of the FET is connected to the read word line (different from the write word line). The write bit line can be used as the top electrode of the MTJ to pass the measuring current.

It has been demonstrated that the write/read cycles of such MRAM circuits can be as low as 5 ns [154], while in laboratory experiments switching of single cells was achieved down to within about 100 ps [155–157]. However, this MRAM design suffers from several limitations. The most obvious ones are:

- Creating high enough magnetic field requires very high currents, for a large cost in wafer area and writing power.
- Moreover, with miniaturization one must increase the anisotropy field of the layers while reducing the field line cross section: a very sharp limit exists due to electromigration around 65 nm node. The magnetic field generated by the word and bit lines can be enhanced by "cladding," which concentrates the flux [158], but this gains only a small factor just able to slightly displace the limit.
- When including all distributions of switching fields, and the crosstalk between adjacent writing lines, the program reliability windows reduces to nothing.

The limitation mentioned in the last point can be overcome by using the toggle switching method [159]. In this method, the free layer of the MTJ consists of two thin magnetic layers coupled anti-ferromagnetically through the exchange coupling, and the easy axis of the free layer is oriented at 45° of the word and bit lines. The free layer magnetization can be reversed by applying a sequence of currents to the word and bit lines as shown in Figure 3.38A. The write phase diagram for toggle switching is shown in Figure 3.38B. One can see from its square shape that the magnetic field on the half-select bits remains quite away from the asteroid boundary.

This clever improvement brought reliability to MRAM [161,162], and made possible the first standalone MRAM product by Everspin [160], mainly targeted for

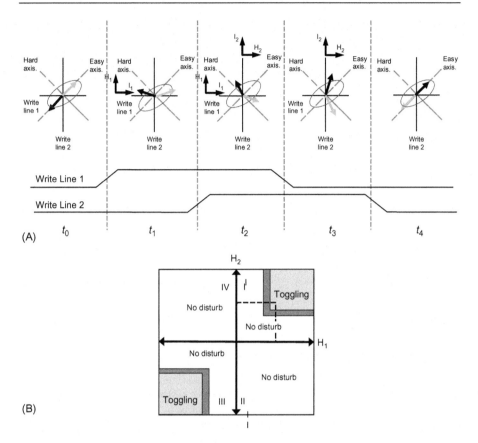

Figure 3.38 (A) The switching sequence of toggle mode MRAM. The free layer is synthetic anti-ferromagnet, shown by dark and light arrows. The switching takes place upon application of two magnetic field pulses along x and y axes. The required timings of the pulses is shown. (B) The asteroid curve for the synthetic anti-ferromagnetic free layer. The shape of the curve becomes square type (compare to Figure 3.37B).
Source: From Ref. [160].

replacing battery backed SRAM. But such MRAM can hardly be miniaturized beyond the 65 nm node of CMOS electronics, and is very limited in density (minimum cell size $\sim 20-40\ F^2$, where F is the usual characteristic dimension of a CMOS node).

The TAS (thermally assisted switching)-MRAM was introduced to overcome all limits [163–167]: The schematic diagram of the MTJ used in the TAS-MRAM is shown in Figure 3.39A. The energy barrier for reversal of the free layer is provided by exchange biasing it to an antiferromagnetic layer with moderate blocking temperature T_{B2}. A short current pulse through the tunnel junctions raises the temperature of the stack above T_{B2}, thus suppressing the energy barrier, and the final direction of the free layer magnetization is then determined by cooling in a moderate magnetic field. The value of this writing field is determined at values around

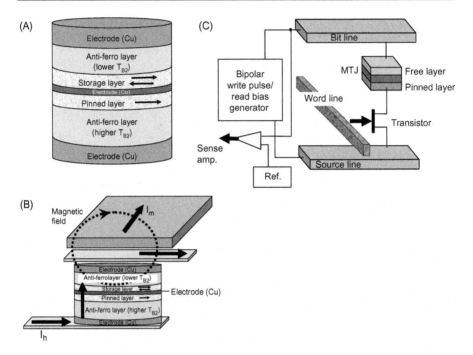

Figure 3.39 (A) Schematic diagram of the MTJ used in TAS-MRAM. The free layer/storage layer is pinned by an anti-ferromagnet with low blocking temperature T_{B2}. (B) Switching process of TAS-MRAM. A current pulse is passed through the MTJ to increase its temperature above T_{B2}. The MTJ is then allowed to cool in a small magnetic field which determines the orientation of the free layer. (C) Schematic diagram of STT-MRAM.
Source: (A) and (C) from Ref. [168] and (B) from Ref. [169].

50−100 Oe only by reliability issues (sensitivity to stray fields from external sources, distribution of anisotropy field in nominally isotropic cells, distribution of blocking temperature or fluctuations in the maximum temperature reached during a writing cycle), and so it is in principle only weakly dependent on the cell size. Besides, the heating process requires a threshold current density, hence it is scalable. It is also possible to store two bits of information on one MTJ by using the large TMR ratio of MgO-based junctions and two writing lines, as shown in Ref. [167].

However, several limitations remain:

- The ultimate speed is high (approximately a few 10 ns) but limited by necessary compromise between fast heating at low current density (obtained by high thermal insulation of cell) and fast cooling (requires good thermal contact between MTJ core and the rest of the circuit).
- Because a nonnegligible magnetic field has to be created by a current line of reducing cross section, the electromigration limit still exists, although it is less sharp and is postponed down to the 30−20 nm nodes.
- The variability and cyclability of exchange bias properties at very small cell dimensions remain a mostly unknown area [170].

3.4.2   The Spin-Transfer Torque MRAM (or STT-RAM, or Spin-RAM)

The development of spin-transfer switching represented a huge step forward for MRAM perspective. A schematic diagram of 1T1MTJ STT-RAM is shown in Figure 3.39C. The word line is connected to the gate of a transistor which is used to select the MTJ which is to be written or read. The writing is done through spin-transfer switching by applying either a positive or a negative voltage pulse between the source line and the bit line. The reading is done by applying a weaker voltage to the bit line to sense the resistance of the MTJ. Thus the reading operation is the same as in field-driven MRAM, but the writing is done differently. The cell architecture no longer requires to create magnetic fields, so the electromigration limit is no more critical, and much higher cell densities should be achievable (with minimum cell size down to $6-10$ F^2).

STT-MRAM has been fully demonstrated in Refs. [169,171], following a previous version of field-assisted STT-RAM [172]. Recently, a 32 Mbit-embedded Spin-RAM demonstrator for SoCs was announced by NEC, with a read/write cycle time of 9 ns (Nikkei Electronics, February 16, 2009). Many studies on the downscaling prospects of STT-RAM [152,173−175] and on failure analysis [176] have been carried out. In short, minimizing the spin-transfer writing threshold of the MTJ is still required by a factor of about 4 or 5 to warrant high density, high speed, and ultimate scalability, together with reducing the resistance area product to better match the CMOS transistor properties, but this maybe come in competition with fast reading as reading voltages near the threshold are likely to induce unwanted program events.

Further improvements in the STT-MRAM can be achieved by combining thermal assistance with spin torque in a TAS-MRAM stack such as Figure 3.39A and B. Indeed, a standard STT-RAM cell already undergoes nonnegligible temperature rise during a write pulse [177].

3.4.3   Toward Magnetic Logic

Logic circuits are now everywhere, from computers to cell phones or cars. An MRAM memory bank can indeed be embedded in a programmable CMOS logic circuit, replacing other nonvolatile memory technologies (such as Flash, ferroelectric RAM, and phase change RAM) to store the logic functions. We will rather consider here more innovative ways to add magnetism to logic. Performing binary logic operation requires a nonlinear behavior at a threshold, represented in CMOS electronic by the switch of the MOS transistor from conducting to insulator, controlled by a gate voltage. Switching of the magnetization of a magnetic nanostructure is indeed providing such a nonlinear behavior, and through an MTJ this switch can be transformed into an electric signal.

However, today's practical tunnel magnetoresistance amplitude is still far from reaching the several orders of magnitude observed in transistors, and the standard MTJ is only a two terminal device. Moreover, CMOS logic is based on a "static" operation mode: assuming no current leakage, energy consumption occurs only during the switch of the transistors at the rising or falling edge of the clock, while the signal is kept available for the other logic gates at zero energy cost as long as the circuit is powered (similar to an SRAM ("Static" RAM) memory for instance). On the contrary, with a standard MTJ a nonzero current is necessary to get a voltage, hence power dissipation through Joule effect.

Last but not least, while in magnetic storage applications a nonnegligible error rate can be compensated by limited redundancy and error correction codes, logic chips require nearly undetectable error rates in their operation.

Here also, the recent breakthroughs in spin electronics, and especially spin-transfer torque effects, should help to bring "magnetic logic" to life. Different propositions can be distributed into three broad classes.

The most mature one is "mixed CMOS/MTJ logic." It consists in finely distributing MTJs above a CMOS logic circuit: the logic functions are still provided by CMOS transistors, but the MTJs provide enhanced functionalities such as instant on/off (allows to turn down power to bring energy cost to zero in standby mode, while recovering full functionalities), or enhanced radiation hardness. In principle, the number of transistors and the wafer area should be only moderately increased (MTJs are integrated above CMOS). In pioneer works, writing of the MRAM cells was made using magnetic field pulses [178,179], but TAS-RAM or spin-transfer technology are rapidly becoming dominant. Typical example is the flip-flop, one of the most present logic function in today's logic circuits, proposed in magnetic non-volatile mode by Zhao et al. in 2006 [180,181] and recently realized by NEC [182] (although with a simplified field-induced writing, as shown in Figure 3.40A). More demonstrations of magnetic logic circuits include TMR-based logic blocks [183], a full nonvolatile adder [184], run time reconfigurable circuits [185] for eventually a 3D-stacked reconfigurable spin processor [186], and nonvolatile FPGA [168,187,188]. Reviews can be found in Refs. [189,190].

A more advanced approach is based on using magnetic interactions between magnetic nanostructures, either isolated, or inserted into magnetoresistive elements, or more generally constituted by nonuniformities of magnetization such a domain wall or a vortex. A first example is the magnetic quantum cellular automata proposed by Cowburn et al. [191], in which lines and patterns of interacting quasi-circular dots would propagate information or perform basic logic operations (NOT, AND, etc.) through competing dipolar interactions, the energy being provided by an oscillating global magnetic field. This first proposition suffered from a too high error rate due to distribution in geometry of nominally identical dots. A. Imre et al. [192] later proposed a way of overcoming this issue by applying a saturating field, supposed to reliably take the dots magnetization to their stable minimum of energy when going back to zero field. However, none of these approaches can really lead to fast, CMOS integrable logic functions, first because magnetic fields are required. A real breakthrough in this class of magnetic logic circuits was recently proposed by Leem and Harris [193]. As shown in Figure 3.40B, interacting magnetic particles are free layers of closely spaced MTJ nanopillars. Three adjacent pillars (two inputs and one central output) can for instance become a NAND or a NOR gate, depending on a choice of current threshold for the spin-transfer switching of the central pillar. At each clock steps, current pulses are sent to write the input dots

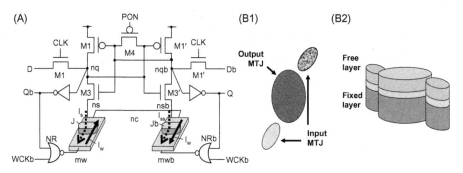

Figure 3.40 (A) Schematic diagram of nonvolatile magnetic flip-flop. A bit of the data and its complement held in the latch circuit can be backed up into the pair of MTJs. Similarly, the data stored in the MTJ pair can also be recalled into the latch. (B) Schematic diagram of spin-torque-based logic device. The energy barrier for magnetization reversal of the middle (output) MTJ can be controlled by the orientation of the free layers of the two side MTJs (inputs).
Source: (A) from Ref. [182]) and (B) from Ref. [201].

and then update the configuration of the output dot. The authors also propose a way of chaining several such magnetic logic gates, applying the idea to the simulation of a ring oscillator [193]. Other recently proposed methods are based on spin waves to transfer and process informations either through inductive coupling between unpatterned ferromagnetic layers [194,195], or in patterned magnetic tracks with control of spin wave phase and spin wave interferences [196,197].

Finally, a new class of logic gates might have been opened by the recently achieved reversible nonlocal switching of the magnetization by Yang et al. [198], toward the realization of logic operation through combination of spin accumulation of spin currents [199]. A logic gate based on spin accumulation effects in ferromagnetic semiconductors was also proposed in Ref. [200], together with means of chining such gates.

Acknowledgments

Y. Suzuki, A. Tulapurkar, and Y. Shiota would like to acknowledge all members in spintronics group in Osaka University, AIST Japan, and Canon ANELVA for experiments and discussions. A. Tulapurkar would also like to acknowledge magnetism group members of SLAC, USA for discussions.

The authors express their gratitude to Ms. Rie Hasegawa for her continuous help to finish the chapter.

References

[1] Johnson M, Silsbee RH. Phys Rev B 1988;37:5326–35.
[2] Johnson M. Phys Rev Lett 1993;70:2142–5.

[3] Jedema FJ, Heersche HB, Filip AT, Baselmans JJA, van Wees BJ. Nature 2001;410:345 2002; 416: 713.

[4] Ohno Y, Young DK, Beschoten B, Matsukura F, Ohno H, Awschalom DD. Nature 1999;402:790.

[5] Fiederling R, Keim M, Reuscher G, Ossau W, Schmidt G, Waag A, et al. Nature 1999;402:787.

[6] Jonker BT, Park YD, Bennett BR, Cheong HD, Kioseoglou G, Petrou A. Phys Rev B 2000;62:8180 Appl Phys Lett 2000; 77: 3989.

[7] Jonker BT, Kioseoglou G, Hanbicki AT, Li CH, Thompson PE. Nat Phys 2007;3:542−6.

[8] Lou X, Adelmann C, Crooker SA, Garlid ES, Zhang J, Reddy KSM, et al. Nat Phys 2007;3:197−202.

[9] Sasaki T, Oikawa T, Suzuki T, Shiraishi M, Suzuki Y, Tagami K. Appl Phys Express 2009;2:053003.

[10] Dash SP, Sharma S, Patel RS, de Jong MP, Jansen R. Nature 2009;462:491.

[11] Suzuki T, Sasaki T, Oikawa T, Shiraishi M, Suzuki Y, Noguchi K. Appl Phys Express 2011;4:023003.

[12] Appelbaum I, Huang B, Monsma DJ. Nature 2007;447:295−8.

[13] Tombros N, van der Molen SJ, van Wees BJ. Phys Rev B 2006;73:233403.

[14] Tombros N, Jozsa C, Popinciuc M, Jonkman HT, van Wees BJ. Nature 2007;448:571−4.

[15] Ohishi M, Shiraishi M, Nouchi R, Nozaki T, Shinjo T, Suzuki Y. Jpn J Appl Phys 2007;46:L605−7.

[16] Shiraishi M, Ohishi M, Nouchi R, Mitoma N, Nozaki T, Shinjo T, et al. Adv Funct Mater 2009;19:3711−6.

[17] Slonczewski JC. J Magn Magn Mater 1999;195:L261−7.

[18] Tsoi M, Jansen AGM, Bass J, Chiang WC, Seck M, Tsoi V, et al. Phys Rev Lett 1998;80:4281 Erratum: 1998; 81: 492

[19] Kiselev SI, Sankey JC, Krivorotov IN, Emley NC, Schoelkopf RJ, Buhrman RA, et al. Nature 2003;425:380−3.

[20] Slonczewski JC. J Magn Magn Mater 1996;159:L1−7.

[21] Berger L. Phys Rev B 1996;54:9353−8.

[22] Myers EB, Ralph DC, Katine JA, Louie RN, Buhrman RA. Science 1999;285:867−70.

[23] Sun JZ. J Magn Magn Mater 1999;202:157−62.

[24] Katine JA, Albert FJ, Buhrman RA, Myers EB, Ralph DC. Phys Rev Lett 2000;84:3149−52.

[25] Huai Y, Albert F, Nguyen P, Pakala M, Valet T. Appl Phys Lett 2004;84:3118−20.

[26] Kubota H, Fukushima A, Ootani Y, Yuasa S, Ando K, Maehara H, et al. Jpn J Appl Phys 2005;44:L1237−40.

[27] Diao Z, Apalkov D, Pakala M, Ding Y, Panchula A, Huai Y. Appl Phys Lett 2005;87:232502.

[28] Mangin S, Ravelosona D, Katine JA, Carey MJ, Terris BD, Fullerton EE. Nat Mater 2006;5:210.

[29] Valet T, Fert A. Phys Rev B 1993;48:7099.

[30] Zutić I, Fabian J, Das Sarma S. Rev Mod Phys 2004;76:1−88.

[31] Schmidt G, Ferrand D, Molenkamp LW, Filip AT, van Wees B. Phys Rev B 2000;62: R4790−3.

[32] Rashba EI. Phys Rev B 2000;62:R16267.

[33] Motsnyi VF, De Boeck J, Das J, Van Roy W, Borghs G, Goovaerts E, et al. Appl Phys Lett 2002;81:265.
[34] Kimura T, Otani Y, Hamrle J. Phys Rev Lett 2006;96:037201.
[35] Mizukami S, Ando Y, Miyazaki T. Phys Rev B 2002;66:104413.
[36] Saitoh E, Ueda M, Miyajima H, Tatara G. Appl Phys Lett 2006;88:182509.
[37] Valenzuela SO, Tinkham M. Nature 2006;442:176.
[38] Kimura T, Otani Y, Sato T, Takahashi S, Maekawa S. Phys Rev Lett 2007;98:156601 Erratum: Phys Rev Lett 2007; 98: 249901.
[39] Seki T, Hasegawa Y, Mitani S, Takahashi S, Imamura H, Maekawa S, et al. Nat Mater 2008;7:125−9.
[40] Kato K, Myers RC, Gossard AC, Awschalom DD. Science 2004;306:1910−3.
[41] Wunderlich J, Kaestner B, Sinova J, Jungwirth T. Phys Rev Lett 2005;94 (047204):1−4.
[42] Liu L, Pai C−F, Li Y, Tseng H−W, Ralph DC, Buhrman RA. Science 2012;336:555.
[43] Murakami S, Nagaosa N, Zhang SC. Science 2003;301:1348.
[44] Slonczewski JC. J Magn Magn Mat 2002;247:324−38.
[45] Slonczewski JC. Phys Rev B 1989;39:6995−7002.
[46] Slonczewski JC. Phys Rev B 2005;71(024411):1−10.
[47] Theodonis I, Kioussis N, Kalitsov A, Chshiev M, Butler WH. Phys Rev Lett 2006;97:237205.
[48] Kubota H, Fukushima A, Yakushiji K, Nagahama T, Yuasa S, Ando K, et al. Nat Phys 2008;4:37−41.
[49] Sankey JC, Cui YT, Sun JZ, Slonczewski JC, Buhrman RA, Ralph DC. Nat Phys 2008;4:67−71.
[50] Wang C, Cui Y−T, Katine JA, Buhrman RA, Ralph DC. Nat Phys 2011;7:496.
[51] Edwards DM, Federici F, Mathon J, Umerski A. Phys Rev B 2005;71:054407.
[52] Zhang S, Li Z. Phys Rev Lett 2004;93:127204.
[53] Thiaville A, Nakatani Y, Miltat J, Suzuki Y. Europhys Lett 2005;69:990.
[54] Barnes SE, Maekawa S. Phys Rev Lett 2005;95:107204.
[55] Tatara G, Kohno H. Phys Rev Lett 2004;92:086601.
[56] Manchon A, Zhang S. Phys Rev B 2009;79:094422.
[57] Miron IM, Gaudin G, Auffret S, Rodmacq B, Schuhl A, Pizzini S, et al. Nat Mater 2010;9:230.
[58] Ohno H, Chiba D, Matsukura F, Omiya T, Abe E, Dietl T, et al. Nature 2000;408:944.
[59] Chiba D, Yamanouchi M, Matsukura F, Ohno H. Science 2003;301:943.
[60] Stolichnov I, Riester SWE, Trodahl HJ, Setter N, Rushforth AW, Edmonds KW, et al. Nat Mater 2008;7:464.
[61] Eerenstein W, Mathur ND, Scott JF. Nature 2006;442:759.
[62] Borisov P, Hochstrat A, Chen X, Kleemann W, Binek C. Phys Rev Lett 2005;94:117203.
[63] Chu Y−H, Martin LW, Holcomb MB, Gajek M, Han S−J, He Q, et al. Nat Mater 2008;7:478.
[64] Novoad V, Ohtani Y, Ohsawa A, Kim SG, Fukamichi K, Koike J, et al. J Appl Phys 2000;87:6400.
[65] Lee J−W, Shin S−C, Kim S−K. Appl Phys Lett 2003;82:2458.
[66] Nie X, Bluegel S. European patent; 2000, 19841034.4.
[67] Duan C−G, Velev JP, Sabirianov RF, Zhu Z, Chu J, Jaswal SS, et al. Phys Rev Lett 2008;101:137201.

[68] Nakamura K, Shimabukuro R, Fujiwara Y, Akiyama T, Ito T, Freeman AJ. Phys Rev
      Lett 2009;102:187201.
[69] Tsujikawa M, Oda T. Phys Rev Lett 2009;102:247203.
[70] Niranjan MK, Duan C-G, Jaswal SS, Tsymbal EY. Appl Phys Lett 2010;96:222503.
[71] Nakamura K, Akiyama T, Ito T, Weinert M, Freeman AJ. Phys Rev B
      2010;81:220409R.
[72] Tsujikawa M, Haraguchi S, Oda T. J Appl Phys 2012;111:083910.
[73] Weisheit M, Fahler S, Marty A, Souche Y, Poinsignon C, Givord D. Science
      2007;315:349.
[74] Maruyama T. Nat Nanotechnol 2009;4:158.
[75] Shiota Y, Maruyama T, Nozaki T, Shinjo T, Shiraishi M, Suzuki Y. Appl Phys
      Express 2009;2:063001.
[76] Nozaki T, Shiota Y, Shiraishi M, Shinjo T, Suzuki Y. Appl Phys Lett 2010;96:022506.
[77] Nozaki T, Shiota Y, Miwa S, Murakami S, Bonell F, Ishibashi S, et al. Nat Phys
      2012;8:491 and its supplementary information.
[78] Zhu J, Katine JA, Rowlands GE, Chen Y-J, Duan Z, Alzate JG, et al. Phys Rev Lett
      2012;108:197203.
[79] Sun JZ. Phys Rev B 2000;62:570.
[80] Stiles MD, Miltat J. Spin dynamics in confined magnetic structures III.
      In: Hillebrands B, Thiaville A, editors. Topics in applied physics, vol. 101. Berlin:
      Springer; 2006. p. 225−308.
[81] Miltat J, Albuquerque G, Thiaville A, Vouille C. J Appl Phys 2001;89:6982−4.
[82] Lee KJ, Deac A, Redon O, Nozières JP, Dieny B. Nat Mater 2004;3:877.
[83] Mizukami S, Ando Y, Miyazaki T. Jpn J Appl Phys 2001;40:580.
[84] Devolder T, Crozat P, Chappert C, Miltat J, Tulapurkar A, Suzuki Y, et al. Phys Rev B
      2005;71:184401.
[85] Stone M. Nucl Phys B 1989;314:557.
[86] Albert FJ, Emley NC, Myers EB, Ralph DC, Buhrman RA. Phys Rev Lett
      2002;89:226802.
[87] Yagami K, Tulapurkar A, Fukushima A, Suzuki Y. Appl Phys Lett 2004;85:5634−6.
[88] Berger L. J Appl Phys 2003;93:7693.
[89] Huai Y, Pakala M, Diao Z, Ding Y. Appl Phys Lett 2005;87:222510.
[90] Nagase T, Nishiyama K, Nakayama M, Shimomura N, Amano M, Kishi T, et al. Abstract:
      C1.00331, American Physics Society. March meeting, 2008, New Orleans, LA.
[91] Sun JZ, Myers EB, Albert FJ, Sankey JC, Bonet E, Buhrman RA, et al. Phys Rev Lett
      2002;89:196801.
[92] Tulapurkar AA, Devolder T, Yagami K, Crozat P, Chappert C, Fukushima A, et al.
      Appl Phys Lett 2004;85:53−8.
[93] Devolder T, Chappert C, Katine JA, Carey MJ, Ito K. Phys Rev B 2007;75:064402.
[94] Krivorotov IN, Emley NC, Buhrman RA, Ralph DC. Phys Rev B 2008;77:054440.
[95] Tomita H, Miwa S, Nozaki T, Yamashita S, Nagase T, Nishiyama K, et al. Appl Phys
      Lett 2013;102:042409.
[96] Koch RH, Katine JA, Sun JZ. Phys Rev Lett 2004;92:088302.
[97] Li Z, Zhang S. Phys Rev B 2004;69:134416.
[98] Coffey W, Kalmykov Yu.P, Waldron JT. With applications to stochastic problems in
      physics, chemistry and electrical engineering. 2nd ed. World Scientific series in con-
      temporary chemical physics. vol. 14. World Scientific, Singapore, 1995.
[99] Brown Jr. WF. Phys Rev 1963;130:1677.

[100] Apalkov DM, Visscher PB. Phys Rev B 2005;72(180405(R)):274−5.
[101] Apalkov DM, Visscher PB. J Magn Magn Mater 2005;286:370.
[102] Taniguchi T, Imamura H. Phys Rev B 2012;85:184403.
[103] Taniguchi T, Shibata M, Marthaler M, Utsumi Y, Imamura H. Appl Phys Exp Lett 2012;5:063009.
[104] Taniguchi T, Utsumi Y, Marthaler M, Golubev DS, Imamura H. Phys Rev B 2013;87:054406.
[105] Krivorotov IN, Emley NC, Sankey JC, Kiselev SI, Ralph DC, Buhrman RA. Science 2005;307:228.
[106] Devolder T, Hayakawa J, Ito K, Takahashi H, Ikeda S, Crozat P, et al. Phys Rev Lett 2008;100:057206.
[107] Tomita H, Konishi K, Nozaki T, Kubota H, Fukushima A, Yakushiji K, et al. Appl Phys Express 2008;1:061303.
[108] Aoki T, Ando Y, Watanabe D, Oogane M, Miyazaki T. J Appl Phys 2008;103:103911.
[109] Acremann Y, Strachan JP, Chembrolu V, Andrews SD, Tyliszczak T, Katine JA, et al. Phys Rev Lett 2006;96:217202.
[110] Strachan JP, Chembrolu V, Acremann Y, Yu XW, Tulapurkar AA, Tyliszczak T, et al. Phys Rev Lett 2008;100:247201.
[111] Tsoi M, Jansen AGM, Bass J, Chiang WC, Tsoi V, Wyder P. Nature 2000;406:46.
[112] Matsumoto R, Fukushima A, Yakushiji K, Yakata S, Nagahama T, Kubota H, et al. Phys Rev B 2009;80:174405.
[113] Kim JV, Mistral Q, Chappert C, Tiberkevich VS, Slavin AN. Phys Rev Lett 2008;100:167201.
[114] Deac AM, Fukushima A, Kubota H, Maehara H, Suzuki Y, Yuasa S, et al. Nat Phys 2008;4:803.
[115] Nazarov AV, Olson HM, Cho H, Nikolaev K, Gao Z, Stokes S, et al. Appl Phys Lett 2006;88:162504.
[116] Houssameddine D, Florez SH, Katine JA, Michel JP, Ebels U, Mauri D, et al. Appl Phys Lett 2008;93:022505.
[117] Devolder T, Bianchini L, Kim J−V, Crozat P, Chappert C, Cornelissen S, et al. J Appl Phys 2009;106:103921.
[118] Cornelissen S, Bianchini L, Hrkac G, Op de Beeck M, Lagae L, Kim J−V, et al. Euro Phys Lett 2009;87:57001.
[119] Wada T, Yamane T, Seki T, Nozaki T, Suzuki Y, Kubota H, et al. Phys Rev B 2010;81:104410.
[120] Maehara H, Kubota H, Seki T, Nishimura K, Tomita H, Nagamine Y, et al. Fifty-fifth magnetism and magnetic materials conference, Atlanta, GA; 2010.
[121] Deac A, Liu Y, Redon O, Petit S, Li M, Wang P, et al. J Phys Condens Matt 2007;19:165208.
[122] Houssameddine D, Ebels U, Delaët B, Rodmacq B, Firastrau I, Ponthenier F, et al. Nat Mater 2007;6:447.
[123] Rippard WH, Pufall MR, Kaka S, Silva TJ, Russek SE. Phys Rev B 2004;70:100406R.
[124] Sankey JC, Krivorotov IN, Kiselev SI, Braganca PM, Emley NC, Buhrman RA, et al. Phys Rev B 2005;72:224427.
[125] Mistral Q, Kim JV, Devolder T, Crozat P, Chappert C, Katine JA, et al. Appl Phys Lett 2006;88:192507.

[126] Petit S, Baraduc C, Thirion C, Ebels U, Liu Y, Li M, et al. Phys Rev Lett 2007;98:077203.
[127] Thadani KV, Finocchio G, Li ZP, Ozatay O, Sankey JC, Krivorotov IN, et al. Phys Rev B 2008;78:024409.
[128] Kim JV. Phys Rev B 2006;73:174412.
[129] Kim JV, Tiberkevich V, Slavin AN. Phys Rev Lett 2008;100:017207.
[130] Tiberkevich VS, Slavin AN, Kim JV. Phys Rev B 2008;78:092401.
[131] Nagamine Y, Maehara H, Tsunekawa K, Djayaprawira DD, Watanabe N, Yuasa S, et al. Appl Phys Lett 2006;89:162507.
[132] Kaka S, Pufall MR, Rippard WH, Silva TJ, Russek SE, Katine JA. Nature 2005;437:389.
[133] Mancoff FB, Rizzo ND, Engel BN, Tehrani S. Nature 2005;437:393.
[134] Georges B, Grollier J, Cros V, Fert A. Appl Phys Lett 2008;92:232504.
[135] Yuasa S, Fukushima A, Nagahama T, Ando K, Suzuki Y. Jpn J Appl Phys 2004;43: L588.
[136] Yuasa S, Nagahama T, Fukushima A, Suzuki Y, Ando K. Nat Mater 2004;3:868.
[137] Parkin SSP, Kaiser C, Panchula A, Rice PM, Hughes B, Samant M, et al. Nat Mater 2004;3:862.
[138] Hayakawa J, Ikeda S, Lee YM, Matsukura F, Ohno H. Appl Phys Lett 2006;89:232510.
[139] Yuasa S, Fukushima A, Kubota H, Suzuki Y, Ando K. Appl Phys Lett 2006;89:042505.
[140] Tulapurkar AA, Suzuki Y, Fukushima A, Kubota H, Maehara H, Tsunekawa K, et al. Nature 2005;438:339.
[141] Sankey JC, Braganca PM, Garcia AGF, Krivorotov IN, Buhrman RA, Ralph DC. Phys Rev Lett 2006;96:227601.
[142] Suzuki Y, Kubota H. J Phys Soc Jpn 2008;77:031002.
[143] Wang C, Cui Y-T, Sun JZ, Katine JA, Buhrman RA, Ralph DC. Phys Rev B 2009;79:224416.
[144] Ishibashi S, Ando K, Seki T, Nozaki T, Kubota H, Yakata S, et al. IEEE Trans Magn 2011;47:3373−6.
[145] Cheng X, Boone CT, Zhu J, Krivorotov IN. Phys Rev Lett 2010;105:047202.
[146] Chiba D, Nakatani Y, Matsukura F, Ohno H. Appl Phys Lett 2010;96:192506.
[147] Shiota Y, Nozaki T, Bonell F, Murakami S, Shinjo T, Suzuki Y. Nat Mater 2012;11:39.
[148] Kim NS, Austin T, Blaauw D, Mudge T, Flautner K, Hu JS, et al. Computer 2003;36:68.
[149] Pohm AV, Beech RS, Bade PA, Chen EY, Daughton JM. IEEE Trans Magn 1994;30:4650.
[150] Tehrani S, Slaughter JM, Chen E, Durlam M, Shi J, DeHerren M. IEEE Trans Magn 1999;35:2814.
[151] Scheuerlein R, Gallagher W, Parkin S, Lee A, Ray S, Robertazzi R, et al. IEEE international solid-state circuits conference; 2000, San Francisco. Digest of technical papers. p. 128−9.
[152] Maffitt TM, DeBrosse JK, Gabric JA, Gow ET, Lamorey MC, Parenteau JS, et al. IBM J Res Dev 2006;50:25.
[153] Gallagher WJ, Parkin SSP. IBM J Res Dev 2006;50:5.
[154] DeBrosse J, Gogl D, Bette A, Hoenigschmid H, Robertazzi R, Arndt C, et al. IEEE J Solid-State Circuits 2004;39:678.

[155] Kaka S, Russek SE. Appl Phys Lett 2002;80:2958.
[156] Gerrits T, van den Berg HAM, Hohlfeld J, Bar L, Rasing T. Nature 2002;418:509.
[157] Schumacher HW, Chappert C, Sousa RC, Freitas PP, Miltat J. Phys Rev Lett 2003;90:017204.
[158] Durlam M, Naji P, Omair A, DeHerrera M, Calder J, Slaughter JM, et al. Symposium on VLSI circuits digest of technical papers; Honolulu, 2002. p. 158−61.
[159] Durlam M, Addie D, Akerman J, Butcher B, Brown P, Chan J, et al. IEEE international electron devices meeting; Washington, DC, 2003. Technical digest. p. 34.6.1−3.
[160] Engel BN, Akerman J, Butcher B, Dave RW, DeHerrera M, Durlam M, et al. IEEE Trans Magn 2005;41:132.
[161] Yamamoto T, Kano H, Higo Y, Ohba K, Mizuguchi T, Hosomi M, et al. J Appl Phys 2005;97:10P503.
[162] Aakerman J, Brown P, Gajewski D, Griswold M, Janesky J, Martin M, et al. Forty-third annual international reliability physics symposium; San Jose, 2005.p. 163.
[163] Prejbeanu IL, Kula W, Ounadjela K, Sousa RC, Redon O, Dieny B, et al. IEEE Trans Magn 2004;40:2625.
[164] Kerekes M, Sousa RC, Prejbeanu IL, Redon O, Ebels U, Baraduc C, et al. Forty-ninth annual conference on magnetism and magnetic materials, Jacksonville, FL; 2005. p. 10P501−3.
[165] Sousa RC, Kerekes M, Prejbeanu IL, Redon O, Dieny B, Nozieres JP, et al. J Appl Phys 2006;99:08N904.
[166] Deak JG, Daughton JM, Pohm AV. IEEE Trans Magn 2006;42:2721.
[167] Leuschner R, Klostermann UK, Park H, Dahmani F, Dittrich R, Grigis C, et al. International electron devices meeting; San Francisco, 2006. p. 1−4.
[168] Zhao W, Belhaire E, Dieny B, Prenat G, Chappert C. International conference on field-programmable technology; Kitakyushu, 2007. p. 153−60.
[169] Hosomi M, Yamagishi H, Yamamoto T, Bessho K, Higo Y, Yamane K, et al. IEEE international electron devices meeting; Washington, DC, 2005. Technical digest. p. 459−62.
[170] Baltz V, Sort J, Rodmacq B, Dieny B, Landis S. Appl Phys Lett 2004;84:4923.
[171] Kawahara T, Takemura R, Miura K, Hayakawa J, Ikeda S, Lee Y, et al. International solid-state circuits conference; San Francisco, 2007. Technical digest. p. 480.
[172] Jeong WC, Park JH, Oh JH, Jeong GT, Jeong HS, Kinam K. Symposium on VLSI circuits; Kyoto,2005. p. 184.
[173] Zhu JG. Proc IEEE 2008;96:1786.
[174] Wang X, Chen Y, Li H, Dimitrov D, Liu H. IEEE Trans Magn 2008;44:2479.
[175] Zhu X, Zhu JG. IEEE Trans Magn 2006;42:2739.
[176] Li J, Augustine C, Salahuddin S, Roy K. Forty-fifth ACM/IEEE design automation conference; Anaheim, 2008. p. 278−83.
[177] Lee DH, Lim SH. Appl Phys Lett 2008;92:233502.
[178] Black Jr. WC, Das B. J Appl Phys 2000;87:6674.
[179] Richter R, Bar L, Wecker J, Reiss G. Appl Phys Lett 2002;80:1291.
[180] Zhao W, Belhaire E, Chappert C. Seventh IEEE conference on nanotechnology; Hong Kong, 2007. p. 399−402.
[181] Zhao W, Belhaire E, Chappert C, Javerliac V, Mazoyer P. Ninth international conference on solid-state and integrated-circuit technology; Beijing, 2008. p. 2136−9.
[182] Sakimura NS, Sugibayashi T, Nebashi R, Kasai N. IEEE custom integrated circuits conference; San Jose, 2008. p. 355−8.
[183] Mochizuki A, Kimura H, Ibuki M, Hanyu T. IEICE Trans Fundam Electron Commun Comput Sci 2005;E88-A:1408−15.

[184] Matsunaga S, Hayakawa J, Ikeda S, Miura K, Hasegawa H, Endoh T, et al. Appl Phys Express 2008;1:091301.
[185] Zhao W, Belhaire E, Chappert C, Mazoyer P. IEEE Trans Magn 2009;45:776.
[186] Sekikawa M, Kiyoyama K, Hasegawa H, Miura K, Fukushima T, Ikeda S, et al. IEEE international electron devices meeting; San Francisco, 2008. p. 1−3.
[187] Paul S, Mukhopadhyay S, Bhunia S. IEEE/ACM international conference on computer-aided design; San Jose, 2008. p. 589−92.
[188] Guillemenet Y, Torres L, Sassatelli G, Bruchon N. Int J Reconfigurable Comput 2008;2008:ID723950.
[189] Wang JP, Yao X. J Nanoelectron Optoelectron 2008;3:12.
[190] Ikeda S, Hayakawa J, Lee YM, Matsukura F, Ohno Y, Hanyu T, et al. IEEE Trans Electron Devices 2007;54:991.
[191] Cowburn RP, Welland ME. Science 2000;287:1466.
[192] Imre A, Csaba G, Ji L, Orlov A, Bernstein GH, Porod W. Science 2006;311:205.
[193] Leem L, Harris JS. IEEE international electron devices meeting; San Francisco, 2008. Technical digest. p. 1.
[194] Khitun A, Bao M, Lee JY, Wang KL, Lee DW, Wang S, et al. J Nanoelectron Optoelectron 2008;3:24.
[195] Khitun A, Mingqiang B, Wang KL. IEEE Trans Magn 2008;44:2141.
[196] Kostylev MP, Serga AA, Schneider T, Leven B, Hillebrands B. Appl Phys Lett 2005;87:153501.
[197] Lee KS, Kim SK. J Appl Phys 2008;104:053909.
[198] Yang T, Kimura T, Otani Y. Nat Phys 2008;4:851.
[199] Brataas A, Bauer GEW, Kelly PJ. Phys Rep 2006;427:157.
[200] Dery H, Dalal P, Cywinski L, Sham LJ. Nature 2007;447:573.
[201] Leem L, Harris JS. J Appl Phys 2009;105:07D102.

4 Dynamics of Magnetic Domain Walls in Nanomagnetic Systems

Teruo Ono and Teruya Shinjo

Institute for Chemical Research, Kyoto University, Uji, Japan

Contents

4.1 Introduction

Weiss pointed out in his famous paper on spontaneous magnetization in 1907 that ferromagnetic materials are not necessarily magnetized to saturation in the absence of an external magnetic field [1]. Instead, they have magnetic domains, within each of which magnetic moments align. The formation of the magnetic domains is energetically favorable because this structure can lower the magnetostatic energy originating from the dipole−dipole interaction. The directions of magnetization of neighboring domains are not parallel. As a result, between two neighboring domains, there is a region in which the direction of magnetic moments gradually changes. This transition region is called a magnetic domain wall (DW).

Nanomagnetism and Spintronics. DOI: http://dx.doi.org/10.1016/B978-0-444-63279-1.00004-6

Recent developments in nanolithography techniques make it possible to prepare nanoscale magnets with simple magnetic domain structure which is suitable for basic studies on the magnetization reversal. For example, in a magnetic wire with submicron width, two important processes in the magnetization reversal, nucleation and propagation of a magnetic DW, can be clearly seen. As shown in Figure 4.1A, in a very narrow ferromagnetic wire, the magnetization is restricted to be directed parallel to the wire axis due to the magnetic shape anisotropy. When an external magnetic field is applied against the magnetization, a magnetic DW nucleates at the end of the wire and the magnetization reversal proceeds by the propagation of this DW through the wire (Figure 4.1B and C).

4.2   Field-Driven DW Motions

4.2.1   Detection of DW Propagation by Using GMR Effect

As described above, a magnetic nanowire is a good candidate for the investigation of nucleation and propagation of a magnetic DW. However, it is very difficult to detect the propagation of the DW, because the change in magnetic moments in this process is very small due to the small volume of the magnetic wire. Here, the principle how to detect the DW propagation in magnetic wires by using the giant magnetoresistance (GMR) effect is described. The GMR effect is the change in electrical resistance caused by the change of the magnetic structure in magnetic multilayers [2]. This means that the magnetic structure of the system can be detected by resistance measurements. Consider the GMR system shown in Figure 4.2, which is composed of magnetic, nonmagnetic, and magnetic layers. The resistance is the largest for the anti-parallel magnetization configuration (Figure 4.2A), and it is the smallest for the parallel configuration (Figure 4.2D). During the magnetization reversal of one of the two magnetizations (Figure 4.2B and C), the total resistance of the system is given by the

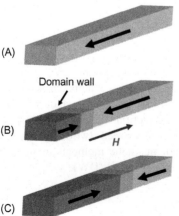

Figure 4.1 Schematic illustration of the magnetization reversal process in a magnetic wire.

sum of the resistances of the parallel magnetization part and the antiparallel magnetization part. Thus, the resistance of this system, R, is given by:

$$R = \frac{x}{L}R_{\uparrow\uparrow} + \frac{L-x}{L}R_{\uparrow\downarrow} \qquad (4.1)$$

where x is the position of the DW, L is the length of the wire, $R_{\uparrow\uparrow}$ is the resistance for parallel configuration, and $R_{\uparrow\downarrow}$ is the resistance for antiparallel configuration. This equation means that we can determine the position of DW by simple resistance measurements.

The above idea has been demonstrated experimentally [3]. The sample is a 500 nm wide wire composed of $Ni_{81}Fe_{19}(20 \text{ nm})/Cu(10 \text{ nm})/Ni_{81}Fe_{19}(5 \text{ nm})$ trilayer structure. Due to the large Cu-layer thickness, the interlayer exchange coupling between the thin and thick NiFe layers is negligible. The magnetoresistance measurements are performed at 300 K under an external magnetic field along the wire axis. As seen in Figure 4.3, the sample has four current–voltage terminals where the voltage is probed over a distance of 20 μm. Furthermore, the sample has an artificial neck (0.35 μm width) introduced at one-third distance from one voltage probe.

Figure 4.4 shows the resistance of the trilayer system as a function of an external magnetic field. Prior to the measurement, a magnetic field of 100 Oe was applied in order to achieve magnetization alignment in one direction. Then, the resistance was measured as the field was swept toward the counter direction. The result of the magnetoresistance measurement displays essentially four very sharp leaps. First and second leaps correspond to the magnetization reversal of the thin NiFe layer, whereas third and fourth leaps correspond to the magnetization reversal of the thick NiFe layer.

How the magnetization reversal takes place in the sample is the following: As long as the counter field is smaller than a critical field, the magnetizations of both thin and thick NiFe layers align parallel, and the resistance shows the lowest value.

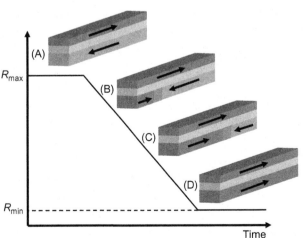

Figure 4.2 Schematic explanation for the detection of DW propagation in magnetic wires by using the magnetoresistance effect.

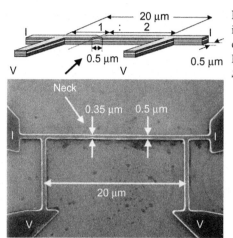

Figure 4.3 SEM image and schematic illustration of the sample. The sample consists of $Ni_{81}Fe_{19}(20\ nm)/Cu(10\ nm)/Ni_{81}Fe_{19}(5\ nm)$.
Source: From Ref. [3].

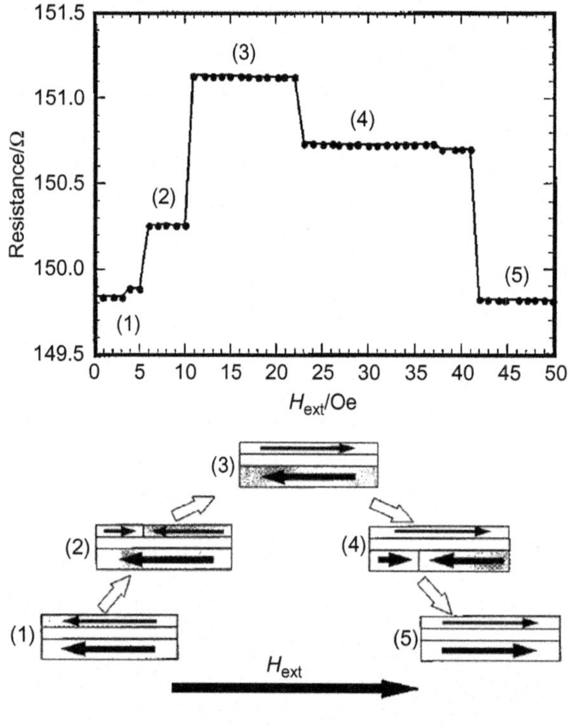

Figure 4.4 Resistance as a function of the external magnetic field at 300 K. The magnetic domain structures inferred from the resistance measurement and the direction of the external field are schematically shown.
Source: From Ref. [3].

As the applied magnetic field exceeds 5 Oe, the resistance abruptly jumps and maintains constant value up to 10 Oe. Then, exceeding 10 Oe, the resistance abruptly jumps again and maintains the largest value up to 22 Oe. The result indicates that the antiparallel magnetization alignment is realized at an external field between 11 and 22 Oe, where the resistance shows the largest value. The ratio of the resistance changes at first and second leaps is 1:2. This means that one-third of the total magnetization of the thin NiFe layer changes its direction at first leap in Figure 4.4, since the GMR change is directly proportional to the switching layer magnetization. The ratio, one-third, corresponds to the ratio of length between one voltage probe and the neck to the overall length of the wire between the voltage probes. Therefore, in this case, a magnetic DW nucleates in the shorter part of the wire (left side of the scanning electron microscope (SEM) image in Figure 4.3) and propagates to the neck, where it is pinned up to 10 Oe. The second leap, upon exceeding 10 Oe, corresponds either to depinning of the magnetic DW from the neck, or to nucleation and propagation of another magnetic DW on the other side of the neck (right side of the SEM image in Figure 4.3). These two possibilities cannot be distinguished from the result shown in Figure 4.4. Though the nucleation position of a DW could not be determined in this experiment, one can inject a DW from one end of a wire by breaking its symmetry [4] or by applying a local magnetic field at the end of a wire [5]. Since the ratio of the resistance changes at third and fourth leaps is also 1:2, the magnetization reversal of the thick NiFe layer takes place in the same manner as in the thin NiFe layer described above.

The above experiment shows that a magnetic DW can be trapped by the artificial neck structure introduced into the wire. This is an example of the control of magnetization reversal process by utilizing the designed structure of mesoscopic magnets. The strength of the trap potential of the artificial neck can be estimated by measuring the temperature dependence of the depinning field of a DW from the neck [6]. It was shown that the trap potential increases with a decrease of the neck width and it reaches 4.3×10^{-19} J (30,000 K) for the 250 nm wide neck in the 20 nm thick and 500 nm wide $Ni_{81}Fe_{19}$ wire. Thus, a DW pinned by the neck is stable against the thermal agitation of room temperature, and this kind of trap potential can be used to define a DW position in the spintronic devices.

It should be noted that the GMR method described here corresponds to a very high sensitive magnetization measurement. For the sample investigated above, the sensitivity is as high as 10^{-13} emu (10^7 spins). The method, in principle, can be applied to smaller samples as far as the resistance of the samples can be measured and the relative sensitivity increases with decreasing sample volume. The GMR method for the detection of the DW position is also applicable to the study of the current-driven DW motion [7].

4.2.2 Ratchet Effect in DW Motions

A ratchet is a mechanism that can limit a motion to one direction, and it is realized on a macroscopic scale by use of a pawl and a wheel with asymmetric-shaped teeth, in which the pawl restricts the wheel to rotate in one direction. Feynman discussed the thermodynamic theory about the ratchet system in his famous lecture. Recently, it has been paid much attention as a model such as molecular motor in living bodies, muscle, and the power of micromachines [8–12].

As described in the previous section, an artificial neck in a magnetic wire works as a pinning potential for a DW motion. Because the energy of a DW is proportional to its area, a DW at wider position in a wire has a larger energy. This energy change of a DW along the wire axis produces the pinning potential for the DW, and this is the reason why we can trap a DW in the artificial neck. The force necessary to move the DW against the potential is given by the derivative of the DW energy with respect to the DW position, which is proportional to the slope of the artificial neck. Thus, we expect different depinning fields depending on the propagation direction of a DW, if we make the artificial structure asymmetric [13−15]. This difference in depinning field between rightward and leftward propagations leads to the unidirectional motion of a DW, i.e., a ratchet effect. The experimental results on the ratchet effect in DW motions are presented in the following text [15].

The samples were fabricated onto thermally oxidized Si substrates by means of e-beam lithography and lift-off method. Figure 4.5 shows a schematic illustration of a top view of the whole sample. A magnetic wire has a trilayered structure consisting of $Ni_{81}Fe_{19}(5\ nm)/Cu(20\ nm)/Ni_{81}Fe_{19}(20\ nm)$. The main body of the magnetic wire has four notches with asymmetric shape. The sample has four current−voltage probes made of nonmagnetic material, Cu. Further, it has two narrow Cu wires crossing the ends of the magnetic wire and a wide Cu wire covering the notched part of the magnetic wire. These Cu wires are electrically insulated from the magnetic wire by SiO_2 layers of 50 nm in thickness. A flow of an electric current in

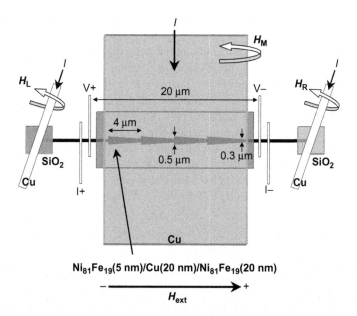

Figure 4.5 Schematic illustration of a top view of the whole sample. The black part is a trilayered magnetic wire consisting of NiFe(5 nm)/Cu(20 nm)/NiFe(20 nm). The main body of the magnetic wire has four asymmetric notches. A flow of electric current in the Cu wire crossing the magnetic wire generates a local magnetic field, which is used to nucleate a magnetic DW.

each Cu wire can generate local magnetic fields, H_L, H_M, or H_R, which act on the left end, the main part with notches, or the right end of the magnetic wire, respectively. H_L (H_R) can trigger the nucleation of a DW at the left (right) end of the magnetic wire. Thus, the propagation direction of a DW can be controlled by H_L (H_R).

Figure 4.6A shows a typical resistance change of the trilayered magnetic wire with asymmetric notches as a function of H_{ext} applied along the wire axis. The resistance increase at 10 Oe corresponds to the magnetization reversal of the NiFe (5 nm) layer, while the decrease in resistance at 160 Oe is due to the magnetization reversal of the NiFe(20 nm) layer. The magnetic DW was not pinned by asymmetric notches during the magnetization reversals of NiFe(20 nm) layer because of its large nucleation field.

In order to nucleate a DW in the NiFe(20 nm) layer at smaller H_{ext} and to pin the DW at the notch, we utilized the method generating a pulsed local magnetic field at the end of the magnetic wire. Figure 4.6B shows the result of the DW injection into the NiFe(20 nm) layer of the magnetic wire by H_L. The measurement procedure is the same as that shown in Figure 4.6A, except applying H_L when $H_{ext} = 60$ Oe. The magnitude and the duration of the pulsed H_L were 200 Oe and 100 ns, respectively. The resistance abruptly decreased after the application of H_L and stayed at a value between the largest and the smallest values. The value of the resistance indicates that a DW injected from the left by H_L was pinned at the first notch. By increasing H_{ext} after the injection of the DW, the resistance abruptly decreased to the smallest value at 98 Oe, which indicates that the DW propagated to the right end of the wire through the asymmetric notches. Thus, the depinning field for the rightward propagation of the DW can be determined to be 98 Oe. On the other hand, the depinning field for the leftward propagation of the DW can be determined from the result shown in Figure 4.6C, which shows the MR measurement when the DW was injected from the right end of the wire by H_R. In this case, the depinning field was 54 Oe. Thus, the depinning field for the rightward propagation is much larger than that for the leftward propagation. If an AC magnetic field between the depinning fields for both directions is applied, a unidirectional DW motion will be induced. This "magnetic ratchet effect" has been recently demonstrated [16].

The magnetic ratchet effect was also demonstrated in the current-driven DW motion; the critical current density necessary for the current-driven DW motion depends on the propagation direction of the DW. The DW moves more easily in the direction along which the slope of the asymmetric notch is smaller [17].

4.2.3 DW Velocity Measurements

Sixtus and Tonks first measured the DW velocity in bulk magnetic wires in 1931 [18]. Figure 4.7 shows a schematic diagram of the circuit for the velocity measurements. Under a homogeneous magnetic field, a magnetic DW is nucleated by adding a local magnetic field, which is produced by an adding coil. The DW traveling along the wire from left to right produces successive voltage surges in two search coils, which are placed around the wire at a known separation. The velocity of the DW can

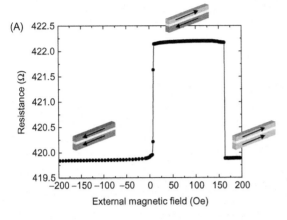

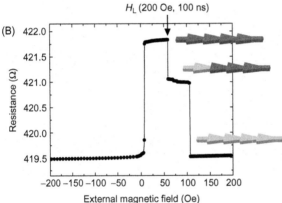

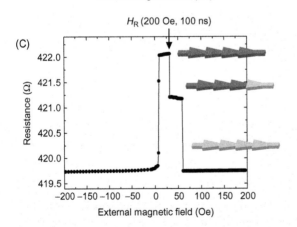

Figure 4.6 Resistance change of the trilayered magnetic wire with asymmetric notches as a function of the external magnetic field. The magnetic domain structures inferred from the resistance measurement are schematically shown. (A) Typical MR curve of the trilayered system. (B) A magnetic DW was injected into the NiFe(20 nm) layer from the left end of the magnetic wire by the pulsed H_L. (C) A DW was injected into the NiFe (20 nm) layer from the right end of the magnetic wire by pulsed H_R.

Source: From Ref. [15].

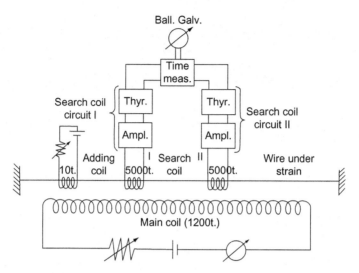

Figure 4.7 Schematic diagram of the circuit for velocity measurements by Sixtus and Tonks.
Source: From Ref. [4].

be calculated from their time interval of the voltage surges and the separation of the coils. The DW velocity, obtained experimentally, was discussed in terms of the dissipation of the magnetic energy by eddy currents.

The GMR detection method described in Section 4.2.1 has an advantage in dynamical measurements because of its simplicity. Here, velocity measurements of a single DW propagating in a magnetic nanowire are presented [19]. Because the GMR detection method provides the information of a DW position, as shown in Eq. (4.1), the DW velocity, $v = dx/dt$, can be determined by the time-domain measurements, and v is given by:

$$v = \frac{dx}{dt} = \frac{L}{R_{\uparrow\downarrow} - R_{\uparrow\uparrow}} \frac{dR}{dt} \tag{4.2}$$

Thus, the GMR method can offer the time variation of the DW velocity. This is an advantage over conventional experimental methods, such as experiments by Sixtus and Tonks and the DW velocity measurements by using the Kerr microscopy with a combination of a pulsed magnetic field, where only the average velocity of a DW can be obtained [18].

The samples for the DW velocity measurements have trilayer structures of $Ni_{81}Fe_{19}$(40 nm)/Cu(20 nm)/$Ni_{81}Fe_{19}$(5 nm). The width of the wire is 0.5 μm and the sample has four current−voltage terminals where the voltage is probed over a distance of 2 mm. The magnetic field was applied along the wire axis. The voltage across two voltage probes was monitored by a differential preamplifier and a digital oscilloscope. The current passing through the electromagnet was also monitored by

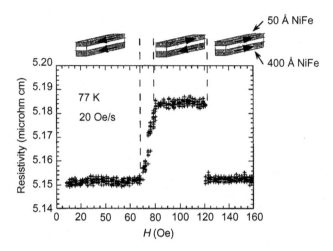

Figure 4.8 Resistance as a function of the external magnetic field at 77 K. The resistance was measured at 10 ms intervals with sweeping the field toward the counter direction at the sweeping rate of 20 Oe/s. The magnetic domain structures inferred from the resistance measurement are schematically shown.
Source: From Ref. [19].

the digital oscilloscope so as to obtain both resistance and applied magnetic field during the magnetization reversal simultaneously.

Figure 4.8 shows the resistance change of the trilayer system at 77 K as a function of an external magnetic field. Prior to the measurement, a magnetic field of 500 Oe was applied in order to align the magnetization in one direction. Then, the resistance was measured at 10 ms intervals with the sweeping field toward the counter direction at the sweeping rate of 20 Oe/s. The result indicates that the antiparallel magnetization alignment is realized in the field range between 80 and 120 Oe, where the resistance shows the largest value. The change in resistance at 80 and 120 Oe is attributed to the magnetization reversals of the 5 nm thick NiFe and 40 nm thick NiFe layers, respectively. Since there is no measured point during the magnetization reversal of the 40 nm thick NiFe in Figure 4.8, it is concluded that the magnetization reversal of the 40 nm thick NiFe is completed within 10 ms. On the other hand, the magnetization reversal of the 5 nm thick NiFe proceeds gradually with increasing an external magnetic field. This indicates that the magnetization reversal of the 5 nm thick NiFe takes place by the successive pinning and depinning of a magnetic DW. Hereafter, we focus on the magnetization reversal of the 40 nm thick NiFe.

Figure 4.9 shows an experimental result on the time variation of the resistance during the magnetization reversal of the 40 nm thick NiFe layer. The data were collected at 40 ns intervals. The linear variation of resistance with time in Figure 4.9 indicates that the propagation velocity of the magnetic DW is constant during the magnetization reversal of the 40 nm thick NiFe layer. The propagation velocity of the magnetic DW at the external magnetic field of 121 Oe is estimated to be

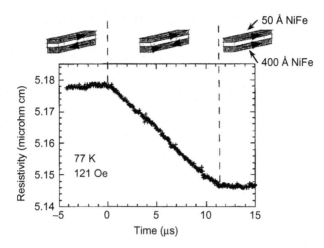

Figure 4.9 Time variation of the resistance during the magnetization reversal of the 40 nm thick NiFe layer at 77 K, which was collected at 40 ns intervals. The applied magnetic field simultaneously monitored by digital oscilloscope was 121 Oe. As the sweeping rate of the applied magnetic field was 20 Oe/s, the variation of the applied magnetic field during magnetization reversal is less than 2×10^{-5} Oe, that is, the applied magnetic field is practically constant during the measurements.
Source: From Ref. [19].

182 m/s, which is calculated from the separation (2 mm) of the two voltage probes and the time (11 μs) the wall traveled across it. Since the sweeping rate of the magnetic field was 20 Oe/s, the variation of the magnetic field during the magnetization reversal is less than 2×10^{-5} Oe, that is, the external magnetic field is regarded as constant during the measurements.

Since the reversal field of the 40 nm thick NiFe varied in every measurement, ranging from 90 to 140 Oe, the wall velocities at various magnetic fields were obtained by repeating the measurements. The result at 100 K is shown in Figure 4.10. The wall velocity depends linearly on the applied magnetic field, and it is described as $v = \mu(H - H_0)$, where v is the wall velocity, H the applied magnetic field, μ so-called wall mobility, and it is obtained that $\mu = 2.6$ (m/s Oe), and $H_0 = 38$ (Oe). Here, we utilized the statistical nature of the magnetization reversal field to obtain the external magnetic field dependence of the DW velocity. A similar but more sophisticated experiment was performed by Himeno et al. [20]. They set two Cu wires crossing the magnetic wire at the ends of the wire, which can produce a pulsed local magnetic field by a flow of a pulsed electric current. This pulsed local magnetic field can nucleate a DW at the end of the magnetic wire under a given external magnetic field. This enables us to determine the DW velocity as a function of the external magnetic field in a controlled manner.

Magneto-optic Kerr effect magnetometer in micron-scale spatial resolution, together with the pulsed magnetic field, offers another approach to measure a DW velocity. It has been reported that the very high DW velocity over 1000 m/s with

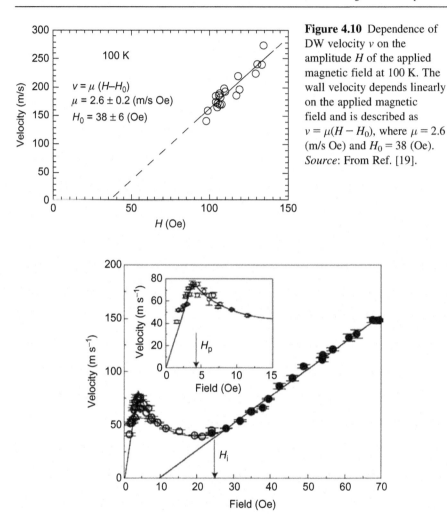

Figure 4.10 Dependence of DW velocity v on the amplitude H of the applied magnetic field at 100 K. The wall velocity depends linearly on the applied magnetic field and is described as $v = \mu(H - H_0)$, where $\mu = 2.6$ (m/s Oe) and $H_0 = 38$ (Oe). *Source*: From Ref. [19].

Figure 4.11 Average DW velocity versus field step amplitude. The inset shows the details around the velocity peak.
Source: From Ref. [22].

high mobility of 30 m/s Oe was realized for a single-layer 5 nm thick $Ni_{80}Fe_{20}$ wire with 200 nm in width [21]. Thus, the mobility in Ref. [21] is ten times larger than that in Ref. [19]. This mystery has been resolved recently by the measurements of the DW velocity in the wide range of an external magnetic field [22]. Figure 4.11 shows the DW velocity as a function of an external magnetic field for the 20 nm thick $Ni_{80}Fe_{20}$ wire with 600 nm in width. One can recognize that there are three regimes: the high DW mobility regime of 25 m/s Oe below 5 Oe, the low DW mobility regime of 2.5 m/s Oe above 25 Oe, and the transition regime with negative differential mobility between 5 and 25 Oe. These results are in accord with

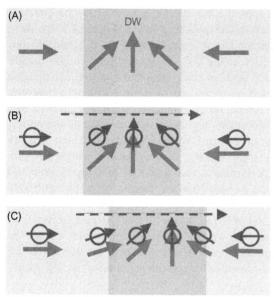

Figure 4.12 Schematic illustration of current-driven DW motion: (A) a DW between two domains in a magnetic wire. The arrows show the direction of magnetic moments. The magnetic DW is a transition region of the magnetic moments between domains, and the direction of moments gradually changes in the DW. (B) The spin of conduction electron follows the direction of local magnetic moments because of the s−d interaction. (C) As a reaction, the local magnetic moments rotate reversely, and in consequence, the electric current displaces the DW.

theoretical predictions that, above a threshold field (so-called Walker field), uniform wall movement gives way to turbulent wall motion, leading to a substantial drop in wall mobility [23−28].

The DW mobilities for low and high fields in Figure 4.11 are very close to those reported in Ref. [21] and in Ref. [19], respectively. Thus, it can be concluded that the measurements in Ref. [19] were performed above the Walker field and those in Ref. [21] were carried out below the Walker field. So, the discrepancy between two experiments was resolved. The high DW mobility in the relatively large external magnetic field in Ref. [21] can be understood by the micromagnetic simulation which suggests that Walker breakdown in nanowires involves the nucleation and motion of Bloch lines (vortices) in the wall, and that finite roughness inhibits this process, pushing the breakdown field higher [23]. In the thinner nanowire of Ref. [21], this effect might become more prominent than in wider nanowires, pushing the breakdown field to the higher value.

4.3 Current-Driven DW Motions

4.3.1 Concept of Current-Driven DW Motion

Figure 4.12A is an illustration of a DW between two domains in a magnetic wire. Here, arrows show the direction of magnetic moments. The magnetic DW is a transition region of the magnetic moments between domains, and the direction of moments gradually changes in the DW. What will happen if an electric current flows through a DW? Suppose a conduction electron passes though the DW from left to right. During this travel, the spin of conduction electron follows the direction of local magnetic moments because of the s−d interaction (Figure 4.12B). As a reaction, the local magnetic moments rotate reversely (Figure 4.12C), and, in consequence, the electric current can displace the DW. This current-driven DW motion was first predicted by Berger [29,30], and his group performed several experiments on magnetic films [31,32]. It needed huge currents to move a DW in a magnetic film due to the large cross section. Recent developments in nanolithography techniques make it possible to prepare nanoscale magnetic wires, resulting in the review of their pioneering works. The current-driven DW motion provides a new strategy to manipulate a magnetic configuration without an assistance of magnetic field, and will improve drastically the performance and functions of recently proposed spintronic devices, whose operation is based on the motion of a magnetic DW [33−36]. Reports on this subject have been increasing in recent years from both theoretical [37−42] and experimental [43−68] points of view because of its scientific and technological importance. However, most of the results cannot be reviewed here due to the limitation of space. The reader may be referred to the chapter by H. Ohno for the current-driven DW experiments in ferromagnetic semiconductor (Ga,Mn)As [50,62].

4.3.2   Magnetic Force Microscopy Direct Observations

In this section, results of the direct observation of the current-driven DW motion in a microfabricated magnetic wire are presented [47]. Magnetic force microscopy (MFM) is used to show that a single DW can be displaced back and forth by positive and negative pulsed currents.

An L-shaped magnetic wire with a round corner, as schematically illustrated in Figure 4.13, was prepared for the experiments. One end of the L-shaped magnetic wire is connected to a diamond-shaped pad, which acts as a DW injector [4], and the other end is sharply pointed to prevent a nucleation of a DW from this end.

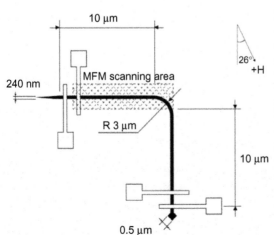

Figure 4.13 Schematic illustration of a top view of the sample. One end of the L-shaped wire is connected to a diamond-shaped pad which acts as a DW injector, and the other end is sharply pointed to prevent a nucleation of a DW from this end. The wire has four electrodes made of Cu. MFM observations were performed for the hatched area at room temperature.
Source: From Ref. [47].

L-shaped magnetic wires of 10 nm thick $Ni_{81}Fe_{19}$ were fabricated onto thermally oxidized Si substrates by means of an e-beam lithography and a lift-off method. The width of the wire is 240 nm.

In order to introduce a DW positioned in the vicinity of the corner, the direction of the external magnetic field was tilted from the wire axis in the substrate plane as shown in Figure 4.13. In the initial stage, a magnetic field of $+1$ kOe was applied in order to align the magnetization in one direction along the wire. Then, a single DW was introduced by applying a magnetic field of -175 Oe. After that, the MFM observations were carried out in the absence of a magnetic field. The existence of the single DW in the vicinity of the corner was confirmed as shown in Figure 4.14A. The DW is imaged as a bright contrast, which corresponds to the stray field from positive magnetic charge. In this case, a head-to-head DW is realized as illustrated schematically in Figure 4.14D. The position and the shape of the DW were unchanged after several MFM scans, indicating that the DW was pinned by a local structural defect, and that a stray field from the probe was too small to change the magnetic structure and the position of the DW.

After the observation of Figure 4.14A, a pulsed current was applied through the wire in the absence of a magnetic field. The current density and the pulse duration were 7.0×10^{11} A/m^2 and 5 μs, respectively. Figure 4.14B shows the MFM image after an application of the pulsed current from left to right. The DW, which had been in the vicinity of the corner (Figure 4.14A), was displaced from right to left

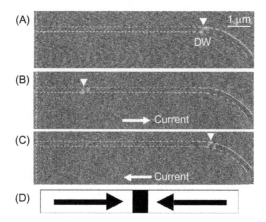

Figure 4.14 (A) MFM image after the introduction of a head-to-head DW. The DW is imaged as a bright contrast, which corresponds to the stray field from positive magnetic charge. (B) MFM image after an application of a pulsed current from left to right. The current density and pulse duration are 7.0×10^{11} A/m^2 and 5 μs, respectively. DW is displaced from right to left by the pulsed current. (C) MFM image after an application of a pulsed current from right to left. The current density and pulse duration are 7.0×10^{11} A/m^2 and 5 μs, respectively. The DW is displaced from left to right by the pulsed current. (D) Schematic illustration of a magnetic domain structure inferred from the MFM image. The DW has a head-to-head structure.
Source: From Ref. [47].

by the application of the pulsed current. Thus, the direction of the DW motion is opposite to the current direction. Furthermore, the direction of the DW motion can be reversed by switching the current polarity as shown in Figure 4.14C. These results are consistent with the spin-transfer mechanism.

The same experiments for a DW with different polarities, a tail-to-tail DW, were performed to examine the effect of a magnetic field generated by the electric current (Oersted field). The tail-to-tail DW was generated by the following procedures. A magnetic field of -1 kOe was applied in order to align the magnetization in the direction opposite to that in the previous experiment. Then, a tail-to-tail DW was introduced by applying a magnetic field of $+175$ Oe. The introduced DW is imaged as a dark contrast in Figure 4.15A, which indicates that a tail-to-tail DW is formed as schematically illustrated in Figure 4.15D. Figure 4.15A−C shows that the direction of the tail-to-tail DW displacement is also opposite to the current direction. The fact that both head-to-head and tail-to-tail DWs are displaced opposite to the current direction indicates clearly that the DW motion is not caused by the Oersted field.

Figure 4.16A−K are successive MFM images with one pulsed current applied between each consecutive image. The current density and the pulse duration were 7.0×10^{11} A/m^2 and 0.5 μs, respectively. Each pulse displaced the DW opposite to the current direction. The difference in the displacement for each pulse is possibly

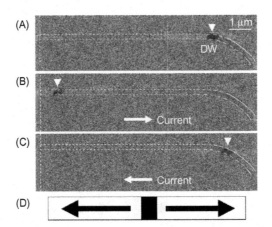

Figure 4.15 (A) MFM image after the introduction of a tail-to-tail DW. The DW is imaged as a dark contrast, which corresponds to the stray field from negative magnetic charge. (B) MFM image after an application of a pulsed current from left to right. The current density and pulse duration are 7.0×10^{11} A/m^2 and 5 μs, respectively. DW is displaced from right to left by the pulsed current. (C) MFM image after an application of a pulsed current from right to left. The current density and pulse duration are 7.0×10^{11} A/m^2 and 5 μs, respectively. The DW is displaced from left to right by the pulsed current. (D) Schematic illustration of a magnetic domain structure inferred from the MFM image. The DW has a tail-to-tail structure.
Source: From Ref. [47].

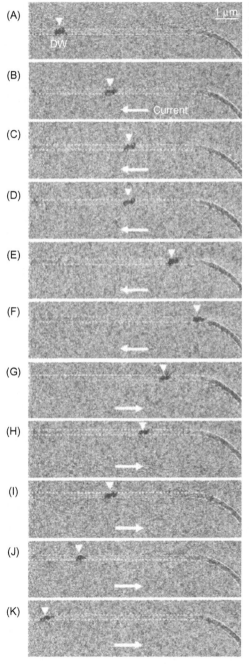

Figure 4.16 Successive MFM images with one pulse applied between each consecutive image. The current density and the pulse duration were 7.0×10^{11} A/m^2 and 0.5 μs, respectively. Note that a tail-to-tail DW is introduced, which is imaged as a dark contrast. *Source*: From Ref. [47].

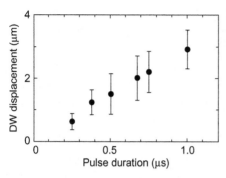

Figure 4.17 Average DW displacement per one pulse as a function of the pulse duration under a condition of constant current density of 7.0×10^{11} A/m^2.
Source: From Ref. [47].

due to the pinning by randomly located defects. The average displacement per one pulse did not depend on the polarity of the pulsed current. The average DW displacement per one pulse as a function of the pulse duration under the condition of constant current density of 7.0×10^{11} A/m^2 is shown in Figure 4.17. The average DW displacement is proportional to the pulse duration, which indicates that the DW has a constant velocity of 3.0 m/s. It was also confirmed that the DW velocity increases with the current density.

4.3.3 Toward Applications of Current-Driven DW Motion

It was shown that a DW position in a wire can be controlled by tuning the intensity, the duration, and the polarity of the pulsed current, and thus the current-driven DW motion has the potentiality for spintronic device applications such as novel memory and storage devises [33−36]. However, there are following issues to be overcome for the practical applications: (1) low threshold current density, (2) high DW velocity, and (3) stability of DW position. These three conditions should be simultaneously satisfied in the operations of real devices.

The experimentally obtained threshold current densities in the absence of a magnetic field are the order of 10^{11}–10^{12} A/m^2 [47,52]. These values are more than an order smaller than the theoretical value and also much smaller than that obtained from the micromagnetic simulation [37,41,42]. Several possibilities to solve this strong discrepancy have been proposed: sample heating by the current, inclusion of new term to the model, and the use of Landau−Lifshitz equation instead of Landau−Lifshitz−Gilbert equation. The reader may be referred to the chapter by H. Kohno and G. Tatara for an approach by an analytical model, and also to the chapter by A. Thiaville and Y. Nakatani for an approach by micromagnetics. In real experiments, the application of the high current density of 10^{11} A/m^2 results in the severe increase of sample temperature up to reaching the Curie temperature [54,65]. Figure 4.18 shows the estimated sample temperatures as a function of current density for two samples with the same dimension prepared on different substrates (SiO$_2$/Si and MgO) [56]. One can recognize at a glance that the Joule heating effect strongly depends on the substrate; the lower thermal conductance of SiO$_2$ leads to the rapid increase of the sample temperature. Thus, the sample

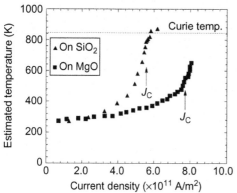

Figure 4.18 Estimated sample temperatures as a function of current density for two samples with the same dimension prepared on different substrates (SiO$_2$/Si and MgO).
Source: From: Ref. [56].

heating depends on the sample architecture (dimension, composition, substrate, etc.). More important message from Figure 4.18 is the large difference of the threshold current densities J_c in two samples with the same dimension. The higher sample temperature of the sample on the thermally oxidized Si substrate results in the lower threshold current density. This clearly suggests the importance of the thermal effect on the current-driven DW motion.

In a simple model in which the transferred spin angular momentum from conduction electrons is fully used to the DW motion, the DW velocity is expected to be $jPg\mu_B/(2eM_s)$, where j is the current density; P, the spin polarization of the current; g, the g-value of an electron; μ_B, the Bohr magneton; e, the electronic charge; and M_s, the saturation magnetization. This predicts about an order larger DW velocity than typical experimental values of several m/s in the absence of a magnetic field at the current density of 10^{11} A/m^2 [47,52]. Thus, the efficiency of the current-driven DW motion is about 10%, and most of the transferred momentum seems to dissipate by local excitations rather than being used to drive a DW. However, recent experiments show that high velocity of about 100 m/s can be obtained with the current density of 1.5×10^{12} A/m^2 for NiFe in the case of very low DW depinning field less than 10 Oe [68]. Thus, the low efficiency in the pioneering studies can be attributable to the pinning effect of a DW due to the defects such as surface and/or edge roughness.

Although the high velocity has been demonstrated for NiFe nanowires as mentioned above and the current-controlled multiple DW motion has been demonstrated by using NiFe nanowire [68], it has been also shown that the threshold current density increases with the DW pinning field. This could become a problem in the application, because DW position should be stabilized by using such as notch structures against the thermal agitation. Recently, very attractive simulation results have been coming out, which suggest that it is possible to reduce the threshold current density, keeping the high thermal stability of the DW position for nanowires with perpendicular magnetic anisotropy [69,70]. The promising experimental results, which support these simulations, have been reported [71−80]. It was shown that the adiabatic spin torque dominates the DW motion in a perpendicularly

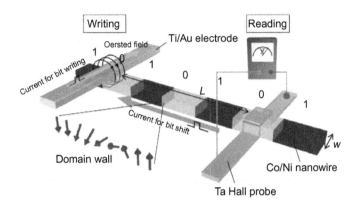

Figure 4.19 The device consists of a Co/Ni nanowire, an electrode (Ti/Au electrode) for writing domain bits, and a local detector (Ta Hall probe) for reading bits.
Source: From Ref. [74]. Copyright 2010, The Japan Society of Applied Physics.

magnetized Co/Ni nanowire and that the threshold current density for the DW motion can be reduced by tuning the wire width [76]. Both threshold current density and DW velocity were found to be almost independent of the external magnetic field in the range of ± 50 Oe [77,78], which shows that the reliable device operation against an external magnetic field disturbance can be achieved using the present system.

All-electrical control and local detection of multiple magnetic DWs in perpendicularly magnetized Co/Ni nanowires have been recently demonstrated [74]. Figure 4.19 is a schematic of the device structure. The device consists of a Co/Ni nanowire, an electrode (Ti/Au electrode) for writing domain bits, and a local detector (Ta Hall probe) for reading bits. The scheme of the demonstration is displayed in Figure 4.20A−G. Before creating the DWs, the wire magnetization was initialized by applying perpendicular positive magnetic field of 4 kOe (Figure 4.20A). The domain with negative magnetization direction was recorded into the wire using the local field (Figure 4.20B), and the created single DW was shifted by 12 multiple current pulses with duration of 11 ns and the current density of 1.3×10^{12} A/m^2 for each single pulse (Figure 4.20C). A subsequent domain with the positive magnetization direction was recorded using the local field (Figure 4.20D), then 12 multiple current pulses were again injected into the wire (Figure 4.20E). Afterward, the above sequence was repeated (Figure 4.20F and G). The Hall resistance was measured after each current pulse to check the magnetization direction under the Hall probe. Figure 4.20H shows the results. The arrows with the labeled numbers (#1, #2,...) show the timing of the DW creation. The alternative switching of the Hall resistance was observed after three DWs were introduced into the wire, which indicates the series of DWs was reproducibly shifted by the application of current pulses. The back and forth shift of multiple DWs was also demonstrated by using the same device.

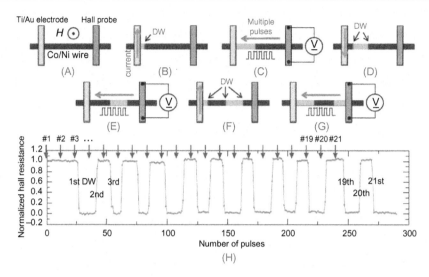

Figure 4.20 (A–G) Schematic illustrations of measurement procedure to create and shift multiple DWs. (H) Normalized Hall resistance as a function of number of pulses when the multiple DWs were introduced into the wire and shifted toward the same direction. The arrows indicate the timing of the DW creation.
Source: From Ref. [74].

4.4 Topics on Nanodot Systems

4.4.1 MFM Studies on Magnetic Vortices in Dot Systems

As mentioned in Introduction, ferromagnetic materials generally form domain structures to reduce their magnetostatic energy. In very small ferromagnetic systems, however, the formation of DWs is not energetically favored. Specifically, in a dot of ferromagnetic material of micrometer or submicrometer size, a curling spin configuration—that is, a magnetization vortex (Figure 4.21)—has been proposed to occur in place of domains. When the dot thickness becomes much smaller than the dot diameter, usually all spins tend to align in-plane. In the curling configuration, the spin directions change gradually in-plane so as not to lose too much exchange energy, but to cancel the total dipole energy. In the vicinity of the dot center, the angle between adjacent spins then becomes increasingly larger when the spin directions remain confined in-plane. Therefore, at the center of the vortex structure, the magnetization within a small spot will turn out of plane and parallel to the plane normal [82].

Cowburn et al. reported magneto-optical measurements on nanoscale superalloy ($Ni_{80}Fe_{15}Mo_5$) dot arrays [83]. From the profiles of the hysteresis loops, they concluded that a collinear-type single domain phase is stabilized in dots with diameters smaller than a critical value of about 100 nm and a vortex phase likely occurs in

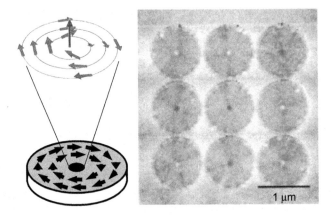

Figure 4.21 MFM image of an array of permalloy dots 1 μm in diameter and 50 nm thick with the schematic spin structure (magnetic vortex and vortex core) in a dot.
Source: From Ref. [81].

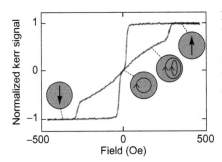

Figure 4.22 Hysteresis loop for the circular dot with a diameter of 300 nm and a thickness of 10 nm under an external magnetic field in the sample plane. The schematic annotation shows the magnetization within a circular dot.
Source: From Ref. [83].

dots with larger diameters. Figure 4.22 shows the hysteresis loop for the dot with a diameter of 300 nm and a thickness of 10 nm under an external magnetic field in the sample plane [83]. The magnetic field increases the area where the magnetization aligns parallel and decreases the area where the magnetization aligns antiparallel to the magnetic field. The gradual increase of magnetization corresponds to the moving of the vortex core from the center to the outside and the jump of magnetization corresponds to the annihilation of the magnetization vortex. The relatively sudden loss in the course of field reducing corresponds to the nucleation of the vortex.

Figure 4.21 is the first proof of such a vortex structure with a nanometer-scale core where the magnetization rises out of the dot plane [81]. The sample is an array of 3 × 3 dots of permalloy ($Ni_{81}Fe_{19}$) with 1 μm in diameter and 50 nm thickness. At the center of each dot, bright or dark contrast is observed, which corresponds to the positive or negative stray field from the vortex core. The direction of the magnetization at the center turns randomly, either up or down, as reflected by the different contrast of the center spots. This is reasonable since up and down

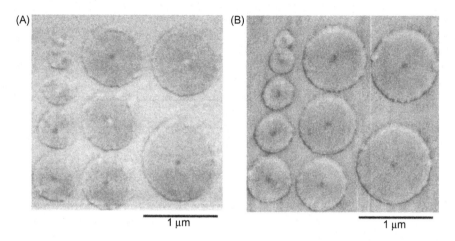

Figure 4.23 MFM images of an ensemble of 50 nm thick permalloy dots with diameters varying from 0.3 to 1 μm: (A) after applying an external field of 1.5 T along an in-plane direction and (B) parallel to the plane normal, respectively.
Source: From Ref. [81].

magnetizations are energetically equivalent without an external applied field and do not depend on the vortex orientation: clockwise or counterclockwise. MFM observations were performed also for an ensemble of permalloy dots with varying diameters, nominally from 0.3 to 1 μm (Figure 4.23). The image in Figure 4.23A was taken after applying an external field of 1.5 T along an in-plane direction. Again, the two types of vortex core with up and down magnetization are observed. In contrast, after applying an external field of 1.5 T normal to the substrate plane, center spots exhibit the same contrast (Figure 4.23B) indicating that all the vortex core magnetizations have been oriented into the field direction. The size of the core cannot be determined from the images since the spatial resolution of MFM is much larger than the theoretical core size. The core size was determined to be 9 nm by using spin-polarized scanning tunneling microscopy which has an atomic scale resolution [84].

Since the size of the perpendicular magnetization at the vortex core is very small, conventional magnetization measurements cannot detect switching of the vortex core under magnetic field. For the measurement of switching field of core magnetization, arrays of circular dots of permalloy with the diameter of 0.2, 0.4, and 1 μm and the thickness of 50 nm were prepared [85]. The sample consists of sets of permalloy dots arrays of square lattice 5×5. First, a strong magnetic field (1.5 T) normal to the sample plane was applied in order to align all the core magnetizations in one direction. The perpendicular magnetization turning to this direction shows dark contrast in MFM image as shown Figure 4.24A. Second, a certain magnetic field H_t was applied to the counter direction for 5 min. Finally, one 5×5 array which was chosen randomly in the sample was observed by MFM in the absence of a magnetic field. The MFM images for the cases of H_t of 3500 Oe are

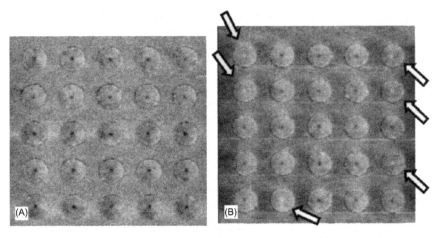

Figure 4.24 MFM images of an array composed of 5×5 circular permalloy dots with the diameter of 0.4 μm: (A) after applying a magnetic field of 1.5 T and (B) after applying a magnetic field $H_t = -0.35$ T.
Source: From Ref. [85].

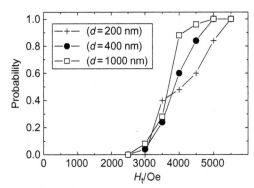

Figure 4.25 Switching probability of core magnetization in circular dots with the diameter of 0.2, 0.4, and 1 μm as a function of magnetic field normal to the sample plane. The average switching fields are 4100, 3900, and 3650 Oe in the samples of 0.2, 0.4, and 1 μm in diameter, respectively.
Source: From Ref. [85].

shown in Figure 4.24B. A bright spot corresponds to core magnetization switched by H_t, while a dark spot represents the core magnetization with its initial direction. The number of bright spots in the MFM images after applying various values of H_t gives the switching probability for each field. The results for the three samples are shown in Figure 4.25. The average switching field is 4100, 3900, and 3650 Oe in the sample of 0.2, 0.4, and 1 μm in diameter, respectively.

The experimental confirmation of the existence of the vortex core by MFM study [81] stimulated the subsequent intensive studies on the dynamics of the vortex core. It has been clarified that a vortex core displaced from the stable position (dot center) exhibits a spiral precession around it during the relaxation process [86–88]. This motion has a characteristic frequency which is determined by the shape of the dot. Thus, the nanodot functions as a resonator for vortex core motion.

4.4.2 Current-Driven Resonant Excitation of Magnetic Vortex

As is clear from Figure 4.12, the underlying physics of the current-driven DW motion is that spin currents can apply a torque on the magnetic moment when the spin direction of the conduction electrons has a relative angle to the local magnetic moment. This leads us to the hypothesis that any type of spin structure with spatial variation can be excited by a spin-polarized current in a ferromagnet.

The ideal example of such a noncollinear spin structure is a curling magnetic structure ("magnetic vortex") realized in a ferromagnetic circular nanodot described in the previous section. In this section, current-induced dynamics of a vortex core in a ferromagnetic dot will be discussed. It is shown that a magnetic vortex core can be resonantly excited by an AC through the dot when the current frequency is tuned to the resonance frequency originating from the confinement of the vortex core in the dot [89]. The core is efficiently excited by the AC due to the resonant nature and the resonance frequency is tunable by the dot shape. It is also demonstrated that the direction of a vortex core can be switched by utilizing the current-driven resonant dynamics of the vortex.

The current-induced dynamics of the vortex core was calculated by the micromagnetic simulations based on the Landau−Lifshitz−Gilbert equation,

$$\frac{\partial \mathbf{m}}{\partial t} = -\gamma_0 \mathbf{m} \times \mathbf{H}_{\text{eff}} + \alpha \mathbf{m} \times \frac{\partial \mathbf{m}}{\partial t} - (\mathbf{u} \cdot \nabla)\mathbf{m} \qquad (4.3)$$

with modifications due to electric/spin current [41,42]. Here, m is a unit vector along the local magnetization, γ_0 the gyromagnetic ratio, $\mathbf{H}_{\text{eff}}$ the effective magnetic field including the exchange and the demagnetizing fields, and α is the Gilbert damping constant. The third term on the right-hand side represents the spin-transfer torque, which describes the effect of spin transfer from conduction electrons to localized spins. This spin-transfer effect is a combined effect of the spatial nonuniformity of magnetization and the current flow. The vector $u = -jPg\mu_B/(2eM_s)$, which has the dimension of velocity, is essentially the spin current associated with the electric current in a ferromagnet, where j is the current density, P the spin polarization of the current, g the g-value of an electron, μ_B the Bohr magneton, e the elementary charge, and M_s is the saturation magnetization.

First, we determined the eigenfrequency f_0 of the vortex core precession in the dot by calculating the free relaxational motion of the vortex core from the off-centered position. The eigenfrequency depends on the aspect ratio h/r (the height h to the radius r) of the dot [86]. Then, the simulations were performed by applying an AC at a given frequency f in the absence of a magnetic field. Figure 4.26A shows the time evolution of the core position when an AC ($f = f_0 = 380$ MHz and $J_0 = 3 \times 10^{11}$ A/m^2) is applied to a dot with $r = 410$ nm and $h = 40$ nm. Once the AC is applied, the vortex core first moves in the direction of the electron flow or spin current. This motion originates from the spin-transfer effect. The off-centered core is then subjected to a restoring force toward the dot center. However, because of the gyroscopic nature of the vortex (i.e., a vortex moves perpendicularly to the force), the core makes a circular precessional motion around the dot center [86]. The precession is amplified by the current to reach a steady orbital motion where the spin transfer from the current is balanced with the damping, as depicted in Figure 4.26A. The direction of the precession depends on the direction of the core magnetization as in the motion induced by the magnetic field [86,90]. It should be noted that the radius of the steady orbital on resonance is larger by more than an order of magnitude as compared to the displacement of the vortex core induced by a DC current of the same amplitude [90]. Thus, the core is efficiently excited by the AC due to resonance.

Figure 4.26B shows the time evolutions of the x position of the vortex core for three different excitation frequencies $f = 250$, 340, and 380 MHz. The steady state appears after around 30 ns on resonance ($f = 380$ MHz). For $f = 340$ MHz slightly off the resonance, the amplitude beats first, and then the steady state with smaller amplitude appears. The vortex core shows only a weak motion for $f = 250$ MHz, which is quite far from the resonance. Figure 4.26C shows the radii of the steady orbitals as a function of the current frequency for the dots with $r = 410$, 530, and

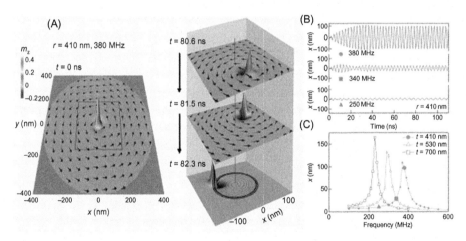

Figure 4.26 (A) Time evolution of the vortex under the AC application. Magnetization direction $\boldsymbol{m} = (m_x, m_y, m_z)$ inside the dot on the xy plane was obtained by micromagnetic simulation. The 3D plots indicate m_z with the $m_x - m_y$ vector plots superimposed. The plot on the left represents the initial state of the vortex core situated at the center of the dot with $r = 410$ nm. The 3D plots on the right show the vortex on the steady orbital at $t = 80.6$, 81.5, and 82.3 ns after applying the AC ($f_0 = 380$ MHz and $J_0 = 3 \times 10^{11}$ A/m²). These plots are close-ups of the square region around the dot center indicated by the black square in the plot on the left. The time evolution of the core orbital from $t = 0$ to 100 ns is superimposed only on the $t = 82.3$ ns plot. (B) Time evolutions of the vortex core displacement (x) for three excitation frequencies $f = 250$, 340, and 380 MHz ($r = 410$ nm and $J_0 = 3 \times 10^{11}$ A/m²). (C) Radius of the steady orbital as a function of the frequency for the dots with $r = 410$, 530, and 700 nm.
Source: From Ref. [89].

700 nm. Each dot exhibits the resonance at the eigenfrequency of the vortex motion.

In order to experimentally detect the resonant excitation of a vortex core predicted by the micromagnetic simulation, we measured the resistance of the dot while an AC excitation current was passed through it at room temperature in the configuration shown in Figure 4.27. An SEM image of the sample is shown in Figure 4.27. Two wide Au electrodes with 50 nm thickness, through which an AC excitation current is supplied, are also seen. The amplitude of the AC excitation current was 3×10^{11} A/m^2. Figure 4.28A shows the resistances as a function of the frequency of the AC excitation current for the dots with three different radii, $r = 410$, 530, and 700 nm. A small but clear dip is observed for each dot; this signifies the resonance. The radius dependence of the resonance frequency is well reproduced by the simulation, as shown in Figure 4.28B.

4.4.3 Switching a Vortex Core by Electric Current

During the study on the current-driven resonant excitation of magnetic vortex, we found that higher excitation currents induce even the switching of the core magnetization during the circular motion [91]. Figures 4.29A–F are successive snapshots of the calculated results for the magnetization distribution during the process of core motion and switching, showing that the reversal of the core magnetization takes place in the course of the circular motion without going out of the dot. Noteworthy is the development of an out-of-plane magnetization (dip) which is opposite to the core magnetization (Figure 4.29A–D). Figure 4.31A displays the vortex core trajectory corresponding to Figure 4.29. One sees that, after switching,

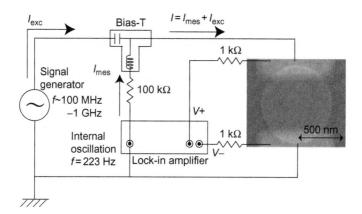

Figure 4.27 SEM image of the sample along with a schematic configuration used for the measurements. The detection of the vortex excitation was performed by resistance measurements with a lock-in technique (223 Hz and current $I_{mes} = 15$ μA) under the application of an AC excitation current $I_{exc} = 3 \times 10^{11}$ A/m^2.
Source: From Ref. [89].

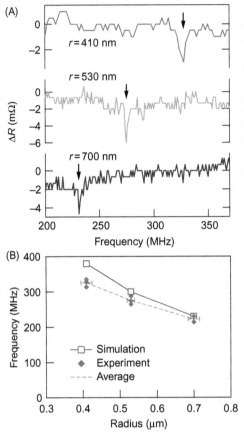

Figure 4.28 (A) Experimental detection of the current-driven resonant excitation of a magnetic vortex core. The resistances are indicated as a function of the frequency of the AC excitation current for the dots with three different radii $r = 410$, 530, and 700 nm. (B) Radius dependence of the resonance frequency. The blue squares and the red diamonds indicate the simulation and the experimental results, respectively. The experimental results for eight samples are plotted. The red dashed line is the averaged value of the experimental data. (For interpretation of the references to color in this figure legend, the reader is referred to the web version of this book.)
Source: From Ref. [89].

the core steps toward the disk center and starts rotating in the opposite sense, again evidencing the core switching.

We confirmed the predicted current-induced switching of the vortex core by the MFM observation as described below [91]. First, the direction of a core magnetization was determined by MFM observation. A dark spot at the center of the disk in Figure 4.30B indicates that the core magnetization directs upward with respect to the paper plane. The core direction was checked again after the application of an AC excitation current of frequency $f = 290$ MHz and amplitude $J_0 = 3.5 \times 10^{11}$ A/m^2 through the disk, with the duration of about 10 s. Here, the current densities were evaluated by dividing current by the disk diameter times thickness. As shown in Figure 4.30C, the dark spot at the center of the disk changed into bright spot after the application of the excitation current, indicating that the core magnetization has been switched. Parts B−L in Figure 4.30 are successive MFM images with an excitation current applied between each consecutive image. It was observed that the direction of the core magnetization after application of the excitation current was changed randomly. This indicates that the switching occurred frequently compared to the

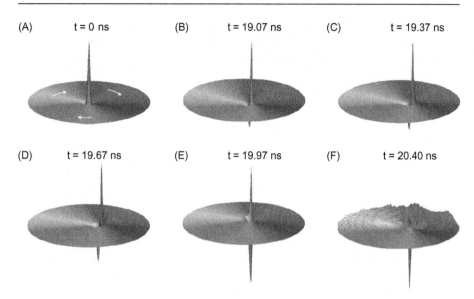

(A)   t = 0 ns   (B)   t = 19.07 ns   (C)   t = 19.37 ns

(D)   t = 19.67 ns   (E)   t = 19.97 ns   (F)   t = 20.40 ns

Figure 4.29 Perspective view of the magnetization with a moving vortex structure. The height is proportional to the out-of-plane (z) magnetization component. Rainbow color indicates the in-plane component as exemplified by the white arrows in (A). (A) Initially, a vortex core magnetized upward is at rest at the disk center. (B) On application of the AC, the core starts to make a circular orbital motion around the disk center. There appears a region with downward magnetization (called "dip" here) on the inner side of the core. (C–E) The dip grows slowly as the core is accelerated. When the dip reaches the minimum, the reversal of the initial core starts. (F) After the completion of the reversal, the stored exchange energy is released to a substantial amount of spin waves. A positive "hump" then starts to build up, which will trigger the next reversal. Calculation with $h = 50$ nm, $R = 500$ nm, and $J_0 = 4 \times 10^{11}$ A/m^2.
Source: From Ref. [91].

duration of the excitation current (about 10 s) and the core direction was determined at the last moment when we turned off the excitation current.

Figure 4.31B shows the core velocity as a function of excitation time which was obtained by the micromagnetic simulation. The sudden decreases of velocity correspond to the repeated core-switching events. Worth noting is that the core switches when its velocity reaches a certain value, $v_{switch} \approx 250$ m/s here, regardless of the value of the excitation current density. This is the crucial key to understand the switching mechanism together with the existence of the dip structure which appears just before the core switching. The rotational motion of the core is accompanied by the magnetization dynamics in the vicinity of the core. This magnetization dynamics in the dot plane produces so-called damping torque perpendicular to the plane according to the second term of Eq. (4.3), which generates the dip structure seen in Figure 4.29B–E. The higher core velocity leads to the stronger damping torque, and eventually the core switching occurs at the threshold core velocity. If the core

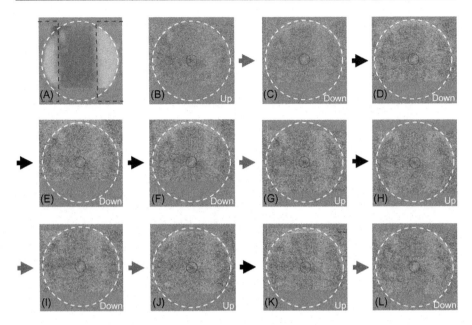

Figure 4.30 MFM observation of electrical switching of vortex core. (A) AFM image of the sample. A permalloy disk fills inside the white circle. The thickness of the disk is 50 nm, and the radius is 500 nm. Two wide Au electrodes with 50 nm thickness, through which an AC excitation current is supplied, are also seen. (B) MFM image before the application of the excitation current. A dark spot at the center of the disk (inside the red circle) indicates that the core magnetization directs upward with respect to the paper plane. (C) MFM image after the application of the AC excitation current at a frequency $f = 290$ MHz and amplitude $J_0 = 3.5 \times 10^{11}$ A/m^2 through the disk with the duration of about 10 s. The dark spot at the center of the disk in (B) has changed into the bright spot, indicating the switching of the core magnetization from up to down. (B–L) Successive MFM images with excitation current applied similarly between consecutive images. The switching of the core magnetization occurs from (B) to (C), (F) to (G), (H) to (I), (I) to (J), and (K) to (L). (For interpretation of the references to color in this figure legend, the reader is referred to the web version of this book.)
Source: From Ref. [91].

switching is governed by the core velocity, the switching should occur regardless of how the core has the threshold velocity. In fact, the core switching was also observed by the resonant excitation with AC magnetic field [92].

It has been shown that an electric current can switch the direction of a vortex core in a magnetic disk. The current necessary for the switching is only several milliamperes, while the core switching by an external magnetic field needs a large magnetic field of several kilooersteds, as described in Section 4.4.1 [85]. Although the repeated vortex core switching by a continuous AC was presented here, it will be possible to control the core direction by a current with an appropriate waveform. The current-induced vortex core switching can be used as an efficient data writing

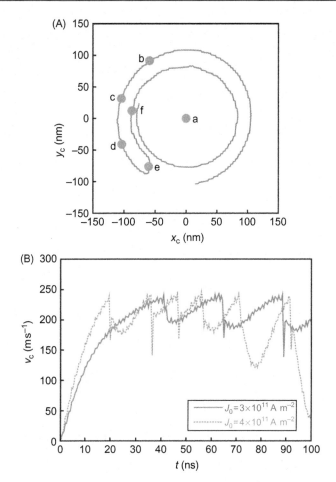

Figure 4.31 Core dynamics under AC spin-polarized current. The calculation was performed for the same disk as in Figure 4.27. (A) In-plane trajectory of the core motion of Figure 4.27 ($J_0 = 4 \times 10^{11}$ A/m^2), from $t = 17.5$ to 23 ns with the snapshot points indicated. (B) Magnitude of the core velocity as a function of time, for two choices of the current density, showing that the maximum velocity is the same.
Source: From Ref. [91].

method for a memory device in which the data are stored in a nanometer size core. It was shown very recently that a nanosecond single current pulse can switch the core magnetization [93]. This method has advantages over the core switching by using the resonance effect described above; it gives short switching time as well as controllability of the core direction which is indispensable in the applications, although the pulse method requires an order higher current density.

All-electrical operation of a magnetic vortex core memory cell has been actually demonstrated recently [94]. The device and measurement setup are shown

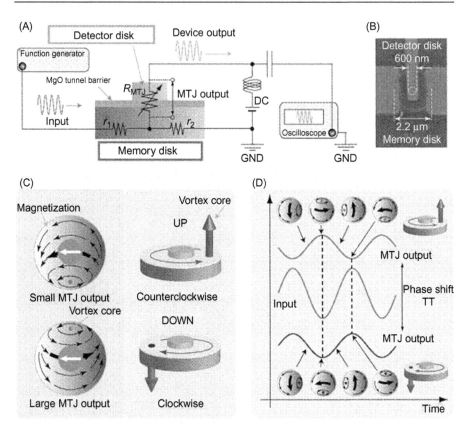

Figure 4.32 Device structure and the measurement. (A) Device and measurement setup.
(B) SEM image of the device. (C) *Left panel*: relation between the core position and the
MTJ output; *right panel*: relation between core direction and core rotation. (D) Device input
and MTJ outputs for the up core and down core.
Source: From Ref. [94]. Reprinted with permission from American Institute of Physics.

schematically in Figure 4.32A. The memory cell consists of a memory disk (MD) and
a detector disk (DD) with a tunneling barrier inserted between them to form an
Magnetic Tunneling Junction (MTJ). It has the right and left electrodes to feed the
input current to the MD and the top electrode for output. The magnetic vortex structure is created in the MD, and the magnetization in the DD is fixed in-plane. When an
input AC (current density j_{AC} and frequency f) is fed from the function generator to
the MD, the gyration motion of the vortex core is excited and the magnetization in the
MD just beneath the DD rotates in synchrony with the motion of the core as shown in
Figure 4.32C. The angle between the magnetization in the MD and the fixed magnetization in the DD thus changes periodically, resulting in time-dependent resistance
change in the MTJ. This resistance change is converted to a voltage change by applying a DC voltage (V_{bias}) to the MTJ. The AC component of the voltage from the cell
is collected by the oscilloscope through a bias tee as the device output.

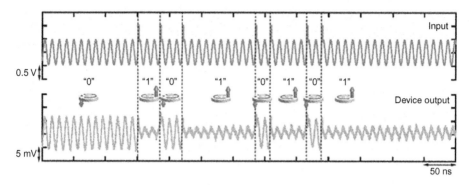

Figure 4.33 Demonstration of all-electrical operation of the vortex core memory. Device output (upper panel) is shown together with the input signals (down panel). Every injection of the input pulse at any time switches the core direction (electrical data writing) and the core direction can be read out as the output amplitude.
Source: From Ref. [94]. Reprinted with permission from American Institute of Physics.

Figure 4.33 shows the demonstration of all-electrical operation of the vortex core memory. As is clear from the schematic circuit in Figure 4.32A, the device output contains not only the MTJ output but also a voltage drop due to the resistance r_2 in the MD. Because there is a phase relation between the input and the MTJ outputs (shown in Figure 4.32D), the MTJ output for the down core adds constructively with the input and the MTJ output for the up core adds destructively with the input, resulting in a large output for the down core and a small output for the up core. The MTJ output shown in the lower panel of Figure 4.33 therefore indicates that we can switch the core direction at any time by applying a pulsed current added to the input signal. It has been shown that the three-terminal device shown in Figure 4.32 works as a vortex core memory cell in which both reading and writing can be done in an electrical way. Binary data corresponding to the core direction can be read out electrically as the amplitude of the output and data can be written electrically by using a pulsed current.

Acknowledgments

This work was partly supported by NEDO, MEXT, and JSPS of Japan. The authors thank S. Kasai, D. Chiba, K. Kobayashi, K. Mibu, N. Nakatani, A. Thiaville, G. Tatara, and H. Kohno for their fruitful discussions.

References

[1] Weiss P. J Phys 1907;6:661.
[2] Baibich MN, Broto JM, Fert A, Nguyen Van Dau F, Petroff F, Etienne P, et al. Phys Rev Lett 1988;61:2472.

[3]   Ono T, Miyajima H, Shigeto K, Shinjo T. Appl Phys Lett 1998;72:1116.
[4]   Shigeto K, Shinjo T, Ono T. Appl Phys Lett 1999;75:2815.
[5]   Himeno A, Ono T, Nasu S, Shigeto K, Mibu K, Shinjo T. J Appl Phys 2003;93:8430.
[6]   Himeno A, Okuno T, Ono T, Mibu K, Nasu S, Shinjo T. J Magn Magn Mater 2005;286:16.
[7]   Grollier J, Boulenc P, Cros V, Hamzic A, Vaures A, Fert A, et al. Appl Phys Lett 2003;83:509.
[8]   Astumian RD. Science 1997;276:917.
[9]   Reimann P. Phys Rep 2002;361:57.
[10] Astumian RD, Hänggi P. Phys Today 2002;55:33.
[11] Linke H, Humphrey TE, Löfgren A, Sushkov AO, Newbury R, Taylor RP, et al. Science 1999;286:2314.
[12] Linke H, editor. Ratchets and Brownian motors: basics, experiments and applications. Appl Phys A 2002;75(2) [special issue].
[13] Hayashi N, Hsu C, Romankiw L, Krongelb S. IEEE Trans Magn 1972;8:16.
[14] Hayashi N. IBM Research, March 1, RC3754, 1972.
[15] Himeno A, Okuno T, Kasai S, Ono T, Nasu S, Mibu K, et al. J Appl Phys 2005;97:066101.
[16] Franken H, Swagten HJM, Koopmans B. Nat Nanotechnol 2012;7:499−503.
[17] Himeno A, Kasai S, Ono T. Appl Phys Lett 2005;87:243108.
[18] Sixtus KJ, Tonks L. Phys Rev 1931;37:930.
[19] Ono T, Miyajima H, Shigeto K, Mibu K, Hosoito N, Shinjo T. Science 1999;284:468−70.
[20] Himeno A, Ono T, Nasu S, Okuno T, Mibu K, Shinjo T. J Magn Magn Mater 2004;272−276:1577.
[21] Atkinson D, Allwood DA, Xiong G, Cooke MD, Faulkner CC, Cowburn RP. Nat Mater 2003;2:85.
[22] Beach GSD, Nistor C, Knutson C, Tsoi M, Erskine JL. Nat Mater 2005;4:741.
[23] Nakatani Y, Thiaville A, Miltat J. Nat Mater 2003;2:521.
[24] Schryer NL, Walker LR. J Appl Phys 1974;45:5406.
[25] Slonczewski JC. Int J Magn 1972;2:85.
[26] Slonczewski JC. J Appl Phys 1974;45:2705.
[27] Malozemoff AP, Slonczewski JC. Magnetic domain walls in bubble materials. New York, NY: Academic Press; 1979.
[28] Redjdal M, Giusti J, Ruane MF, Humphrey FB. J Appl Phys 2002;91:7547.
[29] Berger L. J Appl Phys 1984;55:1954.
[30] Berger L. J Appl Phys 1992;71:2721.
[31] Freitas PP, Berger L. J Appl Phys 1985;57:1266.
[32] Hung CY, Berger L. J Appl Phys 1988;63:4276.
[33] Allwood DA, Xiong G, Cooke MD, Faulkner CC, Atkinson D, Vernier N, et al. Science 2002;296:2003.
[34] Versluijs JJ, Bari MA, Coey JMD. Phys Rev Lett 2001;87:026601.
[35] Parkin SSP. U.S. Patent No. 6834005, 2004.
[36] Numata H, Suzuki T, Ohshima N, Fkami S, Ishiwata K, Kasai N. VLSI Tech Dig 2007:232−3.
[37] Tatara G, Kohno H. Phys Rev Lett 2004;92:086601.
[38] Li Z, Zhang S. Phys Rev Lett 2004;92:207203.
[39] Zhang S, Li Z. Phys Rev Lett 2004;93:127204.
[40] Waintal X, Viret M. Europhys Lett 2004;65:427.

[41] Thiaville A, Nakatani Y, Miltat J, Vernie N. J Appl Phys 2004;95:7049.
[42] Thiaville A, Nakatani Y, Miltat J, Suzuki Y. Europhys Lett 2005;69:990.
[43] Koo H, Krafft C, Gomez RD. Appl Phys Lett 2002;81:862.
[44] Tsoi M, Fontana RE, Parkin SSP. Appl Phys Lett 2003;83:2617.
[45] Klaui M, Vaz CAF, Bland JAC, Wemsdorfer W, Fani G, Cambril E, et al. Appl Phys Lett 2003;83:105.
[46] Kimura T, Otani Y, Tsukagoshi K, Aoyagi Y. J Appl Phys 2003;94:07947.
[47] Yamaguchi A, Ono T, Nasu S, Miyake K, Mibu K, Shinjo T. Phys Rev Lett 2004;92:077205; Phys Rev Lett 2006;96:179904(E).
[48] Lim CK, Devolder T, Chappert C, Grollier J, Cros V, Vaures A, et al. Appl Phys Lett 2004;84:2820.
[49] Vernier N, Allwood DA, Atkinson D, Cooke MD, Cowburn RP. Europhys Lett 2004;65:526.
[50] Yamanouchi M, Chiba D, Matsukura F, Ohno H. Nature 2004;428:539.
[51] Saitoh E, Miyajima H, Yamaoka T, Tatara G. Nature 2004;432:203.
[52] Klaui M, Vaz CAF, Bland JAC, Wernsdorfer W, Faini G, Cambril E, et al. Phys Rev Lett 2005;94:106601.
[53] Klaui M, Jubert P-O, Allenspach R, Bischof A, Bland JAC, Faini G, et al. Phys Rev Lett 2005;95:026601.
[54] Yamaguchi A, Nasu S, Tanigawa H, Ono T, Miyake K, Mibu K, et al. Appl Phys Lett 2005;86:012511.
[55] Ravelosona D, Lacour D, Katine JA, Terris BD, Chapper C. Phys Rev Lett 2005;95:117203.
[56] Yamaguchi A, Yano K, Tanigawa H, Kasai S, Ono T. Jpn J Appl Phys 2006;45:3850.
[57] Beach GSD, Knutson C, Nistor C, Tsoi M, Erskine JL. Phys Rev Lett 2006;97:057203.
[58] Hayashi M, Thomas L, Bazaliy YB, Rettner C, Moriya R, Jiang X, et al. Phys Rev Lett 2006;96:197207.
[59] Fukumoto K, Kuch W, Vogel J, Romanens F, Pizzini S, Camarero J, et al. Phys Rev Lett 2006;96:097204.
[60] Laufenberg M, Bührer W, Bedau D, Melchy P-E, Kläui M, Vila L, et al. Phys Rev Lett 2006;97:046602.
[61] Thomas L, Hayashi M, Jiang X, Moriya R, Rettner C, Parkin SSP. Nature 2006;443:197.
[62] Yamanouchi M, Chiba D, Matsukura F, Dietl T, Ohno H. Phys Rev Lett 2006;96:096601.
[63] Klaui M, Laufenberg M, Heyne L, Backes D, Rüdiger U. Appl Phys Lett 2006;88:232507.
[64] Togawa Y, Kimura T, Harada K, Akashi T, Matsuda T, Tonomura A, et al. Jpn J Appl Phys 2006;45:L683.
[65] Togawa Y, Kimura T, Harada K, Akashi T, Matsuda T, Tonomura A, et al. Jpn J Appl Phys 2006;45:1322.
[66] Hayashi M, Thomas L, Rettner C, Moriya R, Bazaliy YB, Parkin SSP. Phys Rev Lett 2007;98:037204.
[67] Thomas L, Hayashi M, Jiang X, Moriya R, Rettner C, Parkin SSP. Science 2007;315:1553.
[68] Parkin SSP, Hayashi M, Thomas L. Science 2008;320:190.
[69] Fukami S, Suzuki T, Ohshima N, Nagahara K, Ishiwata N. J Appl Phys 2008;103:07E718.
[70] Jung SW, Kim W, Lee TD, Lee KJ, Lee HW. App Phys Lett 2008;92:202508.

[71] Tanigawa H, Kondou K, Koyama T, Nakano K, Kasai S, Ohshima N, et al. Appl Phys Express 2008;1:011301.

[72] Koyama T, Yamada G, Tanigawa H, Kasai S, Ohshima N, Fukami S, et al. Appl Phys Express 2008;1:101303.

[73] Tanigawa H, Koyama T, Yamada G, Chiba D, Kasai S, Fukami S, et al. Appl Phys Express 2009;2:053002.

[74] Chiba D, Yamada G, Koyama T, Ueda K, Tanigawa H, Fukami S, et al. Appl Phys Express 2010;3:073004.

[75] Kim K-J, Lee J-C, Yun S-J, Gim G-H, Lee K-S, Choe S-B, et al. Appl Phys Express 2010;3:083001.

[76] Koyama T, Chiba D, Ueda K, Kondou K, Tanigawa H, Fukami S, et al. Nat Mater 2011;10:194−7.

[77] Koyama T, Chiba D, Ueda K, Tanigawa H, Fukami S, Suzuki T, et al. Appl Phys Lett 2011;98:192509.

[78] Yoshimura Y, Koyama T, Chiba D, Nakatani Y, Fukami S, Yamanouchi M, et al. Appl Phys Express 2012;5:063001.

[79] Ueda K, Koyama T, Hiramatsu R, Chiba D, Fukami S, Tanigawa H, et al. Appl Phys Lett 2012;100:202407.

[80] Koyama T, Ueda K, Kim K-J, Yoshimura Y, Chiba D, Yamada K, et al. Nat Nanotechnol 2012;7:635.

[81] Shinjo T, Okuno T, Hassdorf R, Shigeto K, Ono T. Science 2000;289:930.

[82] Hubert A, Schafer H. Magnetic domains. Berlin: Springer; 1998.

[83] Cowburn RP, Koltsov DK, Adeyeye AO, Welland ME. Phys Rev Lett 1999;83:1042.

[84] Wachowiak A, Wiebe J, Bode M, Pietzsch O, Morgenstern M, Wiesendanger R. Science 2002;298:577.

[85] Okuno T, Shigeto K, Ono T, Mibu K, Shinjo T. J Magn Magn Mater 2002;240:1.

[86] Yu K, Guslienko BA, Ivanov V, Novosad Y, Otani H, Shima K. Fukamichi. J Appl Phys 2002;91:8037.

[87] Park JP, Eames P, Engebretson DM, Berezovsky J, Crowell PA. Phys Rev B 2003;67:020403.

[88] Choe S-B, Acremann Y, Scholl A, Bauer A, Doran A, Stöhr J, et al. Science 2004;304:420.

[89] Kasai S, Nakatani Y, Kobayashi K, Kohno H, Ono T. Phys Rev Lett 2006;97:107204.

[90] Shibata J, Nakatani Y, Tatara G, Kohno H, Otani Y. Phys Rev B 2006;73:020403.

[91] Yamada K, Kasai S, Nakatani Y, Kobayashi K, Kohno H, Thiaville A, et al. Nat Mater 2007;6:269.

[92] Waeyenberge BV, Puzic A, Stoll H, Chou KW, Tyliszczak T, Hertel R, et al. Nature 2006;444:461.

[93] Yamada K, Kasai S, Nakatani Y, Kobayashi K, Ono T. Appl Phys Lett 2008;93:152502.

[94] Nakano K, Chiba D, Ohshima N, Kasai S, Sato T, Nakatani Y, et al. Appl Phys Lett 2011;99:262505.

5 Theoretical Aspects of Current-Driven Magnetization Dynamics

Hiroshi Kohno[1] and Gen Tatara[2]

[1]Graduate School of Engineering Science, Osaka University, Toyonaka, Japan, [2]RIKEN Center for Emergent Matter Science (CEMS), Wako, Saitama, Japan

Contents

Nanomagnetism and Spintronics. DOI: http://dx.doi.org/10.1016/B978-0-444-63279-1.00005-8

5.1   Introduction

Magnetization dynamics driven by electric/spin current in submicron-scale ferromagnets has been a subject of active study for more than a decade in the field of
spintronics [1,2]. It includes magnetization reversal in multilayer systems, displa-
cive and resonant motions of domain walls in thin wires, and those of magnetic
vortices realized in disks. These phenomena are understood as caused by spin torques, the effect that conduction electrons exert on the magnetization through the
microscopic s-d exchange interaction.

Recent activity on this subject was triggered by the introduction of the concept
of "spin-transfer effect" due to Slonczewski [3] and Berger [4]. On the other
hand, it should be remembered that the idea of driving domain walls by electric
current was proposed long ago by Berger. He recognized that the s-d exchange
interaction should be the main origin of the driving in thin samples [5], and proposed the driving mechanisms that are called s-d exchange force (momentum-
transfer effect) [5] and s-d exchange torque (spin-transfer effect) [6–8].
Experimental demonstration was also made by him and collaborators using semimacroscopic samples [9,10].

The continuous progress during this decade of course rests on the experimental achievements. The development of fabrication and observation techniques has enabled us to study submicron-scale magnets having simple domain
structure, which are suitable for scientific studies under well-controlled conditions, notwithstanding for technological applications. On a theoretical side,
modern techniques and viewpoints in solid-state or condensed-matter physics
started to be applied, which will hopefully develop further microscopic
understanding.

In this chapter, we review some basic aspects of the theory of current-driven
magnetization dynamics. We are concerned with magnetization dynamics which
are smoothly varying in space and time compared with the microscopic electronic
scales. Magnetization is thus treated as a continuous function, $n(r, t)$, of space and
time, and the dynamics is described by the Landau-Lifshitz-Gilbert (LLG) equation. Most prototypical phenomenon is the domain-wall motion. In spite of tremendous amount of works on multilayer (pillar) systems [11], they are outside the
scope of this chapter simply because of our lack of ability.

In Section 5.2, a theoretical treatment is given for the domain-wall motion driven by electric current. We start with an elementary introduction to a theoretical description of domain-wall dynamics, by taking the (magnetic-) field-driven motion as an example. The effect of electric current will then be introduced to study the main subject of the current-driven domain wall motion. In Section 5.3, microscopic calculation of spin torques are described in a more general framework of the LLG equation. Generalization of the force to arbitrary magnetization texture, as well as its relation to transport coefficients, is also studied. In Section 5.4, two related topics, current-driven motion of a magnetic vortex and current-induced spin-wave instability, are introduced.

5.2 Dynamics of a Rigid Domain Wall

In this section, we study the dynamics of rigid domain walls by neglecting their deformation. The validity of this approximation will be discussed at the end of this section.

5.2.1 Field-Driven Domain Wall Motion

Suppose we have two nails made of iron. We know that they are ferromagnets at room temperature, but they do not attract each other. The reason is the existence of the magnetic domain structure, which is formed to lower the magnetostatic energy due to long-range dipole-dipole interaction. The typical size of domains is ~1-10 μm, and any semimacroscopic sample inevitably has a multidomain structure with vanishing total moment in zero field (if, e.g., it is prepared by cooling from above the Curie temperature in zero field). On the other hand, nanoscale magnets have simple domain structures suitable for application and basic studies. The transition region between two neighboring domains, as illustrated in Figure 5.1E, is the magnetic domain wall.

If we apply an external magnetic field, magnetic domains with lowered magnetic energy grow at the expense of other domains (Figure 5.1A–D). This magnetization

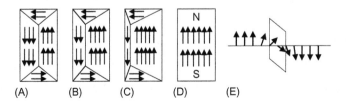

Figure 5.1 (A) Magnetic domain structure with lowest energy in zero external magnetic field. (B) → (C) → (D) illustrates the magnetization process due to domain wall motion. Conventionally, this is induced by applying magnetic field. It can also be induced by applying an electric current, and this is the main subject in this chapter. (E) An example of magnetic domain wall (Bloch wall).

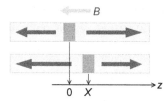

Figure 5.2 A ferromagnetic wire containing a single domain wall. The position of the domain wall is denoted by X. Thick arrows represent magnetization direction rather than spin direction. See footnote 1.

process occurs as the movement of domain walls. In this section, we shall consider how to describe this field-driven domain-wall motion [12–15]. We assume that the magnetization profile of the wall is rigid, and treat it by the method of collective coordinates.

5.2.1.1 Phenomenological Equation of Motion

Consider a ferromagnetic wire containing a single domain wall (Figure 5.2), and let the position of the domain wall be X. If we apply an external magnetic field, B, in the easy-axis direction, the energy of the whole magnet depends on X as:

$$E_B = -2\gamma_0 \hbar S B \frac{AX}{a^3} + \text{const.} \tag{5.1}$$

Here we are considering magnetic moments due to localized spins of magnitude S and gyromagnetic ratio γ_0.[1] A is the cross-sectional area of the wire, and a^3 is the volume per spin. If we regard the domain wall as a particle, whose position is X, Eq. (5.1) represents a potential energy of this particle. This means that this particle feels a force,

$$F_B = 2\gamma_0 \hbar S B \frac{A}{a^3} \tag{5.2}$$

which is proportional to B. One may then expect, naively, that the motion of this particle is described by an equation something like:

$$M_{\mathrm{w}} \ddot{X} = F_B - \eta \dot{X} \tag{5.3}$$

Here the dot represents time derivative, and η is a friction constant. This is indeed a very naive expectation, but it is known that Eq. (5.3) holds under certain conditions. In particular, a domain wall has inertial mass, M_{w}, known as the Döring mass [16]. To see this, and to derive the correct equations of motion for a domain wall, let us next look into the dynamics of individual spins in the domain wall.

---

[1] We consider the case that the spin S and the associated magnetic moment μ are opposite in direction, $\mu = -\gamma_0 \hbar S$ with $\gamma_0 > 0$, as is the case for a free electron.

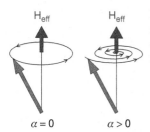

Figure 5.3 Dynamics of a single spin (or precisely, magnetic moment; see footnote 1) in an effective field $\boldsymbol{H}_{\text{eff}}$: (A) without damping and (B) with damping. As for the direction of motion, see footnote 1.

$\alpha = 0$ $\alpha > 0$

5.2.1.2   Equations of Motion from Microscopic Spin Dynamics

To see how the domain wall motion occurs at the microscopic spin level, let us first set up a microscopic model. We assume that the magnetization is carried by localized spins, S_i, located at each site i on a lattice, whose dynamics are governed by the following spin Hamiltonian,

$$H_S = -\tilde{J} \sum_{<i,j>} S_i \cdot S_j + \frac{1}{2} \sum_i (K_\perp S_{i,y}^2 - K S_{i,z}^2) \tag{5.4}$$

Here $\tilde{J}$ is the exchange coupling constant, and K and $K_\perp$ represent easy-axis and hard-axis anisotropy constants, respectively. We have chosen the easy axis to be z-axis, and the hard axis to be y-axis ($K > 0$, $K_\perp \geq 0$). In the ground state of a domain wall, the spins lie in the xz-plane, called easy plane. See Figure 5.4A.

Each spin obeys a simple law of angular-momentum dynamics. It continues precessing around an effective field $\boldsymbol{H}_{\text{eff}}$ if there is no damping (Figure 5.3A). In the presence of damping, the precession will decay and the spin (or, precisely speaking, the associated magnetic moment) eventually points to the direction of $\boldsymbol{H}_{\text{eff}}$ (Figure 5.3B). These are described by the equation[2] [12],

$$\frac{dS}{dt} = -\gamma_0 S \times \boldsymbol{H}_{\text{eff}} - \alpha \frac{S}{S} \times \frac{dS}{dt} \tag{5.5}$$

The first term on the right-hand side describes precession. The second term represents damping, called Gilbert damping, with α being a dimensionless damping constant. Roughly, α is about 0.01 for permalloy, and about 0.001 for YIG. In this subsection, we consider the case of no damping ($\alpha = 0$) for simplicity.

Back to the domain-wall solution of (5.4), suppose a magnetic field $\boldsymbol{B} = (0, 0, -B)$ is applied along the easy axis. It is represented by the additional term:

$$H_B = -\gamma_0 \hbar B \sum_i S_{i,z} \tag{5.6}$$

---

[2] In the equation for magnetization $\boldsymbol{M}$ (instead of spin S; see footnote 1), the damping (α-) term has a plus sign (instead of the minus sign in Eq. (5.5)).

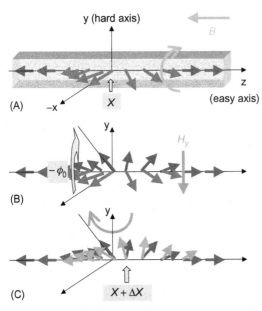

Figure 5.4 Dynamics of spins (actually, arrows shown here are magnetic moments rather than spins; see footnote 1) in the domain wall under applied magnetic field $B = (0, 0, -B)$. Spins in the domain wall rotate around the applied field (A), deviate from the easy plane (B), and rotate around the hard axis (C). As a result, the domain wall makes a translational motion.

in the Hamiltonian. Then, what happens for spins? Since the spins in the domain wall are not parallel to the field B, they will precess around B, that is, around the easy axis (see Figure 5.4B); therefore, they deviate from the easy (xz-) plane and acquire a hard-axis component, S_y. The associated energy increase is given by:

$$E_\perp = \frac{1}{2} K_\perp S_y^2 \tag{5.7}$$

per spin. Then the spins feel an effective field[3]

$$H_y = \frac{1}{\gamma_0 \hbar} \frac{\partial E_\perp}{\partial S_y} = \frac{1}{\gamma_0 \hbar} K_\perp S_y \tag{5.8}$$

in the hard-axis direction, and will precess around this field (Figure 5.4C). Namely, the spins in the domain wall rotate in the easy plane, and this is nothing but the translational motion of the domain wall. The direction of the translation is, as you see, in the correct direction, so that the magnetization occurs in the correct direction.

Let us describe these processes a little more quantitatively. First, the applied B induces a rotation around the easy axis. If we denote the rotation angle by ϕ_0, it obeys the equation:

$$\dot{\phi}_0 = -\gamma_0 B \tag{5.9}$$

---

[3] This may include the demagnetizing field from surface magnetic charges (as in nanowires) or from volume magnetic charges (as in moving Bloch walls [12,17], as well as the effective field due to magnetocrystalline anisotropy.

Now we have finite ϕ_0 (hence finite $S_y = S \sin \phi_0$) and the associated energy increases, $E_\perp = \frac{1}{2} K_\perp S^2 \sin^2 \phi_0$, due to the hard-axis anisotropy. This causes a torque:

$$T_z = -\frac{\partial E_\perp}{\partial \phi_0} = -K_\perp S^2 \sin \phi_0 \cos \phi_0 \tag{5.10}$$

in the easy-axis direction, and this torque induces a change of $S_{tot}^z = \sum_i S_i^z$, which is nothing but the domain wall motion. This process is quantified as:

$$\frac{A\dot{X}}{a^3} 2\hbar S = -N \left(\frac{1}{2} K_\perp S^2 \sin 2\phi_0 \right) \tag{5.11}$$

or

$$\frac{\dot{X}}{\lambda} = -\frac{K_\perp S}{2\hbar} \sin 2\phi_0 \tag{5.12}$$

where $\lambda = a\sqrt{\tilde{J}/K}$ is the domain-wall width (see Eq. (5.22) below), and $N = 2A\lambda/a^3$ is the number of spins in the wall. In Eq. (5.11), the left-hand side represents the change of angular momentum S_{tot}^z per unit time, and the right-hand side represents a torque acting on N spins.[4] The two Eqs. (5.9) and (5.11) (or (5.12)) describe the domain wall motion under B applied in the easy-axis direction. The former Eq. (5.9) corresponds to the x, y components of Eq. (5.5), and the latter Eq. (5.11) corresponds to the z component, with[5]

$$H_{eff} = B + \frac{1}{\gamma_0 \hbar} \frac{\partial E_\perp}{\partial S} \tag{5.13}$$

In Eq. (5.9), the magnetic field B appears on the right-hand side, which is essentially the force in the particle picture (see Eq. (5.2)). According to Newton's equation of motion, it should be equal to the time derivative of "momentum." In this sense, ϕ_0 may be regarded as a momentum of the domain wall.[6] In fact, if $|\phi_0| \ll 1$, we may linearize the sine-term of Eq. (5.12) as $\dot{X} \simeq -(K_\perp S\lambda/\hbar)\phi_0$, and see that ϕ_0

---

[4] We may also estimate the torque as $T_z = -\gamma_0 \hbar (S \times H_{eff})_z = -\gamma_0 \hbar S_x H_y = -K_\perp S_x S_y$ (from Eq. (5.8)), with $S_x = S\cos\phi_0$ and $S_y = S\sin\phi_0$.

[5] Effective fields coming from $\tilde{J}$ and K in the Hamiltonian (5.4) cancel each other, thanks to the stationary property of the domain-wall solution.

[6] If we rewrite Eq. (5.9) as $-(\hbar SN/\lambda)\dot{\phi}_0 = F_B$, with F_B given by (5.2), we see that $P = -(\hbar SN/\lambda)\phi_0$ is the corresponding momentum.

is proportional to $\dot{X}$, the domain wall velocity. From the coefficient, one may read the mass of the domain wall as:[7]

$$M_{\rm w} = \frac{\hbar^2 N}{\lambda^2} \frac{1}{K_\perp} \tag{5.14}$$

This is the inertial mass, known as the Döring mass [16].[8] This shows a remarkable property of a domain wall that it has inertia. Even if there is no force ($B = 0$) applied, there is a moving solution with constant velocity $\dot{X}$ (if there is no damping as assumed here). For $|\phi_0| \ll 1$, the energy increase associated with ϕ_0 (hence $\dot{X}$) is written as:

$$NE_\perp \simeq \frac{1}{2} M_{\rm w} \dot{X}^2 \tag{5.15}$$

and the equation of motion takes the form of Eq. (5.3).

Note that $M_{\rm w}$ is inversely proportional to $K_\perp$; if $K_\perp = 0$, the mass $M_{\rm w}$ is divergent, and the domain wall cannot move even if it feels a force. This is a consequence of the conservation of angular momentum $S_{\rm tot}^z$.[9,10] Thus, $K_\perp$ works as a source of angular momentum, and, in this sense, $K_\perp$ assists the domain-wall motion. This feature is characteristic to the field-driven (or more generally, force-driven) motion. Later, we shall see that $K_\perp$ tends to prevent the domain-wall motion if it is driven by the spin-transfer effect.

5.2.1.3 Lagrangian Formulation

Now we proceed to a little more theoretical treatment and put the analysis into Lagrangian formalism. The Lagrangian formalism is a very convenient method to focus on a few degrees of freedom (such as X and ϕ_0) out of many degrees of freedom. But let us start with a single spin.

The equation of motion of a single spin takes the form of the LLG equation (5.5). Apart from the damping term, it is derived from the Lagrangian (see footnote 1),

$$L = \hbar S \dot{\phi}(\cos\theta - 1) - \gamma_0 \hbar S \cdot B - V_{\rm ani}(\theta, \phi) \tag{5.16}$$

[7] From Eq. (5.12) and footnote 6, we have $P = (\hbar^2 N / \lambda^2 K_\perp)\dot{X}$ for "momentum," which may be identified with $M_{\rm w}\dot{X}$.

[8] For a Bloch wall, the demagnetizing field from volume magnetic charges (due to nonuniform $\phi(x,t)$) contributes to $K_\perp$ by $2\pi M_0^2$, where $M_0 = \gamma_0 \hbar S/a^3$ is the magnetization. Retaining only this contribution leads to the original expression by Döring [17].

[9] The $K_\perp$ breaks spin-rotational symmetry around the z-axis, and hence breaks the conservation of $S_{\rm tot}^z$. Note that the applied field B conserves $S_{\rm tot}^z$.

[10] Introduction of finite damping α also leads to nonconservation of $S_{\rm tot}^z$. In this case, it is possible to exchange angular momentum between X and ϕ_0, as seen from Eqs. (5.26) and (5.27). Then, even if $K_\perp = 0$, there is a solution of a moving domain wall, where the angular momentum necessary for the domain-wall translation (X) is provided by ϕ_0.

with $H_{\text{eff}} = B + (1/\gamma_0\hbar)(\partial V_{\text{ani}}/\partial S)$ in (5.5). We have parameterized the spin orientation as:

$$S(t) = S(\sin\theta\cos\phi, \sin\theta\sin\phi, \cos\theta) \qquad (5.17)$$

where $\theta = \theta(t)$ and $\phi = \phi(t)$ are dynamical variables. (We assume S is constant.) In Eq. (5.16), the first term is the kinetic term; this form is known as the spin Berry phase in quantum mechanics [18,19]; it is important also in classical mechanics in that it leads to the equation of motion characteristic to the angular momentum. (In the present study, we treat the magnetization and domain wall as classical objects.) One can check that the Euler–Lagrange equations

$$\frac{\mathrm{d}}{\mathrm{d}t}\frac{\partial L}{\partial\dot{q}} - \frac{\partial L}{\partial q} = 0 \qquad (5.18)$$

for $q = \theta$ and $q = \phi$, lead to Eq. (5.5) with $\alpha = 0$.

The Euler–Lagrange formalism does not include damping effects: it conserves energy (if L is not explicitly time dependent). The damping can be included by the Rayleigh's method [20]. With a so-called dissipation function:

$$W = \frac{\alpha}{2}\frac{\hbar}{S}\dot{S}^2 = \frac{\alpha}{2}\hbar S(\dot{\theta}^2 + \sin^2\theta\,\dot{\phi}^2) \qquad (5.19)$$

the Euler–Lagrange equation is now modified as:

$$\frac{\mathrm{d}}{\mathrm{d}t}\frac{\partial L}{\partial\dot{q}} - \frac{\partial L}{\partial q} = -\frac{\partial W}{\partial\dot{q}} \qquad (5.20)$$

One can show that $\mathrm{d}H/\mathrm{d}t = -2W$, where $H = \dot{q}(\partial L/\partial\dot{q}) - L$ is the Hamiltonian; therefore, the quantity $2W$ has a physical meaning as the energy dissipation rate.[11] Equation (5.20), with (5.16) and (5.19), then reproduces the full Eq. (5.5).

It is straightforward to extend the above description for a single spin to many-spin systems. Here we restrict ourselves to spin configurations which are slowly varying in space and time, and adopt a continuum description, $S_i(t) \to S(r,t)$, and write it as Eq. (5.17) with $\theta = \theta(r,t)$ and $\phi = \phi(r,t)$ now describing spin configurations in a continuum space r and time t. The Lagrangian is then given by:

$$L_S = \int \frac{\mathrm{d}^3x}{a^3}\left[\hbar S\dot{\phi}(\cos\theta - 1) + \gamma_0\hbar BS\cos\theta\right.$$

$$\left. -\frac{S^2}{2}\{J((\nabla\theta)^2 + \sin^2\theta(\nabla\phi)^2) + \sin^2\theta(K + K_\perp\sin^2\phi)\}\right] \qquad (5.21)$$

where $J = \tilde{J}a^2$.

---

[11] Therefore, W is independent of the choice (parametrization) of dynamical variables, and one can check that Eq. (5.20) is covariant under general variable changes.

We now restrict ourselves to a fixed magnetization profile representing a single domain wall [17,21]. The Euler-Lagrange equation derived from (5.21) has a static domain-wall solution $\theta = \theta_{dw}(x - X)$, $\phi = 0$, where

$$\cos \theta_{dw}(x) = \pm \tanh(x/\lambda) \tag{5.22}$$

or $\sin \theta_{dw}(x) = (\cosh(x/\lambda))^{-1}$ and X is a constant. (We consider an effectively one-dimensional magnet, and neglect transverse variations.) We then "elevate" the domain-wall position X to a dynamical variable and allow its time dependence, $X = X(t)$. Using this domain-wall solution in (5.21), we obtain the Lagrangian for a "rigid" domain wall as:

$$L_{dw} = \pm \frac{\hbar NS}{\lambda} \dot{X} \phi_0 - \frac{1}{2} K_\perp NS^2 \sin^2 \phi_0 \mp F_B X - V_{pin}(X) \tag{5.23}$$

up to total time derivative. Here,

$$\phi_0(t) \equiv \int \frac{dx}{2\lambda} \phi(x, t) \sin^2 \theta_{dw}(x - X(t)) \tag{5.24}$$

is essentially the angle $\phi(x, t)$ at the domain-wall center. We have introduced a pinning potential $V_{pin}(X)$ for the domain wall coming from spatial irregularities of the sample. The first term of Eq. (5.23) means that, except for a proportionality constant, X and ϕ_0 are canonical conjugates to each other; namely, ϕ_0 is the canonical momentum conjugate to X. We have noted this fact before by an intuitive argument, but now have shown it mathematically [21]. Similarly, the dissipation function becomes:

$$W_{dw} = \alpha \hbar S \frac{A}{a^3} \left(\frac{\dot{X}^2}{\lambda} + \lambda \dot{\phi}_0^2 \right) \tag{5.25}$$

Taking variations with respect to X and ϕ_0, respectively, we obtain the equations of motion for a domain wall as follows:

$$\frac{\hbar NS}{\lambda} \left(\pm \dot{\phi}_0 + \alpha \frac{\dot{X}}{\lambda} \right) = \mp F_B + F_{pin} \tag{5.26}$$

$$\frac{\hbar NS}{\lambda} (\pm \dot{X} - \alpha \lambda \dot{\phi}_0) = \frac{NS^2 K_\perp}{2} \sin 2\phi_0 \tag{5.27}$$

These equations (especially, with the lower sign[12]) are equivalent to Eqs. (5.9) and (5.11), but now include damping effects. The magnetic field acts as a force,

---

[12] If we take the upper sign in Eqs. (5.26) and (5.27), and assume that the external field is applied in the positive z-direction, $B = (0, 0, +B)$, they coincide with those of Ref. [22]. Note that Eqs. (5.26) and (5.27) (and Eqs. (5.9) and (5.11)) have been derived for $B = (0, 0, -B)$.

$F_B = \gamma_0\hbar(NS/\lambda)B$ (Eq. (5.2)), and promotes ϕ_0 (i.e., spin rotation around the easy axis), and the hard-axis anisotropy $K_\perp$ acts as a torque and drives X (i.e., translation of the domain wall). If the system is not spatially uniform, we have a pinning force $F_{pin} = -(\partial V_{pin}/\partial X)$. Note that X and ϕ_0 have opposite time-reversal properties, and their mixing (α-terms) means breaking of time-reversal symmetry and leads to damping. For $|\phi_0| \ll 1$, Eqs. (5.26) and (5.27) reduce to Eq. (5.3) with $\eta = \alpha(\hbar NS/\lambda^2)$.[13]

5.2.2 Current-Driven Domain Wall Motion

In the previous section, we have reviewed the domain-wall motion driven by a magnetic field. The contents presented there will be well-known results; in particular, the theoretical descriptions are equivalent to the LLG equation under the constraint that the magnetization profile is fixed to that of a single domain wall (hence a rigid wall). In this section, we proceed to the main topic of current-driven domain-wall motion [23−37]. We introduce the effect of electric current (due to conduction electrons) into the same formalism. We treat conduction-electron and magnetization degrees of freedom as mutually independent.[14] In the following text, "magnetization" means that of localized spins. Also, we refer to conduction electrons simply as "electrons."

5.2.2.1 Interaction of Domain Wall and Electron Flow: An Intuitive Picture

How does the electric current affect the domain-wall motion? The relevant microscopic interaction is the s-d exchange interaction:

$$H_{sd} = -J_{sd} \int d^3x \, S(r) \cdot \sigma(r) \qquad (5.28)$$

between localized spins $S(r)$ and electron spins. Here $\sigma(r) = c^\dagger(r)\sigma c(r)$ represents (twice) the spin density of conduction electrons, with $c^\dagger = (c_\uparrow^\dagger, c_\downarrow^\dagger)$ being the electron creation operator. There are certainly other possible effects, such as hydromagnetic or electromagnetic, but they are ineffective in small or thin systems [6], or can be excluded experimentally.

The s-d interaction affects the domain-wall motion in two different ways. One is the momentum transfer, or force, and the other is the spin transfer, or torque.

Consider a metallic ferromagnet containing a single domain wall, and suppose there is an electron flow from left to right (Figure 5.5). If an electron is reflected

---

[13] Precisely speaking, because of the special form of the damping (α-) terms, the mass and force are renormalized as $M_w = (\hbar^2 N/\lambda^2 K_\perp)(1 + \alpha^2)$ and $F = f + \alpha(\hbar/SK_\perp)\dot{f}$, where $f = F_B + F_{pin}$.

[14] This is not an essential assumption in the following part, but just for simplicity. No essential changes are needed in applying the following results to itinerant ferromagnets where magnetization and conduction are carried by same electrons, as described near the end of Section 5.3.2.

Figure 5.5 Two effects of electric current (electron flow) on a domain wall via the s-d exchange interaction. (A) Electron reflection means the transfer of linear momentum to the domain wall. (B) Adiabatic transmission of electrons means the transfer of spin angular momentum to the domain wall.

by the domain wall, its momentum is changed (Figure 5.5A). This process acts as a force on the domain wall by transferring the linear momentum from the electron to the domain wall. (Note that the s-d interaction (5.28) is translationally invariant and conserves the total momentum of electrons and magnetization.) This force is proportional to the charge current, j, and domain-wall resistivity, ρ_{w}, as we will show later.

On the other hand, if the electron is transmitted through the domain wall *adiabatically*, namely, by keeping its spin direction closely parallel to the local magnetization, the spin angular momentum of the electron is changed (Figure 5.5B). This process acts as a torque on the domain wall by transferring the spin angular momentum from the electron to the domain wall. (Note that the s-d interaction (5.28) also conserves the total spin angular momentum of the electrons and magnetization.) In other words, this change of the electron spin should be absorbed by the magnetization, leading to a translational motion of the domain wall. This torque can be shown to be proportional to the spin current, j_{s}, in the adiabatic limit. This is the spin-transfer effect; a conduction electron exerts a (spin-transfer) torque on the domain wall. These two effects were originally found by Berger long ago [5,6,8]. The spin-transfer effect is now familiar through recent studies on multilayer or pillar systems [3,4].

These two effects enter the equations of motion as follows (see footnote 12):

$$\frac{\hbar NS}{\lambda}\left(\dot{\phi}_0 + \alpha\frac{\dot{X}}{\lambda}\right) = F_{\mathrm{el}} + F_B + F_{\mathrm{pin}} \tag{5.29}$$

$$\frac{\hbar NS}{\lambda}(\dot{X} - \alpha\lambda\dot{\phi}_0) = \frac{NS^2 K_\perp}{2}\sin 2\phi_0 + T_{\mathrm{el},z} \tag{5.30}$$

where F_{el} is the force and $T_{\mathrm{el},z}$ is the torque, both from electrons [8,22]. One might easily convince oneself that the force F_{el} from electrons appears in the equations in exactly the same way as the other forces (F_B and F_{pin}) already considered. As for the torque $T_{\mathrm{el},z}$, we may understand it as follows.

Suppose a spin current I_{s} (in unit of $\hbar/2$, the dimension being, e.g., Ampere) is flowing in the left far region to the domain wall. In the adiabatic case, the spin current flowing in the right far region to the domain wall is given by $-I_{\mathrm{s}}$. Since the change of angular momentum of a single electron is $\hbar$ after a passage through the domain wall, the total angular momentum of electrons is changing at a rate $\hbar I_s/(-e)$, which is transferred to the domain wall, and acts as a torque:

$$T_{el,z} = -\frac{\hbar}{e} I_s \tag{5.31}$$

on the domain wall. This torque is now added to the right-hand side of Eq. (5.11), leading to Eq. (5.30) above.

To derive the expression of the force, consider a situation that the electrons are accelerated by an applied electric field E, and a steady-current state is maintained by reflections from the domain wall. We assume that only the domain wall scatters electrons. Then, the electrons acquire momentum $-eN_{el}E$ per unit time (N_{el} is the total number of electrons and $e > 0$ is the elementary charge), which is nothing but the force, and the whole of this momentum is released to the domain wall under the stationarity condition assumed here. Thus, the domain wall feels a force $F_{el} = -eN_{el}E$ from the electrons. The current density j is related to E as:

$$E = \rho_w j \tag{5.32}$$

via the resistivity ρ_w due to a single domain wall. By eliminating E from these two relations, we obtain:

$$F_{el} = -eN_e \rho_w j \tag{5.33}$$

For thick walls as realized in metallic wires, the reflection probability (or domain-wall resistivity) will be vanishingly small, and the spin-transfer effect will be the dominant driving mechanism. The momentum-transfer effect is considered to be effective only for thin walls, as in nanocontacts and possibly in magnetic semiconductors. However, as described in Section 5.2.2.3, it should also be noted that there is another contribution to the force from spin-relaxation processes, which can be effective even for thick walls.

5.2.2.2 Mathematical Derivation

Mathematically, what is new compared to the field-driven case is the s-d exchange term H_{sd}. Supplementing the electron part by, e.g., that of a free-electron system:

$$L_{el}^0 = \sum_k c_k^\dagger (i\hbar\partial_t - \varepsilon_k)c_k \tag{5.34}$$

with $\varepsilon_k = \hbar^2 k^2/2m$, we consider the total Lagrangian $L_{tot} = L_{dw} + L_{el}^0 - H_{sd}$, and derive the equations of motion by taking variations with respect to X and ϕ_0.

The s-d coupling introduces two new terms into the equations of motion [22]. One comes from the X derivative, which is force:

$$F_{el} \equiv -\left\langle \frac{\partial H_{sd}}{\partial X} \right\rangle = -J_{sd} \int d^3x \, \nabla_x S_{dw}(x - X) \cdot \langle \sigma(x) \rangle_{ne} \tag{5.35}$$

where $S_{dw}(x)$ is the domain-wall spin configuration, and $\langle \sigma(x) \rangle_{ne}$ represents spin density of electrons under the current and/or a moving domain wall. The other one comes from the ϕ_0 derivative, which is torque (force on $-\phi_0$):

$$T_{el,z} \equiv +\left\langle \frac{\partial H_{sd}}{\partial \phi_0} \right\rangle = -J_{sd} \int d^3x \, [S_{dw}(x-X) \times \langle \sigma(x) \rangle_{ne}]_z \tag{5.36}$$

As seen above, the torque $T_{el,z}$ is proportional to the spin current flowing in the bulk (far from the domain wall) in the adiabatic case.

The force comes from the gradient of magnetization; electrons are scattered by this spatial nonuniformity, and there occurs an exchange of linear momentum between the electrons and the domain wall. The torque comes from the mismatch in direction between the domain-wall magnetization and electron spin polarization; they precess around each other and exchange spin angular momentum with each other.

Force and torque both vanish if there is no current (and if the domain wall is at rest). They can be finite in the presence of current (or the domain wall is in motion). In this sense, the calculation of force and torque resembles that of transport coefficients, and we can use the techniques developed for transport coefficients, such as those named after the great physicists, Boltzmann, Kubo, Landauer, Keldysh, and others. We will outline some calculations based on the Kubo formula in the next section.

The equations of motion are obtained as Eqs. (5.29) and (5.30) above. These equations are directly obtained from the effective Lagrangian:

$$L_{eff} = -\frac{\hbar NS}{\lambda} X \dot{\phi}_0 - \frac{1}{2} K_\perp NS^2 \sin^2 \phi_0 + (F_B + F_{el})X - T_{el,z}\phi_0 - V_{pin}(X) \tag{5.37}$$

together with the dissipation function (5.25). Among others, there is a direct linear coupling between ϕ_0 and spin current (or spin-transfer torque, $T_{el,z}$). As noted in Section 5.2.2.3, the spin S in the first term of Eq. (5.37) gets renormalized to $S_{tot} = S + \delta S$ with δS being the contribution from conduction electrons. (See also Section 5.3.1.)

5.2.2.3 Supplementary Remarks

In deriving the expressions, (5.31) and (5.33), for the torque and the force, we have considered an idealized situation that there are no spin- and momentum-relaxation processes in the electron system. This is certainly not the case in real systems.

The effect of spin- and momentum-relaxation processes will be examined in the next section in a more general framework of the LLG equation. According to that, the spin-transfer torque (5.31) is not modified by the normal impurities (momentum-relaxation process) and weak magnetic impurities (spin-relaxation process) (Eq. (5.88)). On the other hand, we will see that spin-relaxation process will induce a new type of current-induced torque, called β-term (see Eq. (5.90) below), which acts as a force:

$$F_\beta = -\beta \frac{\hbar}{2e} j_s \frac{Na^3}{\lambda^2} = \beta \frac{\hbar S N}{\lambda^2} v_s \qquad (5.38)$$

on a rigid wall. Here $j_s = I_s/A$ is the spin-current density, $v_s = -(a^3/2eS) j_s$ is the corresponding drift velocity, and β is a dimensionless parameter characterizing the degree of spin relaxation. The current-induced force thus consists of two contributions, $F_{el} = F_\beta + F_{na}$, where we have written the right-hand side of Eq. (5.33) as F_{na} (the subscript "na" means nonadiabatic). The spin-relaxation process also gives an additional contribution to the Gilbert damping. These issues will be elucidated systematically in Section 5.3.

Actually, even in the absence of spin- and momentum-relaxation processes, the coupling to electrons modifies Eqs. (5.29) and (5.30) in one more way. Namely, it modifies the magnitude of spin from that of the localized spin, S, to the total one, $S_{tot} = S + \delta S$, where δS is the contribution from conduction electrons. This applies to the coefficients of $\dot\phi_0$ and $\dot X$ on the left-hand side of Eqs. (5.29) and (5.30). Let us see their physical implications.

In the above argument that led to Eq. (5.31), we should have been a little more careful about the fact that, when the domain wall is moving, the spin current flowing into the domain wall is modified. To see this, let us write the spin-current density as $-e\sum_\sigma \sigma n_\sigma v_\sigma = -e(n_\uparrow v_\uparrow - n_\downarrow v_\downarrow)$, where n_σ and v_σ are the density and drift velocity of spin-σ electrons, respectively. When the domain wall is moving, v_σ should be measured relative to the domain-wall velocity, $\dot X$; hence, the effective spin-current density seen by the domain wall is given by $-e\sum_\sigma \sigma n_\sigma(v_\sigma - \dot X)$, and the above torque (5.31) is modified as follows:

$$T_{el,z} = -\frac{\hbar}{e} I_s - \frac{\hbar N}{\lambda} \left(\frac{1}{2} a^3 \rho_s \right) \dot X \qquad (5.39)$$

where $\rho_s = \sum_\sigma \sigma n_\sigma = n_\uparrow - n_\downarrow$ is the electron spin density. If the second term of (5.39) is transposed to the left-hand side of Eq. (5.30), it amounts to renormalize the spin S in the coefficient of $\dot X$ to:

$$S_{tot} = S + \frac{1}{2} a^3 \rho_s \qquad (5.40)$$

This effect is expressed as the renormalization $S \rightarrow S_{tot}$ in the first term of the effective Lagrangian (5.37). Therefore, the factor S in front of $\dot\phi_0$ of Eq. (5.29) should also be modified to S_{tot}. It is thus concluded that a similar term should appear also in the force, Eq. (5.33), as:

$$F_{el} = -eN_e\rho_w j - \frac{\hbar N}{\lambda}\left(\frac{1}{2}a^3\rho_s\right)\dot{\phi}_0 \tag{5.41}$$

The second term can be obtained if we replace Eq. (5.32) by:

$$E = \rho_w j + \frac{V'}{L} \tag{5.42}$$

where L is the length of the wire, and

$$V' = \frac{\hbar}{e}\frac{\rho_s}{n}\dot{\phi}_0 \tag{5.43}$$

with $n = n_\uparrow + n_\downarrow$ being the electron density. Equation (5.42) indicates that the domain-wall motion (with $\dot{\phi}_0 \neq 0$) is accompanied by a voltage generation, or the electromotive force, V'. This voltage V' may be considered as originating from the more fundamental "spin motive force" or "spin voltage":

$$V_s = \frac{\hbar}{e}\dot{\phi}_0 \tag{5.44}$$

associated with the domain-wall motion. The two quantities are mutually related via $V' = \langle\sigma\rangle V_s$ with $\langle\sigma\rangle = (n_\uparrow - n_\downarrow)/(n_\uparrow + n_\downarrow) = \rho_s/n$ being the s-electron spin polarization [38].

Berger predicted the existence of motive force $V = (\hbar/e)\dot{\phi}_0$ from the analogy to the Josephson effect in superconductors [7]. He regarded his motive force to be in the charge channel, but now we recognize that it is a spin motive force, Eq. (5.44), acting in the spin channel. Volovik provided a convenient mathematical tool (SU(2) gauge field) to describe spin motive force for general magnetization texture/dynamics [39]. A physical picture of the spin motive force was given by Stern [40] in terms of time-dependent Berry's phase. In the recent context of spintronics, spin motive force was rediscovered by Barnes et al. [38,41], which stimulated further theoretical studies [42−48]. Conceptually, the same phenomena ("spin pumping" and "spin battery") had been proposed in ferromagnet/normal metal junction systems [49].

Experimental detection of spin motive forces has been reported subsequently in domain wall [50,51], magnetic vortex [52], and comb-shaped systems [53]. Similar, but somewhat mysterious, phenomenon has been observed in magnetic tunnel junctions containing magnetic nanoparticles [54].

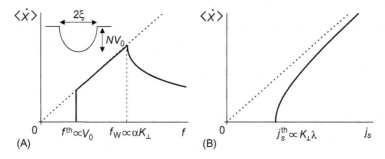

Figure 5.6 Time-averaged domain wall velocity, $\langle \dot{X} \rangle$. (A) *Force-driven case.* As a function of $f = (\lambda/\hbar NS)F$, where $F = F_{el} + F_B$ is the driving force. Inset illustrates model pinning potential. (B) *Current-driven case.* As a function of spin-current density j_s. The dotted line represents $\langle \dot{X} \rangle = v_s$ corresponding to the case of complete spin transfer ($K_\perp = 0$).

5.2.3 Dynamics

In this subsection, we study the dynamics of a domain wall on the basis of Eqs. (5.29) and (5.30). We introduce a pinning potential (see the inset of Figure 5.6A):

$$V_{\text{pin}} = \frac{NV_0}{\xi^2}(X^2 - \xi^2)\Theta(\xi - |X|) \tag{5.45}$$

Here $\Theta(x)$ is the Heaviside step function, V_0 is the pinning strength, and ξ is the pinning range. We rewrite Eqs. (5.29) and (5.30) in the form:

$$\dot{\phi}_0 + \alpha\frac{\dot{X}}{\lambda} = f - \nu_{\text{pin}}\frac{X}{\lambda}\Theta(\xi - |X|) \tag{5.46}$$

$$\frac{\dot{X}}{\lambda} - \alpha\dot{\phi}_0 = \kappa_\perp \sin 2\phi_0 + \nu_s \tag{5.47}$$

where all parameters, except for α, have dimensions of frequency: $\nu_{\text{pin}} = 2V_0/\hbar S$, $\kappa_\perp = SK_\perp/2\hbar$, $\nu_s = T_{el,z}/(\hbar NS) = -I_s/(eNS)$, and $f = (\lambda/\hbar NS)(F_{el} + F_B)$. Explicitly, we have:

$$f = \beta'\frac{v}{\lambda} + \gamma_0 B \tag{5.48}$$

with

$$\beta' = \beta\frac{P}{2S} + \beta_{\text{na}} \tag{5.49}$$

$$\beta_{\text{na}} = \frac{e^2 n\lambda}{2\hbar S}R_w A \tag{5.50}$$

where $P = j_s/j$ is the degree of spin polarization of the current, and we defined $v \equiv -(a^3/e)j$. Note that the effects of (easy-axis) magnetic field, momentum transfer ($\sim \beta_{na}$), and the β-term enter the equation through f defined above.

5.2.3.1 Dynamics Driven by Force

First, we study the domain-wall dynamics driven by force by setting $\nu_s = 0$.

In the absence of a pinning potential ($\nu_{pin} = 0$), Eqs. (5.46) and (5.47) have a stationary solution describing a moving domain wall with constant velocity $\dot{X} = \lambda f/\alpha$ and constant ϕ_0 (i.e., $\dot{\phi}_0 = 0$). The value of ϕ_0 is determined by Eq. (5.27). This stationary solution exists only if $|\dot{X}| < \kappa_\perp \lambda$, namely, $|f| < \alpha \kappa_\perp \equiv f_W$. For $|f| > f_W$, ϕ_0 also becomes time dependent, and the wall motion acquires oscillatory components. This is known as the Walker breakdown.

In the presence of a pinning potential, the domain wall is pinned below the threshold $f^{th} = \nu_{pin}\xi/\lambda$. This corresponds to the threshold field, $\gamma_0 B^{th} = 2V_0\xi/(\hbar S \lambda)$, or the threshold (charge) current,

$$j^{th} = \frac{e}{a^3}\frac{2V_0}{\hbar S}\frac{\xi}{\beta'}, \quad v^{th} = \frac{2V_0}{\hbar S}\frac{\xi}{\beta'} \tag{5.51}$$

The domain-wall velocity as a function of force (B or j) is shown in Figure 5.6A.

5.2.3.2 Dynamics Driven by Spin Transfer

Next, we focus on the spin-transfer effect by setting $f = 0$. We consider only the case of DC current, and assume ν_s is time independent. (This also includes the case of a pulsed current.)

We first consider the case without pinning potential, $\nu_{pin} = 0$. In Eq. (5.47), the spin-transfer torque enters as a source to the domain-wall velocity $\dot{X}$; it tries to drive $\kappa \rightarrow \dot{X}$ directly. However, there is also a hard-axis anisotropy ($\kappa_\perp$) term, which tends to absorb the transferred angular momentum. In fact, if $|\nu_s|$ is smaller than $\kappa_\perp$, spin transfer ν_s is completely absorbed by the $\kappa_\perp$-term, i.e., transferred to the lattice, and is not used for the translational motion ($\dot{X}$) of the domain wall: domain wall is apparently pinned and not driven to a stream motion even in the absence of any extrinsic pinning potential [22]. This is called intrinsic pinning.

The time dependence of X and ϕ_0 are shown in Figure 5.7A. The domain wall approaches a static state with finite displacements of X and ϕ_0. If $|\nu_s|$ exceeds $\kappa_\perp$, the domain wall moves with constant average velocity $\langle \dot{X} \rangle$ with oscillating components superposed (Figure 5.7B). This oscillation is due to the alternating exchange of angular momentum between X and ϕ_0 degrees of freedom. (ϕ_0 also varies with some constant average angular velocity $\langle \dot{\phi}_0 \rangle$, which is smaller than $\langle \dot{X} \rangle$ by a factor of α.) Therefore, for the domain-wall motion driven by the spin-transfer effect, there is a finite threshold spin current,

$$j_s^{th(1)} = \frac{eS^2}{a^3\hbar}K_\perp\lambda, \quad v_s^{th(1)} = \frac{S}{2\hbar}K_\perp\lambda = \kappa_\perp\lambda \tag{5.52}$$

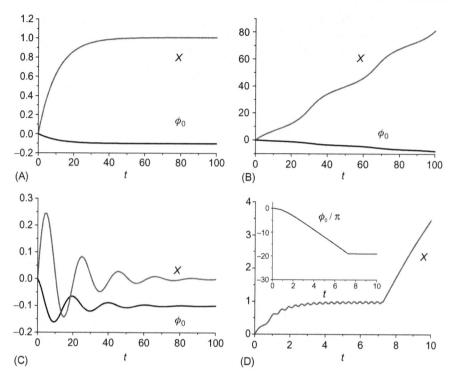

Figure 5.7 Time dependence of X/λ and ϕ_0 under spin current ν_s, without pinning potential $\nu_{pin} = 0$ (a,b) and with pinning potential $\nu_{pin} > 0$ (c,d), for $\kappa_\perp = 0.5$, $\xi = \lambda$, and $\alpha = 0.1$. (A) Below the threshold $j_s^{th(1)}$ ($\nu_s < \kappa_\perp$; $\nu_s = 0.1$). (B) Above the threshold $j_s^{th(1)}$ ($\nu_s > \kappa_\perp$; $\nu_s = 1$). (C) Below $j_s^{th(1)}$, pinned regime ($\nu_s < \kappa_\perp$; $\nu_s = 0.1, \nu_{pin} = 0.1$). (D) Above $j_s^{th(1)}$, and close to but slightly above $j_s^{th(2)}$ ($\nu_s > \kappa_\perp$; $\nu_s = 0.977, \nu_{pin} = 10$), the depinning is seen. For a more realistic value of $\alpha = 0.01$, characteristic features of the domain-wall motion are modified as follows: (A and B) Time scale for both X and ϕ_0 becomes longer in proportion to α^{-1}. Magnitude of X is enhanced ($\propto \alpha^{-1}$) according to $\dot{X}/\lambda = -\alpha^{-1}\dot{\phi}_0$. (C) Decay time becomes long ($\sim \alpha^{-1}$) whereas the oscillation period ($\Delta t \sim 20$) is unchanged; therefore, X and ϕ_0 oscillate many times ($\sim 30-40$) before the decay. (D) The oscillation amplitude of X in the pinned state is much reduced.

essentially determined by $K_\perp$. This threshold is finite even if there is no pinning potential. This is the intrinsic pinning due to $K_\perp$. Here, $K_\perp$ tends to prevent the domain-wall motion by absorbing the transferred spin angular momentum, in contrast to the force-driven case, where, as we have seen before, $K_\perp$ helps the domain-wall motion by supplying angular momentum. Above the threshold, the time-averaged domain-wall velocity is given by (Figure 5.6b):

$$\langle \dot{X} \rangle = \frac{1}{1+\alpha^2} \sqrt{\nu_s^2 - (\nu_s^{th(1)})^2} \qquad (5.53)$$

We next consider the case with extrinsic pinning potential, $\nu_{\text{pin}} > 0$. For $|\nu_s| < \kappa_\perp$, after the initial transient period with oscillatory behavior, the domain wall is eventually pulled back to the pinning center ($X = 0$) and ϕ_0 approaches a constant value (Figure 5.7C). For $|\nu_s| > \kappa_\perp$, the domain wall makes a finite displacement ΔX and oscillates around the mean position (Figure 5.7D). This is the (extrinsically) pinned state. In this pinned state, transferred spin is solely used to drive ϕ_0, and ϕ_0 continues to vary rapidly. Here, AC noise or electromagnetic radiation may be expected due to this ϕ_0 motion. If X happens to "step over" the pinning range, the domain wall is depinned. After depinning, the ϕ_0 motion slows down, and the transferred spin is mainly used for the translational motion of the domain wall.

Let us examine the depinning condition for the latter case (Figure 5.7D). In the pinned state ($\dot{X} \sim 0$), the average displacement may be estimated from Eq. (5.46) as $\Delta X \sim -\lambda\dot{\phi}_0/\nu_{\text{pin}}$, whereas $\dot{\phi}_0 \sim -\nu_s/\alpha$ from Eq. (5.47). The condition for depinning, $|\Delta X| > \xi$, thus leads to the second threshold:

$$
j_s^{\text{th}(2)} = \frac{4e}{a^3\hbar}\alpha V_0\xi, \quad \nu_s^{\text{th}(2)} = \frac{2}{\hbar S}\alpha V_0\xi = \alpha\nu_{\text{pin}}\xi \tag{5.54}
$$

for the spin-current density j_s.

The actual threshold spin-current density j_s^{th} is given by the larger of $j_s^{\text{th}(1)}$ and $j_s^{\text{th}(2)}$. Since $j_s^{\text{th}(2)}$ contains a factor of α, which is usually very small (~ 0.01), j_s^{th} will be determined by the hard-axis anisotropy if the pinning potential is not extremely strong [22]. For this case, the time-averaged domain-wall velocity $\langle \dot{X} \rangle$ is plotted as a function of spin current j_s in Figure 5.6b.

5.2.3.3 Dynamics Driven by both Spin Transfer and Force

We have seen that, even in the absence of extrinsic pinning ($V_0 = 0$), there exists a threshold current density for the domain-wall motion if it is driven solely by the spin-transfer effect. This threshold is due to the hard-axis anisotropy, hence is of intrinsic origin. This intrinsic pinning is expected to be removed by any small amount of force (momentum-transfer effect, or the β-term), but then the extrinsic pinning will play essential roles [55−57].

Threshold current for depinning as a function of pinning strength is shown in Figure 5.8A. It reveals several depinning mechanisms depending on the following parameter regimes:

(I) Weak pinning regime ($\tilde{\Omega} < O(1)$): $\tilde{j}_c$ grows with $\tilde{\Omega}$.
(II) Intermediate regime ($O(1) < \tilde{\Omega} < O(\alpha^{-1/2})$): $\tilde{j}_c \simeq$ constant $\simeq 0.7 - 0.8$.
(III) Strong pinning regime ($\tilde{\Omega} > O(\alpha^{-1/2})$): $\tilde{j}_c \propto \tilde{\Omega}^2 \propto V_0$.

As seen from Figure 5.8A, the β' affects $\tilde{j}_c$ only in regime I, which can be further divided into two subregimes:

(I-a) Small $\beta'/\tilde{\Omega}$: $\tilde{j}_c \propto \tilde{\Omega} \propto \sqrt{V_0}$
(I-b) Large $\beta'/\tilde{\Omega}$: $\tilde{j}_c \propto \tilde{\Omega}^2/\beta' \propto V_0/\beta'$.

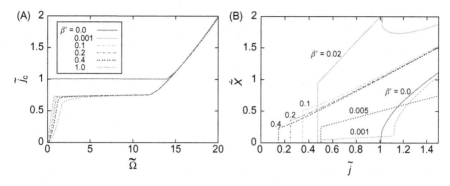

Figure 5.8 (A) Threshold current as a function of pinning frequency. Both quantities are given in dimensionless forms; $\tilde{j} = (a^3/ev_c)j$ for current density j, and $\tilde{\Omega} = \Omega\lambda/v_c$ for pinning frequency Ω defined by $NV_0 = 1/2M_w\Omega^2$. (B) Time-averaged domain-wall velocity as a function of (dimensionless) current density, $\tilde{j}$, for $\tilde{\Omega} = 0.5$ and $\alpha = 0.01$. A jump in wall velocity is seen at the threshold $\tilde{j} = \tilde{j}_c$ (except for $\beta' = 0$). Crossover from regime I-a ($\tilde{j}_c \sim \tilde{\Omega}/\beta'$) to regime I-b ($\tilde{j}_c \sim \tilde{\Omega}$) is seen near $\beta' \simeq \tilde{\Omega}/4 \sim 0.1$.
Source: After Ref. [57].

In regime I, ϕ_0 does not grow because of low current, and the dynamics is described by X. In regime I-a, β' is negligible and the depinning is due to spin transfer. Namely, spin-transfer term gives a finite velocity, $\dot{X} = v_s\lambda = v_s$ (see Eq. (5.47)), to the wall, and if the corresponding kinetic energy $\frac{1}{2}M_w v_s^2$ exceeds NV_0, the domain wall will be depinned. This determines the third threshold:

$$j_s^{\text{th}(3)} = \frac{2eS\lambda}{a^3\hbar}\sqrt{2K_\perp V_0}, \quad v_s^{\text{th}(3)} = \frac{\lambda}{\hbar}\sqrt{2K_\perp V_0} \tag{5.55}$$

In regime I-b, the depinning is governed by the force, and the corresponding threshold current is given by Eq. (5.51).

Dynamics in regimes II and III are determined by ϕ_0. In regime II, the depinning is due to spin transfer, but the terminal velocity is determined by β'. The depinning mechanism here is the same as that discussed in Eq. (5.52), but the lowering of j_c by a factor of $0.7-0.8$ is due to the β' term eliminating the intrinsic pinning. This is the reason why j_c looks different for $\beta' = 0$ and $\beta' \neq 0$ in Figure 5.8A. The regime III corresponds to the strong pinning regime already studied; the threshold is given by Eq. (5.54). In both regimes II and III, the depinning is due to the spin-transfer effect, and β' does not affect the threshold current.

The domain-wall velocity after depinning is shown in Figure 5.8B. It is noted that the wall velocity at the extrinsic threshold is discontinuous. This is because the wall as soon as depinned feels a tilted potential due to β' and has finite velocity. The velocity jump is estimated as:

$$\Delta v^{Ia} = \frac{\beta'}{\alpha} \frac{2S}{P} \frac{\sqrt{2K_\perp V_0}}{\hbar} \lambda \tag{5.56}$$

$$\Delta v^{Ib} = \frac{1}{\alpha} \frac{SV_0}{\hbar} \frac{\lambda^2}{\xi} \tag{5.57}$$

for regime I-a and regime I-b, respectively.

The wall velocity above the threshold is given by:

$$\langle \dot{X} \rangle = \frac{\beta'}{\alpha} v \; (v < v_a) \tag{5.58}$$

$$\langle \dot{X} \rangle = \frac{\beta'}{\alpha} v - \frac{v}{1 + \alpha^2} \sqrt{1 - \left(\frac{v_a}{v}\right)^2} \; (v > v_a) \tag{5.59}$$

where $v = -(a^3/e)j$ (as before) and $v_a = v_s^{\mathrm{th}(1)}/|(P/2S) - (\beta'/\alpha)|$. The v_a corresponds to the Walker breakdown above which ϕ_0 increases with time continuously.

5.2.3.4 Discussion

The present description in terms of only X and ϕ_0 out of many degrees of freedom is called as a "collective-coordinate" description or a "one-dimensional model." This will be a good approximation if the energy scale of the domain-wall motion ($\sim K_\perp, V_0$) is much smaller than the lowest energy of any other modes. The latter may be estimated as the spin-wave gap $\sim S\sqrt{K(K + K_\perp)}$ (see Section 5.4.2). Thus, a necessary condition is $K_\perp \ll K$.

The present analysis assumes a wall structure which is uniform in the width direction, and applies to thin wires (one-dimensional limit). For wider wires, the wall contains a vortex as the most stable structure (called vortex wall [56]), and actually most experiments with permalloy nanowire have been done for such vortex walls [27,31−37]. While the present analysis loses justice for such cases, proper interpretation of the vortex motion in terms of the ϕ_0-motion reveals a surprising usefulness of the present "one-dimensional model," as pointed out in micromagnetic simulations [58] and experiments [33−37].

5.2.3.5 Further References

The analysis presented above applies to absolute zero temperature. Effects of finite temperature (thermally assisted process) are studied in Refs. [59,60]. Analytical treatment of vortex walls is attempted in Ref. [61]. As for spin torques, effects of nonadiabaticity on the spin-transfer torque is studied in Ref. [62]. Conduction electrons are numerically treated with full account of nonadiabaticity in a mesoscopic system in Ref. [63].

5.2.3.6 Experiments

Experimentally, it is important to identify the driving mechanism and the origin of threshold current (pinning/depinning mechanism) in each system. This subject includes the determination of α and β values as material-dependent parameters.

The most commonly used material is the permalloy ($Fe_{19}Ni_{81}$), in which magnetic anisotropy is solely due to shape anisotropy (dipole—dipole interaction). In most cases, the samples are wires patterned from films, and the hard-axis anisotropy is determined by the shape of the cross section of the wire. It is roughly $K_{\perp} = 0.1 \sim 1$ K, and the corresponding threshold current density (due to intrinsic pinning) is estimated as $j_{th} = 10^{13} \sim 10^{14}$ A/m^2. In experiments, the threshold is about 10^{12} A/m^2 and depends on the extrinsic pinning V_0 ($\propto B_{th}$) [64]. These indicate that it is in regime I-a or I-b (weak and extrinsic pinning). Since the domain-wall velocity 110 m/s at $j \sim 1.5 \times 10^{12}$ A/m^2 (in zero magnetic field) is faster than the corresponding spin-transfer velocity, v_s (~ 75 m/s for $P = 0.7$), the IBM group concluded that $\beta/\alpha > 1$[34,35]. From these (and other) results, they deduced the value, $\beta/\alpha = 2 \sim 3$ [64]. Values of $\beta = 0.040 \pm 0.005$ and $P = 0.41 \pm 0.02$ (assuming $\alpha = 0.01$) have been obtained by a different group [65].

Systems of recent particular interest are perpendicularly magnetized multilayer films. These systems have relatively small hard-axis magnetic anisotropy (because of partial cancellation between shape anisotropy and anisotropy arising from interface) and thin domain wall width.

In Ref. [66], the authors study $(Pt/Co)_3/Pt$ multilayers with perpendicular anisotropy. They measured the depinning field under the applied electric current, and concluded that there is a force component in the effects of current. They obtained quite large values, $\beta = 1.45$ ($T = 250$ K) and $\beta = 0.35$ ($T = 300$ K). Since these values are consistent with the resistivity, $\rho_w \sim 10^{-10}$ Ωm, due to the domain wall, they suggest that β in this system comes from electron reflection (not spin-relaxation) process. The domain-wall width is estimated to be 6.3 nm.

In contrast, in Co/Ni multilayers with perpendicular magnetization, Koyama et al. [67] have proved that the spin-transfer mechanism is at work. They observed that the threshold current takes a minimum when the hard-axis magnetic anisotropy (controlled by the wire width) is minimum. They also observed that the threshold current is insensitive to the extrinsic pinning (controlled by external magnetic field). Moreover, the value of the threshold current ($\sim 7.5 \times 10^{11}$ A/m^2) agrees well with the theoretical estimate. These facts clearly indicate that this Co/Ni system is located in regime II, and the driving mechanism is the spin-transfer effect. (Note that this does not necessarily mean $\beta = 0$.)

As for magnetic semiconductors (Ga,Mn)As, also having perpendicular anisotropy, the spin-transfer mechanism seems to well fit the experimental results above the threshold [30], as well as the creep region below the threshold [68], indicating

$\beta = 0$. There is also a recent report [69] demonstrating $\beta \sim 0.25$ ($0.17 < \beta < 0.36$, hence $0.86 < \beta/\alpha < 1.20$). Further investigation will be desired to settle (or reconcile) this discrepancy.

5.3   Microscopic Calculation of Spin Torques

In this section, we describe some basic aspects of the microscopic calculation of spin torques. We focus on magnetization dynamics which is slowly varying in space and time, as described by the LLG equation. Here "slow" means slow compared to the electronic scales, so it is satisfied quite well in most cases for metals and degenerate semiconductors. Throughout this section, we consider general spin texture (magnetic configurations).

5.3.1   General

The LLG equation is given by:

$$\frac{dM}{dt} = \gamma_0 H_{\text{eff}} \times M + \frac{\alpha_0}{M} M \times \frac{dM}{dt} + T_{\text{el}} \tag{5.60}$$

in terms of a magnetization vector M. The first term on the right-hand side represents precessional torque around the effective field H_{eff}, with γ_0 being the gyromagnetic ratio. The effective field includes the ferromagnetic exchange (gradient energy) field, the magnetocrystalline anisotropy field, and the demagnetizing field. The second term represents the Gilbert damping, coming from processes which do not involve conduction electrons, and is thus present even in insulating ferromagnets. The effects of conduction electrons are contained in the third term, T_{el}, called spin torque in particular. This term comes from the s-d exchange coupling H_{sd} to conduction electrons, and is given by:

$$T_{\text{el}} = -M n(r) \times \langle \sigma(r) \rangle_{\text{ne}} \tag{5.61}$$

For notational convenience, we introduce a unit vector, n, whose direction is in the d-spin direction, hence is opposite to magnetization direction,

$$M = -\gamma_0 \frac{\hbar S}{a^3} n \tag{5.62}$$

Note that the magnetization is, by definition, the magnetic moment per unit volume; hence, $|M| = \gamma_0 \hbar S/a^3$. In terms of n, the LLG equation is written as:

$$\dot{n} = \gamma_0 H_{\text{eff}} \times n + \alpha_0 \dot{n} \times n + t'_{\text{el}} \tag{5.63}$$

where we have put $T_{\text{el}}(M) = -(\hbar S/a^3) t'_{\text{el}}(n)$. The dot represents time derivative.

For long-wavelength, low-frequency dynamics, it may be sufficient to consider spin torques which are first order in space/time derivative. Let us call such torques as *adiabatic torques*. In the presence of rotational symmetry in spin space, they are expressed as:

$$\tau_{ad}^{0'} = -(v_s^0 \cdot \nabla)\,n - \beta_{sr}n \times (v_s^0 \cdot \nabla)\,n - \alpha_{sr}(n \times \dot{n}) - \frac{\delta S}{S}\dot{n} \tag{5.64}$$

The first term on the right-hand side is the celebrated spin-transfer torque [70−74], where,

$$v_s^0 = -\frac{a^3}{2eS}j_s \tag{5.65}$$

is the (unrenormalized) "spin-transfer velocity," with j_s being the spin-current density. The second term, sometimes called "β-term," comes from spin-relaxation processes of electrons [55,56,75−79]. Here β_{sr} is a dimensionless constant. The third term is the Gilbert damping, which also results from spin relaxation of electrons. The fourth term contributes as a "renormalization" of spin, as seen below.

If the magnetization varies rapidly, we have in addition a *nonadiabatic torque*, τ_{na}^0, which is oscillatory and nonlocal in space (see Section 5.3.3). The total torque may thus be given by the sum of the two as,

$$t'_{el} = \tau_{ad}^{0'} + \tau_{na}^{0'} \tag{5.66}$$

The LLG equation (5.63) is then written as:

$$\left(1 + \frac{\delta S}{S}\right)\dot{n} = \gamma_0 H_{eff} \times n - (v_s^0 \cdot \nabla)n - \beta_{sr}n \times (v_s^0 \cdot \nabla)n - (\alpha_0 + \alpha_{sr})n \times \dot{n} + \tau_{na}^{0'} \tag{5.67}$$

Here we have transposed the "spin renormalization" term to the left-hand side. We define the total ("renormalized") spin as:

$$S_{tot} = S + \delta S \tag{5.68}$$

with δS being the contribution from conduction electrons, and divide both sides of Eq. (5.67) by S_{tot}/S. Then we arrive at:

$$\dot{n} = \gamma H_{eff} \times n - (v_s \cdot \nabla)\,n - \beta n \times (v_s \cdot \nabla)\,n - \alpha(n \times \dot{n}) + \tau'_{na} \tag{5.69}$$

where $\gamma = (S/S_{tot})\gamma_0$, $\alpha = (S/S_{tot})(\alpha_0 + \alpha_{sr})$, $\beta = \beta_{sr}$, $\tau'_{na} = (S/S_{tot})\tau_{na}^{0'}$, and

$$v_s = \frac{S}{S_{tot}}v_s^0 = -\frac{a^3}{2eS_{tot}}j_s \tag{5.70}$$

is the "renormalized" spin-transfer velocity. Thus, within the LLG equation of Eq. (5.69), the current-driven dynamics is specified by the parameters, v_s, α, and β, and a functional τ'_{na} of n.[15]

In the parameter space of the LLG equation, the manifold of $\alpha = \beta$ (with $\tau'_{na} = 0$) provides a very special case for the dynamics, and there has been a controversy whether the relation $\alpha = \beta$ holds generally or not. If $\alpha = \beta$, the following peculiar dynamics are expected. (i) Any static solution, $n(r)$, in the absence of spin current is used to construct a solution, $n(r - v_s t)$, in the presence of spin current v_s. Namely, the effect of spin current is just a rigid boost for any magnetic texture. (ii) Gilbert damping α does not affect the current-driven motion. (iii) Current-induced spin-wave instability does not occur (see Section 5.4.2) [76].

The relation $\alpha = \beta$ was originally suggested in Ref. [75] based on the Galilean invariance of the system. It may seem natural to assume Galilean invariance in studying long-wavelength and low-frequency dynamics, in which the underlying lattice structure is irrelevant. However, α and β come from spin-relaxation processes [55], which are usually intimately related to the lattice, e.g., through the spin−orbit coupling. The problem is thus subtle, and one has to go beyond the phenomenological argument such as the one based on the Galilean invariance. Instead, a fully microscopic calculation, which starts from a definite microscopic model and does not introduce any phenomenological assumptions, is desired. This section is devoted to outline such theoretical approaches.

Calculations based on a simple model where spin-relaxation processes are introduced by magnetic impurities lead to a result that $\alpha \neq \beta$ in general (even for single-band itinerant ferromagnets). These will be sketched in Section 5.3.2 and Section 5.3.3. More realistic models, especially concerning the spin-relaxation mechanism, can be studied along the same line. Readers who are not interested in theoretical details but in the results can jump to Section 5.3.2.3.

5.3.1.1 Intuitive Picture

Before going into serious calculations, we here present as a preliminary consideration an intuitive picture of the spin-transfer torque and the β-term. This will, in turn, serve to motivate us to do more serious calculations.

A nice picture of the spin-transfer torque was given by Xiao et al. [74]. It is in terms of the Aharonov-Stern effect [81], stating that a magnetic moment, or a spin σ, moving in an inhomogeneous magnetization $n(r)$ (or an external magnetic field, $\gamma \hbar B(r) \equiv M n(r)$), sees an effective field,

$$\gamma \hbar B_{\text{eff}}(t) = M n(r(t)) + \frac{\hbar}{2} n \times (v \cdot \nabla) n \qquad (5.71)$$

---

[15] There will also be contributions from electrons to H_{eff}, such as the exchange field [80]. We do not discuss them here for simplicity.

where $r = r(t)$ is the trajectory of the particle, and $v = \dot{r}(t)$ its velocity. Then the spin σ will tend to align this effective field, and acquires a component $\propto n \times (v \cdot \nabla)n$. The torque, $Mn \times \sigma$, that σ exerts on the magnetization n thus has a component, $\propto (v \cdot \nabla)n$, which is nothing but the spin-transfer torque. They proceeded to argue that the β-term is not induced by the presence of spin relaxation, since the above argument is already based on the relaxation of σ to B_{eff}. However, we note that the field B_{eff} is not a simple static field but contains a field of dynamical origin (second term of Eq. (5.71)), and the effect of spin relaxation needs to be carefully examined.

Here we reformulate the above picture by solving the spin dynamics explicitly. We consider the following two cases.

(a) a spin in a time-dependent magnetization,
(b) a spin moving in a static but spatially varying magnetization.

In both cases, the spin feels a time-dependent field (or magnetization). The essential physics of the spin torque lies in the response of the spin to such time-dependent field (or magnetization). So, let us study the motion of a single spin in a time-dependent field $Mn(t)$ on the basis of the equation

$$\frac{d\sigma}{dt} = \frac{2M}{\hbar} n(t) \times \sigma + \alpha_1 \sigma \times \frac{d\sigma}{dt} \tag{5.72}$$

We have phenomenologically introduced the damping (spin-relaxation) term with strength α_1. Assuming that the time variation of $n(t)$ is very slow, we solve Eq. (5.72) from the adiabatic limit, and calculate the solution as an expansion with respect to the number of time derivative. The result is given by:

$$\sigma = n + \frac{\hbar}{2M}(\dot{n} \times n + \alpha_1 \dot{n}) + \cdots \tag{5.73}$$

On the right-hand side, the first term is the "static" solution in the sense that it is static in the spin frame corotating with n. (Namely, the frame in which n is static. In Section 5.3.3, this frame will be called "adiabatic frame.") The next two terms which are first order in time derivative are "adiabatic" contributions. The ... represents terms which are higher order in time derivative, and are thus "nonadiabatic" contributions. The torque is obtained as $t_{el} = Mn \times \sigma$:

$$t_{el} = \frac{\hbar}{2}(\dot{n} + \alpha_1 n \times \dot{n}) \tag{5.74}$$

Note that the same solution is obtained even if we start from the Landau–Lifshitz damping $(\sim 2M\alpha_1 n \times (n \times \sigma))$ instead of the Gilbert type. The second term of Eq. (5.74) gives a damping torque on n for the situation (a).

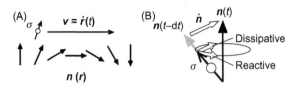

Figure 5.9 Illustration of the response of a spin moving in magnetization texture, $n(r)$.
(A) A moving spin in inhomogeneous magnetization feels a time-dependent magnetization,
$n(t) \equiv n(r(t))$, where $r(t)$ is the real-space trajectory of the spin. (B) Two kinds of response
of the spin to the time-dependent field. One is the reactive response as a precession around
$n(t)$. The other is the dissipative response, tending to orient to $n(t)$. The former gives rise to
the spin-transfer torque and the latter to the β-term.

For case (b), the time dependence of $n(r(t))$ arises from that of $r(t)$. Hence, the
result in this case may be obtained from Eq. (5.74) by the replacement,

$$\partial/\partial t \to v \cdot \nabla \qquad (?) \tag{5.75}$$

as

$$t_{\mathrm{el}} = \frac{\hbar}{2} \left[(v \cdot \nabla) \, n + \beta_1 n \times (v \cdot \nabla) \, n \right] \tag{5.76}$$

with $\beta_1 = \alpha_1$. The first term represents the spin-transfer torque, and the second
term corresponds to the β-term. As noted above, both torques may be classified as
adiabatic torques. Torques of higher order (in space-time derivative) would be
called nonadiabatic. For example, electrons scattered (reflected) by some magnetization texture feel a sudden change of the field, and respond in a way to produce
nonadiabatic torques.

In a similar way that we "derived" $\beta_1 = \alpha_1$ here, one may "derive" a relation
$\beta = \alpha$ quite generally. However, we should note that an implicit assumption was
made there that the damping constant α_1 is unchanged (Galilean invariance) under
the replacement (5.75). In reality, the spin-relaxation processes are often related to
the underlying lattice (such as spin-orbit coupling), and the Galilean invariance
cannot be expected in general. This problem therefore requires careful study. One
way is to perform microscopic calculations based on some definite microscopic
model (especially concerning spin relaxation) without introducing any phenomenological assumptions. Such calculations will be outlined in the following two subsec-
tions [77,82].

5.3.2   Small-Amplitude Method

5.3.2.1   Microscopic Model

Let us first set up a microscopic model. We take a localized picture for ferromagnetism, and consider the so-called s-d model. It consists of localized d spins, S, and

conducting s electrons, which are coupled via the s-d exchange interaction. The total Lagrangian is given by $L_{tot} = L_S + L_{el} - H_{sd}$, where L_S is the Lagrangian for d spins (Eq. (5.21)),

$$L_{el} = \int d^3x \, c^\dagger \left[i\hbar \frac{\partial}{\partial t} + \frac{\hbar^2}{2m} \nabla^2 + \varepsilon_F - V_{imp} \right] c \tag{5.77}$$

is the Lagrangian for s electrons, and

$$H_{sd} = -M \int d^3x \, n(r) \cdot \sigma(r) \tag{5.78}$$

is the s–d exchange coupling (Eq. (5.28)). Here, $c^\dagger = (c_\uparrow^\dagger, c_\downarrow^\dagger)$ is the spinor of electron creation operators, $\sigma(r) = c^\dagger(r)\sigma c(r)$ represents (twice) the s-electron spin density, with σ being a vector of Pauli spin matrices. We have put $S = Sn$ with the magnitude of spin, S, and a unit vector n,[16] and $M = J_{sd}S$ with J_{sd} being the s-d exchange coupling constant.

The s electrons are treated as a free-electron gas in three dimensions subject to the impurity potential:

$$V_{imp} = u \sum_i \delta(r - R_i) + u_s \sum_j S_j \cdot \sigma \delta(r - R_j') \tag{5.79}$$

The first term describes potential scattering. The second term represents quenched magnetic impurities, which is aimed at introducing spin-relaxation processes. The averaging over the impurity spin direction is taken as $\overline{S_i^\alpha} = 0$ and

$$\overline{S_i^\alpha S_j^\beta} = \frac{1}{3} S_{imp}^2 \delta_{ij} \delta^{\alpha\beta} \tag{5.80}$$

The damping rate of s electrons is then given by:

$$\gamma_\sigma = \frac{\hbar}{2\tau_\sigma} = \pi n_i u^2 \nu_\sigma + \frac{\pi}{3} n_s u_s^2 S_{imp}^2 (2\nu_{\bar\sigma} + \nu_\sigma) \tag{5.81}$$

Here n_i (n_s) is the concentration of normal (magnetic) impurities, and $\nu_\sigma = mk_{F\sigma}/2\pi^2\hbar^2$ (with $\hbar k_{F\sigma} = \sqrt{2m\varepsilon_{F\sigma}}$) is the density of states at energy $\varepsilon_{F\sigma} \equiv \varepsilon_F + \sigma M$. (The subscript $\sigma = \uparrow, \downarrow$ corresponds, respectively, to $\sigma = +1, -1$ in the formula, and to $\bar\sigma = \downarrow, \uparrow$ or $-1, +1$.) We assume that $\gamma_\sigma \ll \varepsilon_{F\sigma}$ and $\gamma_\sigma \ll M$, and calculate the torques in the lowest nontrivial order in $\gamma_\sigma/\varepsilon_{F\sigma}$ and γ_σ/M.

---

[16] We define n to be a unit vector in the direction of spin, which is opposite to the direction of magnetization.

5.3.2.2 General Framework

The spin torque from H_{sd} is given by:

$$t_{el}(r) \equiv Mn(r) \times \langle \sigma(r) \rangle_{ne} \tag{5.82}$$

This is related to T_{el} of Eq. (5.61) via $t_{el}[n] = -T_{el}[M]$, and to t'_{el} of Eq. (5.66) via

$$t_{el} = \frac{\hbar S}{a^3} t'_{el} \tag{5.83}$$

with a^3 being the volume per d spin.

The calculation of spin torque is thus equivalent to that of s-electron spin polarization, $\langle \sigma(r) \rangle_{ne}$, or precisely speaking, its orthogonal projection[17] $\langle \sigma_\perp(r) \rangle_{ne}$ to n. The expectation value $\langle \cdots \rangle_{ne}$ is taken in the following nonequilibrium states depending on the type of the torque.

(a) Nonequilibrium states under the influence of uniform but *time-dependent magnetization*. This leads to torques with time derivative of n, namely, Gilbert damping and spin renormalization.

(b) Nonequilibrium states with *current flow* under static but *spatially varying magnetization*. This leads to current-induced torques, namely, spin-transfer torque and the β-term.

In the presence of spin-rotational symmetry for electrons, adiabatic spin torques, which are first order in space−time derivative, are expressed as:

$$\tau_{ad}^0 = a_0 \dot{n} + (a \cdot \nabla) n + b_0 (n \times \dot{n}) + n \times (b \cdot \nabla) n \tag{5.84}$$

The corresponding s-electron spin polarization is given by (see footnote 17):

$$\langle \sigma_\perp \rangle_{ne} = \frac{1}{M} [b_0 \dot{n} + (b \cdot \nabla) n - a_0 (n \times \dot{n}) - n \times (a \cdot \nabla) n] \tag{5.85}$$

To calculate the coefficients a_μ and b_μ microscopically, it is sufficient to consider small transverse fluctuations, $u = (u^x, u^y, 0)$, $|u| \ll 1$, around a uniformly magnetized state, $n = \hat{z}$, such that $n = \hat{z} + u + O(u^2)$ [76,83]. Then, up to $O(u)$, Eq. (5.85) becomes:

$$\langle \sigma_\perp \rangle_{ne} = \frac{1}{M} [b_0 \dot{u} + (b \cdot \nabla) u - a_0 (\hat{z} \times \dot{u}) - \hat{z} \times (a \cdot \nabla) u] \tag{5.86}$$

This equation can be regarded as a linear response of $\sigma_\perp$ to u, and the coefficients, a_μ and b_μ, are obtained as linear-response coefficients. (Precisely speaking, $\langle \sigma_\perp \rangle_{ne}$ due to current is calculated as a linear response to the applied electric field.)

---

[17] We define $\hat{\sigma}_\perp = \hat{\sigma} - n(n \cdot \hat{\sigma})$ and $\hat{\sigma}_{\perp'} = \hat{\sigma} - \hat{z}(\hat{z} \cdot \hat{\sigma})$. Note that $\langle \hat{\sigma} \rangle_{ne}$ in Eq. (5.82) can be replaced by $\langle \hat{\sigma}_\perp \rangle_{ne}$.

5.3.2.3 Results

The results are given by:

$$\delta S = \frac{1}{2}\rho_s a^3 \tag{5.87}$$

$$v_s = -\frac{a^3}{2e(S + \delta S)}j_s \tag{5.88}$$

$$\alpha = \frac{a^3 v_+}{4(S + \delta S)}\frac{\hbar}{\tau_s} + \frac{S}{S + \delta S}\alpha_0 \tag{5.89}$$

$$\beta = \frac{\hbar}{2M\tau_s} \tag{5.90}$$

where $\rho_s = n_\uparrow - n_\downarrow$, $v_+ = v_\uparrow + v_\downarrow$, and $j_s = \sigma_s E = j_\uparrow - j_\downarrow$ is the spin current, with $\sigma_s = (e^2/m)(n_\uparrow\tau_\uparrow - n_\downarrow\tau_\downarrow)$ being the "spin conductivity." (n_σ is the density of spin-σ electrons.) We have defined the spin-relaxation time τ_s by:

$$\frac{\hbar}{\tau_s} = \frac{4\pi}{3}n_s u_s^2 S_{imp}^2 v_+ \tag{5.91}$$

As expected, only the spin scattering ($\sim\tau_s^{-1}$) contributes to α and β, and the potential scattering ($\sim n_i u^2$) does not. (For α, the second term on the right-hand side of Eq. (5.89) comes from processes which do not involve s-electrons.)

The ratio β/α cannot be unity in general for the two-component s-d model, since it contains mutually independent quantities, e.g., S for d electrons and δS for s electrons. For a single-band itinerant ferromagnet (as described by, e.g., the Stoner model), the results are obtained by simply putting $S = 0$ and $\alpha_0 = 0$ in Eqs. (5.87)–(5.90), and by using the spin polarization of itinerant electrons for δS [77]. We still see that $\alpha \neq \beta$, but it was pointed out that the ratio

$$\frac{\beta}{\alpha} = \frac{\rho_s}{Mv_+} \simeq 1 + \frac{1}{12}\left(\frac{M}{\varepsilon_F}\right)^2 \tag{5.92}$$

is very close to unity [76]. Even so, if we generalize Eq. (5.80) to the anisotropic one,

$$\overline{S_i^\alpha S_j^\beta} = \delta_{ij}\delta_{\alpha\beta} \times \begin{cases} \overline{S_\perp^2} & (\alpha, \beta = x, y) \\ \overline{S_z^2} & (\alpha, \beta = z) \end{cases} \tag{5.93}$$

we have

$$\frac{\beta}{\alpha} = \frac{3\overline{S_\perp^2} + \overline{S_z^2}}{2(\overline{S_\perp^2} + \overline{S_z^2})} \tag{5.94}$$

which ranges from $\frac{1}{2}$ (for $\overline{S_\perp^2} \ll \overline{S_z^2}$) to $\frac{3}{2}$ (for $\overline{S_\perp^2} \gg \overline{S_z^2}$). Therefore, we conclude that $\alpha \neq \beta$ in general, and that the value β/α is very sensitive to the details of the spin-relaxation mechanism.

The results obtained by Zhang and Li [55] based on the phenomenological spin-diffusion equations can be written in the form:

$$\alpha_{ZL} = \frac{\delta S}{S + \delta S} \cdot \frac{\hbar}{2M\tau_s} \tag{5.95}$$

whereas $\beta_{ZL} = \beta$ is the same as Eq. (5.90). Thus, it predicts $\alpha = \beta$ for single-band itinerant ferromagnets, $S = 0$. This disagrees with the present microscopic calculation, however. So far, all phenomenological theories predict $\alpha = \beta$, in contrast to the present microscopic results showing $\alpha \neq \beta$ in general.

5.3.3 Gauge-Field Method

In the previous section, we considered small-amplitude fluctuations of magnetization, and calculated the torques in the first order with respect to these small fluctuations. In this sense, the spin torques calculated there are limited to small-amplitude dynamics. (Only for systems with spin-rotational symmetry, where the form of the torque is known as Eq. (5.84), this small-amplitude method is sufficient to determine the coefficients, hence the torque.) In this section, we describe a theoretical formalism which is not restricted to small-amplitude dynamics, but can treat finite-amplitude (arbitrary) dynamics directly [82].

5.3.3.1 Adiabatic Spin Frame and Gauge Field

To treat finite-amplitude dynamics of magnetization, we work with a local/instantaneous spin frame (called "adiabatic frame" in the following) for s electrons whose spin quantization axis is taken to be the local/instantaneous d-spin direction, n [80,84,85]. The electron spinor $a(x)$ in the new frame is related to the original spinor $c(x)$ as $c(x) = U(x)a(x)$, where U is a 2×2 unitary matrix satisfying $c^\dagger(n \cdot \sigma)c = a^\dagger \sigma^z a$. It is convenient to choose U satisfying $U^2 = 1$.

Since $\partial_\mu c = U(\partial_\mu + U^\dagger \partial_\mu U)a \equiv U(\partial_\mu + iA_\mu)a$, the a-electrons satisfy the Schrödinger equation,

$$i\hbar\left(\frac{\partial}{\partial t} + iA_0\right)a(x) = \left[-\frac{\hbar^2}{2m}(\nabla_i + iA_i)^2 - M\sigma_z + \tilde{V}_{imp}\right]a(x) \tag{5.96}$$

The original electrons moving in time-dependent/inhomogeneous magnetization is thus mapped to new electrons moving in a uniform and static magnetization $M\sigma_z$ but there arises a coupling to an SU(2) gauge field:[18]

$$A_\mu = -iU^\dagger(\partial_\mu U) = A_\mu^\alpha \sigma^\alpha \equiv A_\mu \cdot \sigma \tag{5.97}$$

---

[18] We use the vector (bold italic) notation for the spin component. The space–time components are indicated by subscripts such as μ, ν (= 0, 1, 2, 3) or i, j (= 1, 2, 3).

Here A_μ is a measure of temporal ($\mu = 0$) or spatial ($\mu = 1, 2, 3$) variation of magnetization.

Let us introduce a 3×3 orthogonal matrix R, representing the same rotation as U but in a three-dimensional vector space, and satisfying $\det R = 1$. Note that $R\hat{z} = n$, $Rn = \hat{z}$, $c^\dagger \sigma c = R(a^\dagger \sigma a)$, and that $R(a \times b) = (Ra) \times (Rb)$ for arbitrary vectors a and b. Then, the spin-torque density, Eq. (5.82), is written as:

$$t_{\mathrm{el}}(x) = MR(\hat{z} \times \langle \tilde{\sigma}(x) \rangle_{\mathrm{ne}}) \tag{5.98}$$

where $\tilde{\sigma}(x) = (a^\dagger \sigma a)_x$.

Since the gauge field A_μ contains a space$-$time derivative of magnetization, one may naturally formulate a gradient expansion in terms of A_μ to calculate, e.g., the torque (or spin polarization). In particular, the adiabatic torques are obtained as the first-order terms in A_μ:

$$\langle \tilde{\sigma}_\perp \rangle_{\mathrm{ne}} = \frac{2}{M} \left[a_\mu A_\mu^\perp + b_\mu (\hat{z} \times A_\mu^\perp) \right] \tag{5.99}$$

Here $\tilde{\sigma}_\perp = \tilde{\sigma} - \hat{z}(\hat{z} \cdot \tilde{\sigma})$ and $A_\mu^\perp = A_\mu - \hat{z}(\hat{z} \cdot A_\mu)$ are the respective transverse components (see footnote 17), and the sums over $\mu = 0, 1, 2, 3$ are understood. From the identities,

$$RA_\mu^\perp = -\frac{1}{2} n \times (\partial_\mu n), \quad R(\hat{z} \times A_\mu^\perp) = \frac{1}{2} \partial_\mu n \tag{5.100}$$

together with Eq. (5.98), we see that Eq. (5.99) leads to the adiabatic torque density τ_{ad}^0 of Eq. (5.84).

5.3.3.2 Results

If we regard Eq. (5.99) as a linear response to the gauge field A_μ appearing in Eq. (5.96), the coefficients a_μ and b_μ are calculated as linear-response coefficients, and $\delta S, v_s$, and β thus obtained coincide with those obtained by the small-amplitude method (Eqs. (5.87), (5.88), and (5.90)). However, it predicts $\alpha_{\mathrm{sr}} = 0$, and fails to reproduce the Gilbert damping.

5.3.3.3 Gilbert Damping

The above difficulty that the gauge-field method apparently fails to reproduce the Gilbert damping term has been resolved in Ref. [82] by noting that the impurity spins, which are static (quenched) in the original frame, become time dependent in the adiabatic frame (Figure 5.10). Namely, the spin part of V_{imp} is expressed as:

$$S_j \cdot c^\dagger \sigma c = \tilde{S}_j(t) \cdot a^\dagger \sigma a \tag{5.101}$$

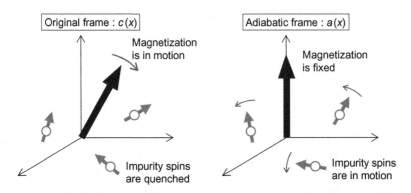

Figure 5.10 Left: magnetization vector $n(t)$ and impurity spins S_j in the original frame. Right: magnetization vector $\hat{z}$ and impurity spins $\tilde{S}_j(t)$ in the adiabatic frame.

where

$$\tilde{S}_j(t) = R(t)S_j \qquad\qquad (5.102)$$

is the impurity spin in the adiabatic frame, which is time dependent. (This fact is expressed by $\tilde{V}_{\text{imp}}$ in Eq. (5.96).) Actually, we can obtain the gauge field from this time dependence as:

$$[R(t)\dot{R}(t)]^{\alpha\beta} = 2\varepsilon^{\alpha\beta\gamma}A_0^{\gamma} \qquad\qquad (5.103)$$

Explicit calculation of $\langle\tilde{\sigma}_\perp\rangle_{\text{ne}}$ in the second order in $\tilde{S}_j(t)$ (nonlinear response) gives:

$$\langle\tilde{\sigma}_\perp(t)\rangle_{\text{ne}} = -\frac{2\pi\hbar}{3M}n_s u_s^2 S_{\text{imp}}^2 \nu_+^2(\hat{z}\times A_0^\perp(t)) \qquad\qquad (5.104)$$

leading to the Gilbert damping, with damping constant exactly the same as Eq. (5.89).

The present calculation provides us a new picture of Gilbert damping. While the s-electron spin tends to follow the instantaneous d-spin direction $n(t)$, it is at the same time pinned by the quenched impurity spins. These two competing effects are expressed by the time dependence of $\tilde{S}_j(t)$ in the adiabatic frame, and this effect causes Gilbert damping. Namely, the Gilbert damping arises since spins of s electrons are "dragged" by impurity spins.

Generally, any terms in the Hamiltonian leading to spin relaxation break spin-rotational symmetry of s electrons, and thus acquire time dependence in the adiabatic frame. Therefore, the same scenario as presented here is expected to apply to other type of spin-relaxation processes quite generally.

5.3.3.4 Nonadiabatic Torque

The nonadiabatic torques come from large-momentum processes, so let us put $V_{imp} = 0$ for simplicity. If we retain the full q-dependence in the response function, we have, among others, a contribution:

$$\langle \tilde{\sigma}_{\perp}(q) \rangle_{ne} = \chi_{ij}(q) E_i(\hat{z} \times A_{q,j}^{\perp}) + \cdots \tag{5.105}$$

where the function $\chi_{ij}(q)$ is nonzero only when $|k_{F\uparrow} - k_{F\downarrow}| < q < k_{F\uparrow} + k_{F\downarrow}$. In particular, $\chi_{ij}(q) = 0$ in the vicinity of $q = 0$, and hence the long-wavelength approximation cannot be applied. The resulting torque

$$\tau_{na}^0(x) = -ME_i \int d^3x' \chi_{ij}(r - r') R(x) A_j^{\perp}(x') \tag{5.106}$$

is characterized by an oscillatory function $\chi_{ij}(r - r')$ in real space, and is essentially nonlocal [22,74,86–88]. This is the nonadiabatic, momentum-transfer torque due to electron reflection [22], generalized here to arbitrary magnetization texture. For details, see Ref. [87].

5.3.4 Force

We have seen in Section 5.2.2 that electrons reflected by a domain wall exert a force on the (rigid) domain wall as a back reaction. It was also recognized that the adiabatic β-term also acts as a force on a domain wall [55,56]. Here, we consider a generalization of the concept of force to a fixed but arbitrary magnetization texture, and derive a general expression of the force acting on the texture.

From $t_{el} = Mn \times \langle \sigma_{\perp} \rangle_{ne}$ and $F_{el,i} = -M \int d^3x \, (\partial_i n) \cdot \langle \sigma_{\perp} \rangle_{ne}$ (Eq. (5.35)) [22,89], one can eliminate $\langle \sigma_{\perp} \rangle_{ne}$ to obtain the following relation:

$$F_{el,i} = -\int d^3x \, (n \times \partial_i n) \cdot t_{el} \tag{5.107}$$

This is a general formula relating the force to the torque [90].

Each term of the spin torque (5.84) corresponds to a different type of force. The spin-transfer torque exerts a "transverse" force:

$$F_{ST,i} = -\frac{\hbar}{2e} j_{s,\ell} \int d^3x \, (\partial_i n \times \partial_\ell n) \cdot n \tag{5.108}$$

for a magnetic configuration n subtending a finite solid angle (spin chirality). An example is given by the magnetic vortex in a ferromagnetic dot, whose dynamics driven by electric/spin current will be studied in Section 5.4.2. The β-term leads to the force:

$$F_{\beta,i} = -\beta \frac{\hbar}{2e} j_{s,\ell} \int d^3x \, (\partial_i n) \cdot (\partial_\ell n) \tag{5.109}$$

The nonadiabatic torque, τ_{na}, gives a force,

$$F_{na,i} = -2ME_i \int d^3x \int d^3x' \chi_{ij}(r - r')A_i^{\perp}(r) \cdot A_j^{\perp}(r') \tag{5.110}$$

This force is due to the reaction to the electron reflection, and is closely related to the resistivity as noted in the domain-wall case [22], and also shown below.

5.3.4.1 Relation to Transport Coefficients

Since the force is, by definition, the momentum transfer from s electrons to magnetization, its back reaction will affect the s-electrons' orbital motion, hence their transport. Therefore, it is expected that the force and the transport coefficients are closely related with each other. In fact, we have observed a relation, Eq. (5.33) or Eq. (5.41), for a domain wall in an idealized situation that there are no spin- and momentum-dissipation processes (other than the domain wall) for s electrons. Explicit calculation shows that these relations are not modified by the presence of normal impurities (momentum-dissipation processes) [22]. Here we extend such relations to more general magnetization texture, and to the case that the s electron transport is diffusive [87,88]. The texture is assumed to be static here.

For simplicity, we neglect the spin-relaxation processes, and calculate the transport coefficients by using Mori formula (memory function formalism) [91,92]. The electrical resistivity due to (static) magnetization texture is obtained as:

$$\rho_{ij} = \left(\frac{2M}{enV}\right)^2 \sum_q \left[\chi^{(0)}(q)(A_{q,i}^{\perp} \cdot A_{-q,j}^{\perp}) + \chi^{(1)}(q)\left(A_{q,i}^{\perp} \times A_{-q,j}^{\perp}\right)^z\right] \tag{5.111}$$

$$\equiv \rho_{ij}^{refl} + \rho_{ij}^{Hall} \tag{5.112}$$

where

$$\chi^{(0)}(q) = \frac{2\hbar}{V}\sum_k \delta(\xi_{k+q/2}^{\sigma})\delta(\xi_{k-q/2}^{\bar{\sigma}}) = \frac{m^2}{2\pi\hbar^3}\frac{1}{q}\Theta_{st}(q) \tag{5.113}$$

$$\chi^{(1)}(q) = \frac{\hbar}{V}\sum_\sigma \sigma \sum_k \frac{f_{k+q/2}^{\sigma} - f_{k-q/2}^{\bar{\sigma}}}{(\xi_{k+q/2}^{\sigma} - \xi_{k-q/2}^{\bar{\sigma}})^2} = \frac{\hbar}{2M^2}\rho_s + O(q^2) \tag{5.114}$$

with $\xi_k^{\sigma} = \varepsilon_k - \varepsilon_{F\sigma} = \hbar^2(k^2 - k_{F\sigma}^2)/2m$ and $\Theta_{st}(q) = 1$ for $|k_{F\uparrow} - k_{F\downarrow}| < q < k_{F\uparrow} + k_{F\downarrow}$, and $\Theta_{st}(q) = 0$ otherwise. In deriving Eq. (5.111), we have used the same model as used in Section 5.3.3, and neglected the difference, $\tau_{\uparrow} - \tau_{\downarrow} \simeq 0$, in the s-electron lifetimes. We have denoted the symmetric part as ρ_{ij}^{refl} and the antisymmetric part as ρ_{ij}^{Hall}.

Let us first consider the antisymmetric part, ρ_{ij}^{Hall}. Since the kernel $\chi^{(1)}(q)$ is nonzero in the adiabatic limit, $q \to 0$, one may make a long-wavelength approximation as:

$$\rho_{ij}^{\text{Hall}} = \frac{2\hbar\rho_s}{e^2 n^2 V} \int d^3x \, [A_i^{\perp}(r) \times A_j^{\perp}(r)]^z \tag{5.115}$$

$$= \frac{\hbar\rho_s}{2e^2 n^2 V} \int d^3x \, \boldsymbol{n} \cdot (\partial_i \boldsymbol{n} \times \partial_j \boldsymbol{n}) \tag{5.116}$$

In the second line, we have used Eq. (5.100). By comparing with the spin-transfer force of Eq. (5.108), we see the relation as:

$$F_{\text{ST},i} = -\frac{en^2 V}{\rho_s} \sum_{\ell} \rho_{i\ell}^{\text{Hall}} j_{s,\ell}$$

$$\simeq -eN_e \sum_{\ell} \rho_{i\ell}^{\text{Hall}} j_{\ell} \tag{5.117}$$

Here, $n = n_{\uparrow} + n_{\downarrow}$ is the s-electron density, $N_e = nV$ is the total number of s electrons, and $\boldsymbol{j}$ is the electric (charge) current. In the second line, we have used the same approximation ($\tau_{\uparrow} \simeq \tau_{\downarrow}$) as used above, and put $(n/\rho_s)\boldsymbol{j}_s \simeq \boldsymbol{j}$.

For the symmetric part, ρ_{ij}^{refl}, the kernel $\chi^{(0)}(q)$ is finite only for $|k_{F\uparrow} - k_{F\downarrow}| \leq q \leq k_{F\uparrow} + k_{F\downarrow}$, and vanishes near $q = 0$, hence one cannot make a long-wavelength approximation. This term is due to the electron reflection. The force due to nonadiabatic torque is obtained as:

$$F_{\text{na},i} = -\sum_{\ell} R_{i\ell} j_{\ell} \tag{5.118}$$

where

$$R_{i\ell} = \frac{(2M)^2}{enV} \sum_{q,j} \frac{q_j q_{\ell}}{q^2} \chi^{(0)}(q)(A_{q,i}^{\perp} \cdot A_{-q,j}) \tag{5.119}$$

There seems no exact relation between ρ_{ij}^{refl} and R_{ij} in general. However, if the magnetization texture is one-dimensional, namely, if it depends on only one Cartesian coordinate, say x, the gauge field A_i is nonzero only for this component, $i = x$, and $R_{i\ell}$ is nonzero only for $i = \ell = x$. In this case, we see that $R_{xx} = enV\rho_{xx}^{\text{refl}} = eN_e\rho_{xx}^{\text{refl}}$, and hence

$$F_{\text{na},x} = -eN_e \rho_{xx}^{\text{refl}} j_x \tag{5.120}$$

Namely, the x-component of the current exerts a force in the x direction. (The relative minus sign is due to the fact that the direction of electron flow is opposite to that of the electric current.)

5.3.4.2 Further References

The spin-transfer torque in the form of Eq. (5.64) was first derived by Bazaliy et al. [70]. The β-term was first derived by Zhang and Li [55] based on a spin-diffusion equation with a spin-relaxation term included, and by Thiaville et al. [56] as a continuum limit of a special type of torque known in multilayer systems [93].

Duine et al. put the present microscopic (small-amplitude) calculation into the Keldysh formalism and developed a functional description of spin torques, giving a theoretical basis of finite-temperature and/or fluctuation dynamics [79]. Some attempts with the Boltzmann equation are done in Ref. [78]. Some developments in the treatment of Gilbert damping can be seen in Refs. [83,94,95]. Phenomenological theory [96,97] based on irreversible thermodynamics is also developed, which, however, needs to be reconciled with the microscopic theory.

The nonadiabatic torque is studied in Refs. [22,74,86,87]. The domain-wall resistivity is studied in Refs. [85,98]. The effects of spin-orbit coupling is studied on domain-wall resistance [99] and domain-wall mobility [100] in a model of ferromagnetic semiconductors.

5.4 Related Topics

In this section, we present two topics on the current-driven magnetization dynamics.

The first topic is the current-driven motion of a magnetic vortex. It is mainly governed by the spin-transfer torque, and nicely exemplifies the "spin-transfer force," $\mathbf{F}_{ST}$, of Eq. (5.108).

The second topic is related to another aspect of the spin-transfer torque. Namely, the spin-transfer torque favors and even induces a topologically nontrivial spin texture along the spin current [70−73], in quite a similar way that the spin Berry phase does in the time direction, namely, precession. It causes the Doppler shift and the softening of spin-wave frequency, and eventually leads to an instability of the uniformly magnetized state.

5.4.1 Current-Driven Motion of Magnetic Vortices

Just as the magnetic domain wall is a topologically stable structure in easy-axis magnets, such as thin wires of soft ferromagnets, a magnetic vortex is a topologically stable structure in easy-plane magnets such as films. In thin disks in particular, it is realized as the most stable structure for a certain range of parameters

[101]. The current-driven dynamics of such magnetic vortices is the subject of this subsection [102−105].

A magnetic vortex realized in a disk is illustrated in Figure 5.11A and B. It is characterized by circulating in-plane magnetization in the region away from the center ($|r| \gg \delta_v$), and out-of-plane magnetization in the core (center) region ($|r| \leq \delta_v$), where δ_v is the size of the core. Such a vortex configuration is characterized by two integers, p and C. The first one is the polarity, $p = \pm 1$, which specifies the direction of the out-of-plane component at the vortex center. The second one is the chirality, C, which specifies the sense of circulation, namely, counterclockwise ($C = 1$) or clockwise ($C = -1$).

More generally, magnetic configurations illustrated in Figure 5.11C are also called vortices. These new members are distinguished by introducing a third integer, $q = \pm 1, \pm 2, \ldots$, called vorticity. It represents the number of circulations of the in-plane magnetization component. A vortex constrained by the disk boundary condition corresponds to $q = 1$, but in other situations, vortices with $q \neq 1$ can also appear in general. In the latter case, the value of C may not be restricted to integers, but can take general real numbers. Magnetization pattern for a vortex specified by (p, q, C) is expressed as:

$$-n(r) = \begin{cases} \hat{x}\cos(q\varphi + \pi C/2) + \hat{y}\sin(q\varphi + \pi C/2) & (|r| \gg \delta_v) \\ p\hat{z} & (r \to 0) \end{cases} \tag{5.121}$$

where $\varphi = \tan^{-1}(y/x)$ with $r = (x, y, z)$. (The minus sign on the left-hand side indicates that we define the numbers (p, q, C) with respect to the direction of magnetization, not spin.)

In the following text, we mainly consider a vortex realized in a disk, hence with $q = 1$. However, some of the equations have more generality, so we leave q to be a general integer in the equations.

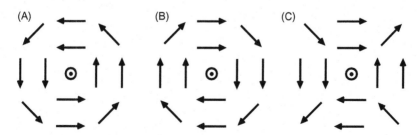

Figure 5.11 Schematic illustration of magnetic vortices. (A) A vortex with $q = 1$, $p = 1$, and $C = 1$. (B) A vortex with $q = 1$, $p = 1$, and $C = -1$. (C) An example of antivortex ($q = -1$). In each case, the magnetization in the center (core) region rises out of the plane, either upward ($p = 1$) or downward ($p = -1$). In (A–C), a vortex with $p = 1$ is shown, as indicated by a circle with a dot. In a ferromagnetic dot with suitable size, either (A) or (B) is realized, whereas in a film, free vortices and antivortices can appear.

5.4.1.1 Equation of Motion

We focus on the position of the vortex center, $X = \begin{pmatrix} X \\ Y \end{pmatrix}$. The equation of motion for X is known as the Thiele equation [89,102,106],

$$-G \times \dot{X} - \alpha D \dot{X} + F = 0 \tag{5.122}$$

which is obtained as a force balance equation. The first term on the left-hand side is known as a gyroforce. A vector

$$G = \hat{z} \frac{M_s d}{\gamma_0} \int_{\text{disk}} \mathrm{d}x\,\mathrm{d}y\,(\partial_x n \times \partial_y n) \cdot n \tag{5.123}$$

is called as the gyrovector, which is perpendicular to the disk (xy-) plane. Here the magnetization is assumed constant in the thickness (z-) direction, and the film thickness d is simply multiplied. The integral in Eq. (5.123), which is thus over the two-dimensional disk region, is a topological quantity equal to the solid angle subtended by the magnetization vector n over the disk, which is estimated to be $2\pi pq$ using Eq. (5.121). Thus,

$$G = 2\pi pq \frac{M_s d}{\gamma_0} \hat{z} \tag{5.124}$$

The second term of Eq. (5.122) represents the damping force, where α is the Gilbert damping constant and D is given by:

$$D = \frac{M_s d}{\gamma_0} \int_{\text{disk}} \mathrm{d}x\,\mathrm{d}y\,(\nabla n)^2 \tag{5.125}$$

The third term, F, represents any other force acting on the vortex center. The force due to electric/spin current is given by:

$$F_{\text{el}} = G \times v_s + \beta' D v_s \tag{5.126}$$

as seen in Section 5.3.4. Here β' includes contributions from the β-term due to spin relaxation, and also from the nonadiabatic torque due to electron reflection. Note that, in the absence of applied field, the chirality does not enter the equation of motion, and only p and q (contained in G) are relevant for the current-driven dynamics.

The value of D will actually depend on the vortex position owing to the deformation of the vortex profile. On the other hand, the value of G is quite robust since it is a topological invariant; as far as the magnetization at the disk perimeter is along the surface ("disk boundary condition"), and the continuum approximation is valid, it is quantized as seen above.

5.4.1.2 Current-Driven Dynamics

Let us study a current-driven motion of a vortex confined in a disk. The equation of motion is given by:

$$G \times (\dot{X} - v_s) + \alpha D\dot{X} - \beta' Dv_s = -kX \tag{5.127}$$

Here the restoring force due to the confinement is approximated by $-kX$ with a force constant k. We assume k and D to be constant for simplicity.

We first consider the case that a steady DC current, represented by $v_s = (v_s, 0)$, is applied. The solution is easily obtained in terms of a complex coordinate, $Z = X + iY$, as

$$Z(t) = e^{i\zeta t}Z(0) + i\frac{v_s}{\Omega}(1 - i\tilde{\beta})(1 - e^{i\zeta t}) \tag{5.128}$$

where $\Omega = k/G$, $\zeta = \Omega/(1 - i\tilde{\alpha}) = k/(G - i\alpha D)$, $\tilde{\alpha} = \alpha D/G$, and $\tilde{\beta} = \beta' D/G$. In a vector notation, it reads as:

$$X(t) = R(\zeta' t)[X(0) - X(\infty)] e^{-\zeta'' t} + X(\infty) \tag{5.129}$$

where $\zeta' = \operatorname{Re} \zeta$, $\zeta'' = \operatorname{Im} \zeta > 0$, and

$$R(\theta) = \begin{pmatrix} \cos\theta & -\sin\theta \\ \sin\theta & \cos\theta \end{pmatrix} \tag{5.130}$$

is the rotation matrix. Thus, the vortex center exhibits a spiral motion in clockwise (for $\zeta' < 0$, i.e., $pq = -1$) or counterclockwise (for $\zeta' > 0$, i.e., $pq = 1$) sense, and eventually approaches the equilibrium position:

$$X(\infty) = \frac{v_s}{\Omega}\begin{pmatrix} \tilde{\beta} \\ 1 \end{pmatrix} = \frac{v_s}{k}\begin{pmatrix} \beta' D \\ G \end{pmatrix} \tag{5.131}$$

as $t \to \infty$. Note that the vortex configuration is defined here in terms of magnetization direction, not spin direction, in contrast to Ref. [102].

For AC current, $v_s = (v_s \cos\omega t, 0)$, the solution is obtained as [105]:

$$Z(t) = e^{i\zeta t}Z(0) + \frac{i}{2}v_s(1 - i\tilde{\beta})\left[\frac{e^{-i\omega t} - e^{i\zeta t}}{\Omega + (1 - i\tilde{\alpha})\omega} + \frac{e^{i\omega t} - e^{i\zeta t}}{\Omega - (1 - i\tilde{\alpha})\omega}\right] \tag{5.132}$$

in the complex notation, or as

$$X(t) = R(\zeta' t)[X(0) - X_\omega - X_{-\omega}] e^{-\zeta'' t} + R(\omega t)X_\omega + R(-\omega t)X_{-\omega} \tag{5.133}$$

in the vector notation, where

$$X_{\pm \omega} = \frac{1}{2} \frac{v_s}{(\Omega \mp \omega)^2 + (\tilde{\alpha}\omega)^2} \begin{pmatrix} \tilde{\beta}\Omega \pm (\tilde{\alpha} - \tilde{\beta})\omega \\ \Omega \mp (1 + \tilde{\alpha}\tilde{\beta})\omega \end{pmatrix} \tag{5.134}$$

The first term of Eq. (5.133) represents eigen rotation [107], which will eventually decay, and the second and third terms represent forced rotation. Near the resonance, $\omega \simeq \pm \Omega$, it exhibits a steady circular motion,

$$X(t) \simeq R(\pm \omega t)X_{\pm \omega} \simeq R(\Omega t)X_{\Omega} \tag{5.135}$$

for $\zeta'' t \gg 1$. The amplitude just on resonance is given by:

$$X_{\Omega} = \frac{v_s}{2\tilde{\alpha}\Omega} \begin{pmatrix} 1 \\ -\tilde{\beta} \end{pmatrix} \tag{5.136}$$

which is enhanced by a factor of $\tilde{\alpha}^{-1}$ compared with the displacement (5.131) due to DC current.

5.4.1.3 Lagrangian Formalism

The Lagrangian formalism for vortex dynamics with the spin-transfer effect has been developed by Shibata et al. [102]. They obtained the Lagrangian

$$L_v = \frac{1}{2} G \cdot (\dot{X} \times X) - G \cdot (v_s \times X) - U(X) \tag{5.137}$$

which, together with the dissipation function

$$W_v = \frac{1}{2} \alpha D \dot{X}^2 \tag{5.138}$$

leads to Eq. (5.127) with $\beta' = 0$. The first term of Eq. (5.137), namely, $G(\dot{X}Y - \dot{Y}X)/2$, shows that X and Y are canonically conjugates with each other. Note that these variables for the vortex are quite symmetric with each other, in contrast to X and ϕ_0 for the domain-wall case.

5.4.2 Current-Induced Spin-Wave Instability

The spin-transfer torque is expressed by the term

$$H_{ST} = \frac{\hbar}{2e} \int d^3x \, (j_s \cdot \nabla\phi)(1 - \cos\theta) \tag{5.139}$$

in the spin Hamiltonian [73]. The spin current couples to a gradient of magnetization. It thus favors spatial variation of magnetization. This leads to the softening, and even to instability, of spin waves [70–73]. It also lowers the domain-wall energy under spin current [73].

It has been known that a sufficiently large spin current leads to a spin-wave instability of the uniformly magnetized state. This was first shown by Bazaliy et al. [70] for a layered system with an interface between a normal metal and a ferromagnet, and later emphasized by Fernández-Rossier et al. [72] to be a bulk phenomenon. Indeed, Eq. (5.139) can be seen as a spatial-derivative version of the kinetic (Berry phase) term, and a kind of "precession" along the spin current may be expected.

Let us consider the LLG equation,

$$\dot{\boldsymbol{n}} = S'(-J\nabla^2\boldsymbol{n} - Kn_z\hat{z} + K_\perp n_y\hat{y}) \times \boldsymbol{n} + \alpha\,\dot{\boldsymbol{n}} \times \boldsymbol{n} + (\boldsymbol{v}_\mathrm{s} \cdot \nabla)\,\boldsymbol{n} \qquad (5.140)$$

where $S' = S^2/\hbar S_{\mathrm{tot}}$. The ground state is given by $\boldsymbol{n} = \hat{z}$, namely, uniformly magnetized in z-direction, which actually satisfies Eq. (5.140). Next, as an excited state, we consider a small fluctuation $\delta\boldsymbol{n}$ from the ground state, $\boldsymbol{n} = \hat{z}$, and put $\boldsymbol{n}(\boldsymbol{r}, t) = \hat{z} + \delta\boldsymbol{n}(\boldsymbol{r}, t)$. The spin-wave dispersion is obtained by assuming a solution of the form, $\delta\boldsymbol{n}(\boldsymbol{r}, t) = (\delta n_x, \delta n_y, 0)e^{i(\boldsymbol{q}\cdot\boldsymbol{r}-\omega t)}$. For $\alpha = 0$, it is given by:

$$\Omega_{\boldsymbol{q}} = KS'\sqrt{\omega_q(\omega_q + \kappa)} + \boldsymbol{q} \cdot \boldsymbol{v}_\mathrm{s} \qquad (5.141)$$

where $\omega_q = q^2\lambda^2 + 1$, $\lambda = \sqrt{J/K}$, and $\kappa = K_\perp/K$. The first term represents the spin-wave energy in the absence of spin current, and has a spectral gap $S'\sqrt{K(K + K_\perp)}$ due to magnetic anisotropy. The second term represents the effect of spin current, called Doppler shift [108]. It reduces the spin-wave gap and shifts the energy-minimum point from $\boldsymbol{q} = \boldsymbol{0}$ to some finite wave vector. If the spin-current density exceeds a certain critical value, $v_\mathrm{s} > v_\mathrm{s}^{\mathrm{cr}}$, the spin-wave energy becomes negative for a range of $\boldsymbol{q}$ (spin-wave instability). This means that the uniformly magnetized state is not the ground state any more, and is energetically unstable.

In the presence of a domain wall, the spin-wave spectrum is modified to be [73]:

$$\Omega_{\boldsymbol{q}} = KS'\sqrt{\omega_q(\omega_q + \kappa) + 2\kappa^2\sin^2\phi_0\cos 2\phi_0} + q(v_\mathrm{s} - \dot{X}) \qquad (5.142)$$

The spin-transfer velocity appears relative to the domain-wall velocity, and the above instability is reduced. There is a range of the spin current where the uniformly magnetized state is unstable whereas the domain-wall state is stable (at least locally in configuration space). It was then proposed that in such case the domain walls are created spontaneously. Indeed, for large $K_\perp$, the state with a domain wall is shown to have lower energy than that of the uniform ferromagnetic state under large current [73]. Current-induced formation of domain walls was observed in one-dimensional micromagnetic simulations [109]. For films or wide wires having easy-plane magnetic anisotropy, micromagnetic simulations found that the spin-wave instability is followed by the formation of magnetic vortices [110].

Acknowledgment

We would like to express our deep thanks to J. Shibata for his fruitful contributions at every stage of our collaboration. We would also like to thank G.E.W. Bauer, A. Brataas, D. Chiba, R. Duine, H. Fukuyama, M. Hayashi, J. Ieda, J. Inoue, S. Kasai, M. Kläui, K.-J. Lee, S. Maekawa, A.H. MacDonald, Y. Nakatani, H. Ohno, T. Ono, S.S.P. Parkin, E. Saitoh, J. Sinova, M. Stiles, Y. Suzuki, A. Thiaville, Y. Tserkovnyak, A. Yamaguchi, M. Yamanouchi, and S. Zhang for valuable discussions. H.K. is indebted to K. Miyake for continual encouragement.

References

[1] Maekawa S, editor. Concepts in spin electronics. Oxford: Oxford University Press; 2006.

[2] Xu YB, Thompson SM, editors. Spintronic materials and technology. Taylor & Francis; 2007.

[3] Slonczewski JC. J Magn Magn Mater 1996;159:L1.

[4] Berger L. Phys Rev B 1996;54:9353.

[5] Berger L. J Appl Phys 1984;55:1954.

[6] Berger L. J Appl Phys 1978;49:2156.

[7] Berger L. Phys Rev B 1986;33:1572.

[8] Berger L. J Appl Phys 1992;71:2721.

[9] Freitas PP, Berger L. J Appl Phys 1985;57:1266.

[10] Salhi E, Berger L. J Appl Phys 1993;73:6405.

[11] Tserkovnyak Y, Brataas A, Bauer GEW, Halperin BI. Rev Mod Phys 2005;77:1375.

[12] Chikazumi S. Physics of ferromagnetism. 2nd ed. Oxford University Press; 1997.

[13] Hubert A, Schäfer R. Magnetic domains. Springer-Verlag; 1998.

[14] de Leeuw FH, van den Doel R, Enz U. Rep. Prog. Phys. 1980;43:689.

[15] Slonczewski JC. Int J Magn 1972;2:85 J Appl Phys 1974;45:2705

[16] Döring W. Z Naturforsch 1948;3A:373.

[17] Braun H-B, Loss D. Phys Rev B 1996;54:3237.Braun H-B, Kyriakidis J, Loss D. Phys Rev B 1997;56:8129.

[18] Auerbach A. Interacting electrons and quantum magnetism. Springer Verlag; 1994 [Chapter 10]

[19] Berry MV. Proc R Soc London 1984;A392:45.

[20] Goldstein H, Poole C, Safko J. Classical mechanics. 3rd ed. Addison Wesley; 2002 [Chapter 1, Section 5]

[21] Takagi S, Tatara G. Phys Rev B 1996;54:9920.

[22] Tatara G, Kohno H. Phys Rev Lett 2004;92: 086601; 2006;96:189702

[23] Grollier J, Lacour D, Cros V, Hamzic A, Vaurés A, Fert A, et al. J Appl Phys 2002;92:4825.Grollier J, Boulenc P, Cros V, Hamzic A, Vaurés A, Fert A, et al. Appl Phys Lett 2003;83:509.

[24] Tsoi M, Fontana RE, Parkin SSP. Appl Phys Lett 2003;83:2617.

[25] Kläui M, Vaz CAF, Bland JAC, Wernsdorfer W, Faini G, Cambriland E, et al. Appl Phys Lett 2003;83:105.

[26] Vernier N, Allwood DA, Atkinson D, Cooke MD, Cowburn RP. Europhys Lett 2004;65:526.

[27] Yamaguchi A, Ono T, Nasu S, Miyake K, Mibu K, Shinjo T. Phys Rev Lett 2004;92:077205.

[28] Saitoh E, Miyajima H, Yamaoka T, Tatara G. Nature 2004;432:203.
[29] Yamanouchi M, Chiba D, Matsukura F, Ohno H. Nature 2004;428:539.
[30] Yamanouchi M, Chiba D, Matsukura F, Dietl T, Ohno H. Phys Rev Lett 2006;96:096601.
[31] Kläui M, Vaz CAF, Bland JAC, Wernsdorfer W, Faini G, Cambril E, et al. Phys Rev Lett 2005;94:106601.
[32] Kläui M, Jubert P-O, Allenspach R, Bischof A, Bland JAC, Faini G, et al. Phys Rev Lett 2005;95:026601.
[33] Hayashi M, Thomas L, Bazaliy YaB, Rettner C, Moriya R, Jiang X, et al. Phys Rev Lett 2006;96:197207.
[34] Hayashi M, Thomas L, Rettner C, Moriya R, Jiang X, Parkin SSP. Phys Rev Lett 2006;97:207205.
[35] Hayashi M, Thomas L, Rettner C, Moriya R, Parkin SSP. Nat Physics 2007;3:21.
[36] Thomas L, Hayashi M, Jiang X, Moriya R, Rettner C, Parkin SSP. Nature 2006;443:197.
[37] Thomas L, Hayashi M, Jiang X, Moriya R, Rettner C, Parkin SSP. Science 2007;315:1553.
[38] Barnes SE, Maekawa S, Chapter 7 of Ref. [1].
[39] Volovik GE. J Phy C 1987;20:L83.
[40] Stern A. Phys Rev Lett 1992;68:1022.
[41] Barnes SE, Maekawa S, Ieda J. Appl Phys Lett 2006;89:122507.
[42] Barnes SE, Maekawa S. Phys Rev Lett 2007;98:246601.
[43] Duine RA. Phys Rev B 2008;77:014409.
[44] Stamenova M, Todorov TN, Sanvito S. Phys Rev B 2008;77:054439.
[45] Yang SA, Xiao D, Niu Q, cond-mat/0709.1117.
[46] Tserkovnyak Y, Mecklenburg M. Phys Rev B 2008;77:134407.
[47] Kim K-W, Moon J-H, Lee K-J, Lee H-W. Phys Rev B 2011;84:054462.
[48] Shibata J, Kohno H. Phys Rev B 2011;84:184408.
[49] Brataas A, Tserkovnyak Y, Bauer GEW, Halperin BI. Phys Rev B 2002;66:060404.
[50] Yang SA, Beach GSD, Knutson C, Xiao D, Niu Q, Tsoi M, et al. Phys Rev Lett 2009;102:067201.
[51] Hayashi M, Ieda J, Yamane Y, Takahashi YK, Mitani S, Maekawa S. Phys Rev Lett 2012;108:147202.
[52] Tanabe K, Chiba D, Ohe J, Kasai S, Kohno H, Barnes SE, et al. Nat Commun 2012;3:845.
[53] Yamane Y, Sasage K, An T, Harii K, Ohe J, Ieda J, et al. Phys Rev Lett 2011;107:236602.
[54] Hai PN, Ohya S, Tanaka M, Barnes SE, Maekawa S. Nature 2009;458:489−93.
[55] Zhang S, Li Z. Phys Rev Lett 2004;93:127204.
[56] Thiaville A, Nakatani Y, Miltat J, Suzuki Y. Europhys Lett 2005;69:990.
[57] Tatara G, Takayama T, Kohno H, Shibata J, Nakatani Y, Fukuyama H. J Phys Soc Jpn 2006;75:064708.
[58] Nakatani Y, unpublished results.
[59] Tatara G, Vernier N, Ferre J. Appl Phys Lett 2005;86:252509.
[60] Duine RA, Núñez AS, MacDonald AH. Phys Rev Lett 2007;98:056605.
[61] He J, Li Z, Zhang S. Phys Rev B 2006;73:184408.
[62] Falloon PE, Jalabert RA, Weinmann D, Stamps RL. Phys Rev B 2004;70:174424.
[63] Ohe J, Kramer B. Phys Rev Lett 2006;96:027204.
[64] Parkin SSP, Hayashi M, Thomas L. Science 2008;320:190.

[65] Lepadatu S, Hickey MC, Potenza A, Marchetto H, Charlton TR, Langridge S, et al. Phys Rev B 2009;79:094402.

[66] Boulle O, Kimling J, Warnicke P, Kläui M, Rüdiger U, Malinowski G, et al. Phys Rev Lett 2008;101:216601.

[67] Koyama T, Chiba D, Ueda K, Kondou K, Tanigawa H, Fukami S, et al. Nat Mater 2011;10:194.

[68] Yamanouchi M, Ieda J, Matsukura F, Barnes SE, Maekawa S, Ohno H. Science 2007;317:1726.

[69] Adam J-P, Vernier N, Ferre J, Thiaville A, Jeudy V, Lemaitre A, et al. Phys Rev B 2009;80:193204.

[70] Bazaliy YaB, Jones BA, Zhang S-C. Phys Rev B 1998;57:R3213.

[71] Ansermet J-Ph. IEEE Trans Magn 2004;40:358.

[72] Fernández-Rossier J, Braun M, Núñez AS, MacDonald AH. Phys Rev B 2004;69:174412.

[73] Shibata J, Tatara G, Kohno H. Phys Rev Lett 2005;94:076601.

[74] Xiao J, Zangwill A, Stiles MD. Phys Rev B 2006;73:054428.

[75] Barnes SE, Maekawa S. Phys Rev Lett 2005;95:107204.Barnes SE. Phys Rev Lett 2006;96:189701.

[76] Tserkovnyak Y, Skadsem HJ, Brataas A, Bauer GEW. Phys Rev B 2006;74:144405.

[77] Kohno H, Tatara G, Shibata J. J Phys Soc Jpn 2006;75:113706.

[78] Piéchon F, Thiaville A. Phys Rev B 2007;75:174414.

[79] Duine RA, Núñez AS, Sinova J, MacDonald AH. Phys Rev B 2007;75:214420.

[80] Tatara G, Fukuyama H. Phys Rev Lett 1994;72:772 J Phys Soc Jpn 1994;63:2538

[81] Aharonov Y, Stern A. Phys Rev Lett 1992;69:3593.

[82] Kohno H, Shibata J. J Phys Soc Jpn 2007;76:063710.

[83] Tserkovnyak Y, Fiete GA, Halperin BI. Appl Phys Lett 2004;84:5234.

[84] Korenman V, Murray JL, Prange RE. Phys Rev B 1977;16:4032.

[85] Tatara G, Fukuyama H. Phys Rev Lett 1997;78:3773.

[86] Waintal X, Viret M. Europhys Lett 2004;65:427.

[87] Tatara G, Kohno H, Shibata J, Lemaho Y, Lee K-J. J Phys Soc Jpn 2007;76:054707.

[88] Kohno H, Tatara G, Shibata J, unpublished note.

[89] Thiele AA. Phys Rev Lett 1973;30:230−3.

[90] Kohno H, Tatara G, Shibata J, Suzuki Y. J Magn Magn Mat 2007;310:2020.

[91] Mori H. Prog Theor Phys 1965;33:423 1965;34:399

[92] Götze W, Wölfle P. Phys Rev B 1972;6:1226.

[93] Heide C, Zilberman PE, Elliott RJ. Phys Rev B 2001;63:064424.

[94] Sakuma A, unpublished (cond-mat/0602075).

[95] Skadsem HJ, Tserkovnyak Y, Brataas A, Bauer GEW. Phys Rev B 2007;75:094416.

[96] Stiles MD, Saslow WM, Donahue MJ, Zhangwill A. Phys Rev B 2007;75:214423.

[97] Saslow WM. Phys Rev B 2007;76:184434.

[98] Simánek E, Rebei A. Phys Rev B 2005;71:172405.

[99] Nguyen AK, Shchelushkin RV, Brataas A. Phys Rev Lett 2006;97:136603.

[100] Nguyen AK, Skadsem HJ, Brataas A. Phys Rev Lett 2007;98:146602.

[101] Shinjo T, Okuno T, Hassdorf R, Shigeto K, Ono T. Science 2000;289:930.

[102] Shibata J, Nakatani Y, Tatara G, Kohno H, Otani Y. Phys Rev B 2006;73:020403.

[103] Ishida T, Kimura T, Otani Y. Phys Rev B 2006;74:014424.

[104] Kasai S, Nakatani Y, Kobayashi K, Kohno H, Ono T. Phys Rev Lett 2006;97:107204.

[105] Yamada K, Kasai S, Nakatani Y, Kobayashi K, Kohno H, Thiaville A, et al. Nat Mater 2007;6:269.

[106] Huber DL. J Appl Phys 1982;53:1899—900.
[107] Guslienko KYu, Ivanov BA, Novosad V, Otani Y, Shima H, Fukamichi K. J Appl Phys 2002;91:8037.
[108] Lederer P, Mills DL. Phys Rev 1966;148:542.
[109] Nakatani Y, unpublished results.
[110] Nakatani Y, Shibata J, Tatara G, Kohno H, Thiaville A, Miltat J. Phys Rev B 2008;77:014439.

6 Micromagnetics of Domain Wall Dynamics in Soft Nanostrips

André Thiaville[1] and Yoshinobu Nakatani[2]

[1]Laboratoire de Physique des Solides, Université Paris-Sud, Orsay, France,
[2]Department of Computer Science, University of Electro-communications,
Chofu, Tokyo, Japan

Contents

Nanomagnetism and Spintronics. DOI: http://dx.doi.org/10.1016/B978-0-444-63279-1.00006-X

6.1   Introduction

6.1.1   *Micromagnetics*

The micromagnetic theory was introduced progressively (1930−1960) as a continuous theory of the magnetic structures, with William F. Brown, Jr. as the main contributor to its formalization. By structures here, we mean those that spread across many atomic distances, so that the atomic nature of magnetism (ferromagnetism or ferrimagnetism, localized or itinerant magnetism) is not relevant. The preceding theory was the so-called domain theory, that considers uniformly magnetized domains separated by magnetic walls of zero thickness, with a number of rules concerning the orientation of the magnetization in the domains and of the domain walls (DWs hereafter). In micromagnetics, domains and walls are all described by a continuous function—the local magnetization—a classical vector that is the local average over a small volume of the magnetization density. Several hypotheses underlie the standard formulation of micromagnetics.

- The small volume should contain many atoms, yet be smaller in size than any magnetic structure to be described.
- Temperature has to be low enough so that the thermodynamic equilibrium value of local magnetization $M_s(T)$ is attained everywhere, with negligible thermodynamic fluctuations, using this averaging volume.
- The time variation of magnetization should be slow enough to use the magnetostatic approximation for the computation of the electromagnetic interactions. This last hypothesis is not essential. However, the requirement that the local average of magnetization is the thermodynamic value limits dynamic phenomena to timescales much larger than femtoseconds.

The above points imply that the local magnetization density can then be written $M(r, t) = M_s m(r, t)$ where m is the unit vector indicating the local orientation of the magnetization (in this chapter, bold symbols will be used for denoting vectors). The evolution in space and time of this vector is ruled by an energy functional and a dynamic equation.

Micromagnetic energy is defined as the integral over the sample volume of a density E composed of several terms (we use SI units throughout). The exchange term, written $A\Sigma_{i,j}(\partial m_j / \partial r_i)^2$ (where i and j refer to the spatial components), is the continuous form of the Dirac–Heisenberg exchange energy. The Zeeman term, written as $-\mu_0 M_s H_{app} \cdot m$, expresses the action of an external field H_{app}. The anisotropy term, accounting for preferred orientations of the magnetization, is written generally as $K\, G(m)$ where G is an angular function and K the anisotropy constant. The magnetostatic interaction results in a term reading $-(1/2)\mu_0 M_s H_d \cdot m$, where H_d is the demagnetizing field given by the solution of the magnetostatic problem.

The magnetization dynamics is governed by the Landau–Lifshitz–Gilbert (LLG) equation:

$$\frac{\partial m}{\partial t} = \gamma_0 H_{eff} \times m + \alpha m \times \frac{\partial m}{\partial t}, \tag{6.1}$$

where the effective field is defined by the functional derivative of the energy density with respect to magnetization orientation:

$$H_{eff} = -\frac{1}{\mu_0 M_s}\frac{\delta E}{\delta m}, \tag{6.2}$$

$\gamma_0 = \mu_0|\gamma|$ is the gyromagnetic ratio and α the Gilbert damping constant (a functional derivative is required as the exchange energy density depends on the gradients of m). When solved, the LLG Eq. (6.1) becomes of the form put forward by Landau and Lifshitz (LL):

$$\frac{\partial m}{\partial t} = \frac{\gamma_0}{1 + \alpha^2} H_{eff} \times m + \frac{\alpha\gamma_0}{1 + \alpha^2} m \times (H_{eff} \times m). \tag{6.3}$$

For recent reviews about micromagnetics, the reader should consult Refs. [1–4].

6.1.2 Numerical Micromagnetics

Exact solutions of the equations of micromagnetics are very rare, as the problem is nonlinear ($|m| = 1$) and integro-differential (magnetostatic and exchange energy terms). For finite samples in particular, the magnetostatic problem cannot be circumvented. With the increase of computers power and the recent availability of micromagnetic codes,[1,2,3,4,5,6] numerical micromagnetics has become a widely spread tool for the analysis of the magnetic structures in samples as nanofabricated nowadays.

Most codes discretize the sample volume into equal cells, with a finite differences implementation of the equations. This works perfectly and very efficiently for idealized samples with straight edges. For complex sample shapes, the finite elements formulation is better suited in principle as it allows any sample shape [5]. For both methods, the cell size should be smaller than the characteristic relevant length of the material [1]. Quite generally, one defines the exchange length:

$$\Lambda = \sqrt{2A/\mu_0 M_s^2} \tag{6.4}$$

and, in cases where anisotropy is large, the Bloch wall width parameter:

$$\Delta = \sqrt{A/K}. \tag{6.5}$$

The typical value for Λ is 5 nm, whereas Δ can vary between 1 and 100 nm depending on the anisotropy value.

A large part of the numerical calculation time is devoted to the solution of the magnetostatic problem. The demagnetizing field obeys the same equations as the electric field in electrostatics, with "magnetic" charges spread in the sample volume (density $-M_s \, \mathrm{div} \, \boldsymbol{m}$) and on the sample surface (density $M_s \boldsymbol{m} \cdot \boldsymbol{n}$, with $\boldsymbol{n}$ the outward-oriented normal to the surface). Thus, the demagnetizing field can be computed directly as a convolution product of the charges with a kernel function. In the finite differences method where all cells are identical, this operation is a discrete convolution product that can be evaluated through fast Fourier transformation, much more rapidly than by direct summation (the cost being $N\log N$ instead of N^2, where N is the number of mesh cells). This is the main advantage of the finite differences method. For the finite elements technique, the calculation is more complex

[1] OOMMF is a free software (in fact, an open framework for micromagnetics routines) developed by M.J. Donahue and D. Porter mainly, from NIST. It is available at <http://math.nist.gov/oommf>.

[2] The commercial LLG Micromagnetics Simulator software is developed by M.R. Scheinfein. See the website <http://llgmicro.home.mindspring.com>.

[3] The commercial MicroMagus software is developed by D.V. Berkov, Innovent Technology Development e.V., Jena, Germany. See the website <http://math.micromagus.de>.

[4] The commercial finite elements micromagnetic simulator FEMME, from Vienna University of Technology, is being developed by D. Süss. See the website <http://www.suessco.com>.

[5] The finite difference micromagnetic simulator MuMax, written for GPUs, was developed by Arne Vansteenkiste and Ben Van de Wiele at Ghent University. It is available at <http://code.google.com/p/mumax>.

[6] The finite elements micromagnetic simulator Nmag was developed at Southampton University by Hans Fangohr, Thomas Fischbacher, and Matteo Franchin, mainly. It is available at <http://nmag.soton.ac.uk>.

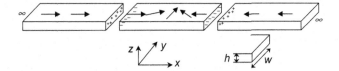

Figure 6.1 Sketch of a calculation region in an infinite nanostrip made out of a material with no anisotropy. The magnetic charges at the wire ends are drawn ($+$ and $-$ signs). The separation between the calculation region and the rest of the sample was increased for the sake of drawing. The coordinates system is also defined, as well as the sample dimensions.

as the magnetostatic problem needs to be solved on the same sample mesh, but involves the field ouside the sample (see the review [6] for more details).

For samples shaped as nanowires where the dynamics of one domain wall is investigated, some additional points need to be considered:

• Nanowires may be very long, thus impossible to mesh entirely. Moreover, far from the DW and from the wire ends, the magnetization becomes invariant along the wire length. In such a case, a restricted calculation region can be used, centered on the DW and moving with it (Figure 6.1). An infinitely long wire can be mimicked by canceling the magnetostatic charges at the ends of the calculation region (this applies to soft nanostrips as considered in this chapter; the charges are different for domains magnetized vertically (z) or transversally (y)). The length of the calculation region along the wire has of course to be well above the static DW width, and for dynamics one should always check that no structure propagates to its ends.
• The total energy of an isolated DW in the wire is not that of the calculation region alone, because the stray field of the DW extends to infinity. Indeed, the DW in nanowires considered here bear a nonzero magnetostatic charge, contrarily to most DW in bulk samples (so that the prohibition of charged DW was one of the rules of the old domain theory). Thus, small deviations of the magnetization exist in the domains, far from the DW.

The representative calculations to be discussed below were performed with parameters characteristic of the soft alloy $Ni_{80}Fe_{20}$ that has been used in many experiments, namely $M_s = 8 \times 10^5$ A/m, $A = 10^{-11}$ J/m, K = 0, $\alpha = 0.02$, and $\gamma_0 = 2.21 \times 10^5$ m/(As). They were performed using a 2D mesh, with mesh size 4 nm (x) $\times$ 4 nm (y) $\times$ h (z), where $h = 5$ nm, mostly.

6.1.3   Domain Wall Structures in Soft Nanostrips

As all experiments performed up to now use samples patterned by lithography techniques out of thin magnetic films, we shall consider flat nanowires of rectangular cross section (width w and thickness h, the area of the cross section will be called $S = wh$) that are referred to as nanostrips. Starting from a schematic DW configuration, the numerical calculations (no applied field, damping set to $\alpha = 0.5$ in order to reach equilibrium rapidly) show the existence of several stable structures [7,8] that are displayed in Figure 6.2.

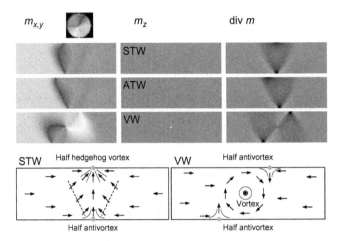

Figure 6.2 Computed domain wall structures in a permalloy nanostrip of cross section $w = 300$ nm wide and $h = 5$ nm thick. Each structure is shown by the in-plane (x, y) magnetization components (left) with circular color coding (map shown on top), by the perpendicular (z) component in black and white (center), and finally by the volume charge (div m), in black and white (right). Schematic drawings of the two basic structures are also included. The ATW can exist in two variants with left and right inclination, only the left one being shown here. Images are 1 μm wide, one half the width of the calculation region. All walls are of head-to-head type, with a positive total magnetostatic charge (coded as black on the right column of images). For tail-to-tail walls, the magnetostatic charge is reversed, but the structures are the same ($m \rightarrow -m$ transformation). (For interpretation of the references to color in this figure legend, the reader is referred to the web version of this book.)

6.1.3.1 The Symmetric Transverse Wall

In this structure, closest to what is expected for a one-dimensional magnetic wire, the DW average magnetic moment is oriented along the wire transverse axis. To quantify the DW transverse moments, we define for $i = y, z$:

$$\mu_i = \frac{1}{\pi S} \int m_i d^3 r \tag{6.6}$$

that has the dimensions of a length (for a Bloch wall profile with $m_i = 1/\cosh(x/\Delta)$ one gets $\mu_i = \Delta$). As the cross section is very flat ($w \gg h$), this moment is oriented in the plane of the magnetic film. The DW region has the shape of a triangle. In the framework of the domain theory, the avoidance of magnetic charges inside the strip would indeed lead to 45° inclined boundaries between the "domains" and "wall" blocks. The structure is also characterized by two topological objects, located on the edges of the nanostrip [9], as depicted by the schematic drawing included in Figure 6.2. At the triangle apex, a half antivortex is present. A full antivortex (entirely contained inside the sample) would show a magnetization perpendicular to the film plane at his core. Here, as shown in the picture coded with the z component, there is no trace of a perpendicular magnetization in statics. However, the in-plane magnetization component conforms that of a half antivortex. Thus, the topological charge (winding number of the in-plane magnetization) of this object is $q = -1/2$ [9].

At the strip edge opposite to the apex, another topological structure exists. It is a half hedgehog vortex, again with no core, that is topologically equivalent to a half vortex hence with a topological charge $q = +1/2$. The half hedgehog vortex bears, contrarily to the half vortex, a magnetostatic charge. In the block structure discussed above, all the DW charge would be transferred to this edge, at the base of the triangle. In the magnetostatic limit (no role of exchange energy), all the DW charge would be concentrated on the half hedgehog vortex. The numerical simulations show that, depending on the sizes of the wire cross section, the charge is spread all over the DW region (Figure 6.2). The charge image, in a first approximation [10], corresponds to the image of the DW structure obtained by magnetic force microscopy.

6.1.3.2 The Vortex Wall

This DW structure does not show any transverse magnetic moment. However, due to the presence of the vortex core, a perpendicular magnetization component exists locally. As the vortex core is small (a few Λ [1]), the average perpendicular moment is much smaller than the transverse moment of the STW.

Topologically, the structure consists in a vortex ($q = 1$) with two half antivortices, one at each edge of the nanostrip, as drawn in the schematic drawing included in Figure 6.2.

In terms of magnetic charges, this structure spreads the DW charge more than the STW and thus reduces the magnetostatic energy, at the expense of a larger exchange energy.

6.1.3.3 The Asymmetric Transverse Wall

This structure is a variant of the STW, that becomes stable at larger film thickness h and/or width w. The half hedgehog vortex is no longer aligned with the half antivortex along the transverse direction, but shows some offset (mainly visible on the volume charge pictures in Figure 6.2).

6.1.3.4 Energies and Phase Diagram

These three DW structures can be absolutely stable or metastable, depending on the nanostrip width w and thickness h.

The equal energy line between vortex wall (VW) and TW was found to lie at $wh \approx 61.3\Lambda^2$ (1500 nm^2 for NiFe) [7,8]. However, as shown experimentally and by micromagnetic calculations, in the case where the DW is nucleated by application of a field transverse to the wire, the TW can be found well above this line [11]. This metastability is due to the existence of an energy barrier. Indeed, in order to transform a VW into a TW, the vortex has to be expelled at one of the strip edges, leaving the center where its energy is lowest. Similarly, in order to transform a TW into a VW, the half hedgehog vortex has to leave the strip edge, thus creating the vortex core with perpendicular magnetization that makes most of the vortex energy. If it were possible to vary continuously the film thickness h, one would say that the TW−VW transition is first order.

In addition, above some thickness, the STW transforms to an ATW. An energy barrier for this process has not been seen in the numerical calculations. Indeed, this transition can be realized continuously (a shift of DW center along the strip length that depends on position across the strip width). Thus, the STW−ATW transition can be described as second order, and we shall use the symbol TW to denote both structures.

Figure 6.3A plots these two equal energy lines, as obtained from the numerical calculations. One should note that, as these calculations were performed with one cell in the film thickness (2D calculations), the "phase diagram" thus obtained is limited to low thicknesses. A deformation of the transition lines is anticipated at larger thicknesses. This phase diagram was confirmed by experiments where the most stable DW structure was obtained by thermal annealing [12]. This study also revealed the role of sample imperfections, that may hinder the DW conversion and even lead to a reentrant VW phase at very low thickness. The variation with film thickness of the energies of the three DW types presented here is plotted in Figure 6.3B, for one value of the wire width, namely $w = 300$ nm.

These are the simplest DW structures, corresponding to the small sizes of the nanostrips. The micromagnetic parameters that characterize these three structures are shown in the Table 6.1.

For much larger thicknesses, comparable to the nanostrip width, it has been shown for example that another type of DW exists, called Bloch point wall [13,14]. At large width and small thickness, cross-tie walls and diamond structures appear [12]. Eventually, the benefit of the reduced dimensions disappears and we recover the full complexity of the bulk domain structures.

6.2 Field Dynamics of Domain Walls

Domain wall dynamics under an applied field is a rather old subject [15]. The early studies however concerned bulk samples, and only more recently thin films. We will address here DW dynamics in soft nanostrips, where the DWs are confined and their dynamics simpler. This research field is rather recent, as samples of high quality have been fabricated in the last 15 years only.

6.2.1 Panorama

Let a field be applied along the nanostrip axis (x direction; coordinates are defined in Figure 6.1) so as to displace the DW. Far away from the DW, the field exerts no torque as m and H_{app} are collinear, so that the magnetization in

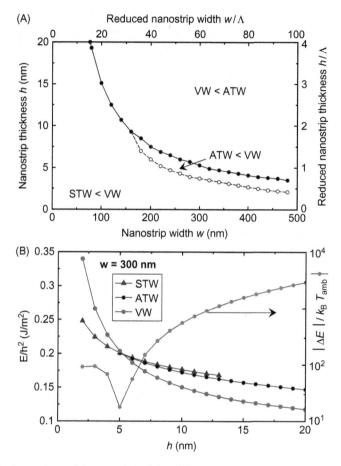

Figure 6.3 Comparison of the energies of the different DW structures upon variation of the nanostrip dimensions. (A) "Phase diagram" mapping the lowest energy structures [8]. (B) Energies of the three DW structures in permalloy nanostrips of cross section $w = 300$ nm wide and varying thickness, summed over the $2\,\mu$m calculation region. VW and TW have very close energies at the thickness $h = 5$ nm, a value that applies to all numerical results shown in this chapter. Because of its symmetry, the STW structure can be obtained above the thickness for transition to ATW, but in an unstable state. In addition, the energy difference (absolute value) beween VW and TW is compared to the thermal energy at room temperature (note the log scale). These calculations were performed with a 2D mesh.

the left and right domains does not change. This mathematical evidence is in fact correct only for low fields: in real samples some deviations of magnetization are always present (the ripple structure [1]) that become exacerbated under a reverse applied field, eventually leading to the nucleation of domains of magnetization aligned with the field before sufficient DW motion may take place. Thus the dynamics of one DW can be properly calculated (and measured) only under low fields.

Table 6.1 Main Micromagnetic Parameters for the Three Simplest DW Structures in a 300 nm Wide and 5 nm Thick NiFe Nanostrip

Parameter	STW	ATW	VW
DW energy (10^{-18} J)	5.014	5.001	5.071
$\Delta_T(v = 0)$(nm)	52.2	49.5	25.4
DW y moment μ_y(nm)	53.06	53.62	-0.53
DW z moment μ_z(nm)	0	0	0.19
$\alpha\tau_{DW}$(ps)	6.3		400
v_{max}(m/s)	644.7		61.1
$\Delta_T(v_{max})$(nm)	31.8		26.1

Symbols are Δ_T the Thiele domain wall width (Eq. 6.8), v_{max} the DW maximum velocity, τ_{DW} the characteristic time for DW structure relaxation, and μ_i the DW transverse magnetization moments (Eq. 6.6). DW energies were evaluated in a 2-μm long calculation box. Note that the dynamic parameters (lower half) are identical for the STW and ATW as the STW is unstable at this nanostrip size.

It is not possible here to describe what is obtained for all points in the phase diagram of Figure 6.3A. The reader is referred to Chapter 5 by H. Kohno and G. Tatara for an approach of this physics by the collective coordinates model. In addition, a chapter devoted to field dynamics of DW in soft nanowires and nanostrips was written recently by us [13], where this model, also called 1D model,[7] of DW dynamics is discussed in detail and compared to numerical calculations. We shall here recall the main features of these dynamics as revealed by numerical calculations, keeping the nanostrip size constant (width $w = 300$ nm and thickness $h = 5$ nm). One interest of this size is that the TW and VW have very similar energies, as just shown.

Figure 6.4 shows the velocity versus field curves for this sample, for an (A)TW and a VW. The values were obtained from a collection of time-dependent calculations where a DW at rest is submitted at time $t = 0$ to an applied field. Two very distinct regions can be seen, that correspond to different regimes of motion.

6.2.2 Steady-State Regime

For all DW structures, a first regime, beginning at zero field, shows a linear—or slightly curved—increase of velocity with field. This behavior is to be expected as the field supplies energy and the damping term in LLG is akin to a viscous friction.

This regime of motion is called steady state. Were the DW width in motion the same as at rest, the $v(H)$ curve would be perfectly linear. Indeed, Thiele [17] has shown that for a stationary DW motion (a motion where the DW structure does not change when seen in the moving frame so that $m(r, t) = m_0(r - q(t))$), the DW velocity v obeys the famous Walker relation:

[7] More appropriately, this model should be called 1D collective coordinates model. Indeed, it does describe a 1D domain wall motion, but on the other hand it is not a full 1D micromagnetic model that treats the entire function $m(x)$, as it is sometimes necessary to consider [16].

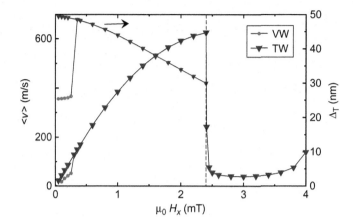

Figure 6.4 Computed velocity v of domain walls, under a field applied along the axis of a nanostrip with a cross section $w = 300$ nm wide and $h = 5$ nm thick. The damping constant is $\alpha = 0.02$. A transverse wall (TW, in fact an asymmetric transverse wall) and a vortex wall (VW) are compared, the latter existing only up to a velocity of about 60 m/s. In the steady-state regime (up to the dashed line), the Thiele domain wall width Δ_T is also plotted. The velocity in the nonsteady-state is a time-averaged value.

$$v = \frac{\gamma_0 \Delta_T}{\alpha} H, \tag{6.7}$$

where the Thiele DW width Δ_T is defined [13] as an integral over the moving DW structure m_0:

$$\frac{2}{\Delta_T} = \frac{1}{S} \int \left(\frac{\partial m_0}{\partial x} \right)^2 d^3 r. \tag{6.8}$$

This relation results from the evaluation of the Thiele dissipation matrix, Eq. (6.7) being just a reexpression of the balance of forces in the Thiele equation.

The corresponding time-dependent evolution for a TW is plotted in Figure 6.5A for a representative field applied abruptly at time $t = 0$. It shows that the DW reaches an invariant structure after some time, and that this structure differs from the one at rest. The exponential relaxation of DW structure involves the characteristic time τ_{DW}. The deformation of the DW structure expresses magnetization rotation around the applied field (an m_z component appears for the TW, and an m_y component for the VW, see Figure 6.5D).

Figure 6.5A shows the initial decrease of Δ_T, more pronounced as H becomes larger (see Figure 6.4), that explains the downward curvature of the $v(H)$ curve for the TW. The change of the Thiele DW width reflects the higher energy of the moving DW structure when compared to the structure at rest. As for small field (hence velocity) this energy increase scales as v^2, the factor in front of v^2 may be called a

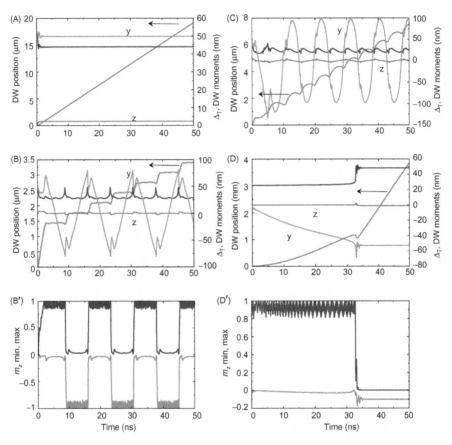

Figure 6.5 Computed motion of domain walls, under a field applied along the axis of a
nanostrip with a cross section $w = 300$ nm wide and $h = 5$ nm thick. The damping constant
is $\alpha = 0.02$. The starting structure is either a TW, with applied fields $\mu_0 H_x = 1, 2.6$, and
4 mT (panels (A), (B), and (C), respectively) or a VW (0.4 mT, panel (D)). The figures show
the DW position (left axis) and, on the right axis, the Thiele DW width (Δ_T) as well as the
two transverse moments, along y and z—see labels, of the DW magnetization (Eq. 6.6). For
cases (B) and (D), supplementary graphs (B′, D′) plot the time evolution of the maximum
and minimum perpendicular magnetization component. When these last values reach the
vicinity of ± 1, a localized structure (V, AV) is present. The fluctuations seen below $+1$ or
above -1 are due to the discreteness of the mesh. The applied field for the VW is just
above the first Walker threshold, so that the VW is seen to transform to a TW during the
calculation.

mass, known as the Döring mass [13,18]. The existence of this mass explains the
transient dynamics of the DW when the field is applied abruptly.

Figure 6.6 shows a series of snapshots of DW structures in steady-state motion,
under increasing fields. For the TW (Figure 6.6A), all structures are ATW.
Magnetization rotation (as shown by the charge image) peaks at the leading edge

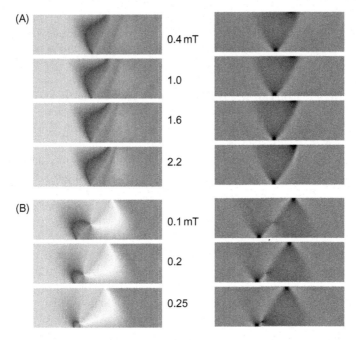

Figure 6.6 Snapshots of DW structures in steady-state motion (toward the right of the pictures) under different fields (values of $\mu_0 H_x$ are indicated), starting from a TW (A) and a VW (B). The nanostrip size is $w = 300$ nm and $h = 5$ nm. The left images display the in-plane magnetization orientation with the same color code as in Figure 6.2, and the right images map the magnetization in-plane divergence (the magnetic volume charge). Images are 1 μm wide, one half of the calculation region. The corresponding velocities are 169, 385, 532, and 612 m/s for the TW and 19, 41, and 52 m/s for the VW. (For interpretation of the references to color in this figure legend, the reader is referred to the web version of this book.)

of the DW as well as at the half hedgehog vortex and half antivortex on the sample edges, increasing with DW velocity. The m_z component also increases in this region as field and velocity get larger (not shown). For the VW (Figure 6.6B), the main deformation under motion is the progressive shift of the vortex core in the transverse y direction, that causes the change of the y moment of the VW (Figure 6.5D).

The comparison of the results for the TW and VW (Figure 6.4) reveals that the VW moves more slowly than the TW, at the same field. The Thiele DW width is indeed smaller for the VW, due to the larger magnetization gradients that occur because of the existence of a vortex core (Table 6.1 and Eq. (6.8)). Note that this result is opposite to the naive appreciation of the DW width from the magnetization pictures (Figure 6.2). More quantitatively, if the magnetization profile averaged over the transverse variables y and z were fitted by the profile of a Bloch wall ($m_x = \pm\tanh(x/\Delta_x)$), the resulting Δ_x would indeed be larger for the VW. The reason for this difference is that the Thiele DW width derives from the energy dissipation in LLG, and one should rather think of the DW width that governs DW dynamics in the steady-state regime as a measure of energy dissipation. The existence of this "dynamic" DW width shows that a straightforward application of the Bloch wall dynamics solution to DW in nanostrips may turn out to be very far from reality.

Starting from a VW, two steady-state regimes can be seen (Figure 6.4), the second steady-state regime being identical to that of a TW. The reason is that the breakdown marking the end of the first steady-state regime leads to a TW structure, that can move in steady-state conditions up to higher fields. Indeed, in the first regime, magnetization precession inside the VW leads to the appearance of an m_y component, linked to the displacement of the vortex core in the y direction from the nanostrip center (Figure 6.6B). This displacement increases with field (thus, and more appropriately in fact, with DW velocity, see later) up to a point where the vortex core is expelled from the sample. This transforms the VW into a TW (Figure 6.5D,D').

In all cases, the end of the (last) steady-state regime is characterized by a velocity that reaches a maximum. More precisely, due to the curvature of the $v(H)$ curve, the maximum may be even reached slightly before the end the steady-state regime (the collective coordinates model predicts that this is possible [13]).

All this phenomenology was uncovered by Walker in the case of the 1D Bloch wall in an infinite sample with uniaxial anisotropy [19], so that the end of the steady-state regime may be described as Walker field and Walker velocity. These terms are kept here, even if the walls existing in the nanostrips have nothing to do with a Bloch wall.

6.2.3   Motion Above the Walker Field

When the steady-state regime ends, the average velocity drops abruptly as shown in Figure 6.4 (this being called the Walker breakdown). The time traces for this case (Figure 6.5B) show that some regularity in time however exists, where the DW magnetization components oscillate as well as DW width and velocity. Thus, an average velocity can be defined, that has been plotted in Figure 6.4. The snapshots of the DW structures (Figure 6.7), as well as the plot of the maximum and minimum of the perpendicular component of magnetization (Figure 6.5B'), reveal the now well-known continuous transformation of the DW structures, by the injection, displacement across the nanostrip width and expulsion of single antivortices (AV) and vortices (V). Note that the breakdown mechanism with the appearance of these very localized structures is rather different from the predictions of the 1D collective coordinates model, where a precession of the DW magnetization is expected (this is also reflected in the very different oscillation amplitudes of the y and z DW moments, whereas in the 1D collective coordinates model they would be equal). The topological analysis presented above is useful for understanding which structures do appear and where [20]. Note that the above-described mechanism is specific to soft materials.

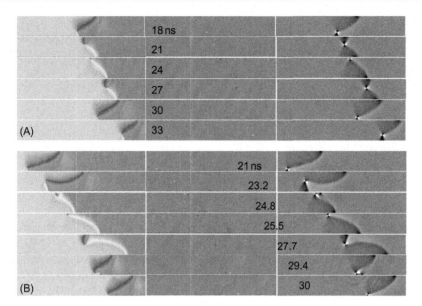

Figure 6.7 Snapshots of a TW set into motion by fields larger than the Walker value, namely $\mu_0 H_x = 2.6$ mT (A) and 4.0 mT (B). The nanostrip size is $w = 300$ nm and $h = 5$ nm , and the same part of the nanostrip is shown in each series. The left images display the magnetization in-plane component, the out-of-plane component is displayed in the center images and the right images map the magnetization divergence. For a field weakly larger than the Walker value (A), the DW structure changes by continuous injection of an AV at one strip edge, that crosses the sample width to disappear at the other edge, followed by the injection of an AV with opposite core magnetization at that edge and so on. For a larger field (B), the AV annihilates with a V of opposite core magnetization that is injected before it reaches the other edge of the sample. Note the increasing DW x extension as field gets larger, and the "anchoring" role played by the AV core. (For interpretation of the references to color in this figure legend, the reader is referred to the web version of this book.)

At still larger fields the regularity progressively disappears (Figure 6.5C and 6.7B) because more than one AV or V is present. Consequently, the evaluation of an average velocity is not so easy, and the numerical data show some scatter. In this regime, the velocity appears to increase again with field. Thus, the $v(H)$ curve shows a minimum value slightly above the Walker field, that we will denote by v_{min}. As the processes become rather complex, calculations should be checked for mesh independence and robustness to disorder (see next section) before attempting to compare these results to experiments. In this regime, it is obvious that the collective coordinates model will not work. Too many degrees of freedom exist in the nanostrip considered in this example (the stripe width is about 60Λ), that can be excited.

The control of the Walker breakdown has been the subject of several studies, with the obvious goal of keeping a DW velocity close to its maximum value.

They involve either an exchange coupling to a perpendicularly magnetized film [21], or the application of a large enough transverse field that leads to a separation of the nucleated AV from the DW [22].

6.2.4 Imperfect Samples

6.2.4.1 Propagation Field

One important characteristic of the DW dynamics in the perfect samples considered up to now is the existence of a linear regime at small applied fields, down to zero. In real samples however, a minimum field has to be applied in order to displace an existing DW, that is therefore called propagation field, H_P. This field is different from the coercive field of the sample, H_c, that can be measured by magnetometry, as the latter involves in general DW nucleation, propagation, and disappearance. One advantage of the nanowires is to allow separating propagation of DWs from their nucleation, as recognized early [23]. The propagation field may originate from many types of defects, so that its faithful micromagnetic modeling is difficult. The defect type most considered for nanostrips is edge roughness [24–26], but local variation of the micromagnetic parameters (A, M_s, or even h) has also been considered. The edge roughness mechanism is anticipated to be dominant among the microfabrication-induced defects, and microfabrication plays indeed a big role as, very often, $H_P > H_c$. This is also consistent with the experimental observation that H_P increases as the nanostrip width decreases. Values of H_P are rarely given in the literature, the published numbers ranging from ≈ 1.5 Oe [27] to 100 Oe [28] ($\mu_0 H_P = 0.15$ to 10 mT). For comparison, a good NiFe film has $H_c \approx 1$ Oe ($\mu_0 H_c = 0.1$ mT). We will therefore discuss now the effects of edge roughness on DW dynamics by field.

Periodic roughness was first tried [24], but is not realistic at least for cases where roughness is not intentional. Random roughness is therefore preferred. Its generation is inspired by the existence of crystallites inside the sample. A random grain pattern can be generated as the Voronoi cells for a random distribution of points [26]. The rough edge is then obtained as the boundary of the grains contained within a nominal nanostrip width with, if necessary, some regularization to eliminate excessive roughness. The parameter of such a roughness is the point density or, equivalently, the mean grain diameter $<D>$. In order to implement the moving calculation region technique, a sufficiently long rough nanostrip has to be generated first. As the magnetization in a rough nanostrip deviates from strict uniformity, this should be taken into account at the x boundaries of the calculation region. However, as the roughness considered here is small, of the order of the exchange length, these deviations are neglected for simplicity.

The calculated DW dynamics with such a roughness is shown in Figure 6.8, to be compared with Figure 6.4 in the perfect case. The figure depicts one of the problem of the numerical micromagnetics of imperfect samples. For a given field, there is some probability that a DW stops at some place where the pinning potential induced by roughness is too large. This occurs especially as these calculations were

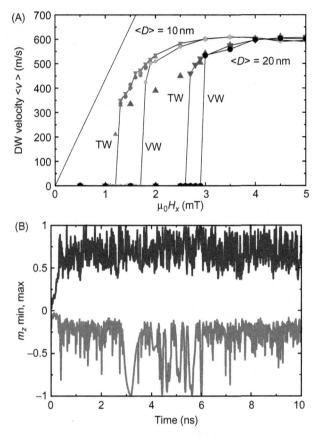

Figure 6.8 Motion of domain walls by field in a nanostrip with rough edges, of cross section $w = 300$ nm wide in average and $h = 5$ nm thick. The damping constant is $\alpha = 0.02$. (A) Computed velocity of domain walls. A transverse wall (TW, in fact an asymmetric transverse wall, triangular symbols) and a vortex wall (VW, circles) are compared. Two values of the average size of grains that are used to define the nanostrip edges are shown, $<D> = 10$ (smaller symbols) and 20 nm (larger symbols). For each case, the velocity in linear motion and the final velocity are indicated (calculations last 50 ns), showing the field window in which the DW motion is stochastic. The straight line shows the linear mobility of the TW extracted from Figure 6.4. Lines link the data points with the lower velocities. (B) Time evolution of the maximum and minimum m_z magnetization components for a TW, under field $\mu_0 H_x = 4$ mT for the smaller edge roughness $<D> = 10$ nm.

performed without thermal fluctuations (a systematic study of the effect of including the thermal fluctuations [29] showed a jerky DW motion with pinning and depinning close to H_P, with an exponential variation of velocity with field). In the absence of a better statistical analysis, criteria have to be adopted for deciding whether the DW moves or not, so as to compute its velocity. In the case of Figure 6.8, motion was declared to take place when the DW moved for more than

1 μm, and in cases it eventually stopped the velocity during the motion phase was evaluated, together with a zero value indicating DW stopping. Thus, in a certain field range (that depends on the calculation total time and on the 1 μm criterion), DW motion and stop can occur. This corresponds to the experimental behavior.

Figure 6.8 shows that propagation fields $H_P \approx 10 - 30$ Oe ($\mu_0 H_P = 1-3$ mT) can be simulated by this technique. The grain size $<D>$ is however limited by the numerical mesh size, so that for lower H_P smaller $<D>$, hence smaller mesh sizes, are required. As a consequence, small velocities of DW motion are not observed when using standard mesh sizes of micromagnetics (below, but comparable to the exchange length Λ). The approach to zero velocity in the vicinity of H_P is also not computed, as it would require the introduction of thermal fluctuations and extremely long calculation times [30].

In the case chosen for Figure 6.8, it appears that a VW is more pinned than a TW. This is however not a general conclusion, as for other sizes the opposite was seen. The DW pinning is also quite sensitive to the type of defects. For example, film thickness reduction (up to zero) for randomly chosen cells is very effective on a VW, as this DW structure bears a large energy localized at the vortex core [1].

6.2.4.2 Walker Breakdown

Apart from the appearance of a propagation field, disorder as introduced here has a striking consequence: the Walker breakdown is pushed to much higher fields (not even visible in the field range of this plot). Therefore, already little above the propagation field the DW travels at an average velocity clipped to the maximum velocity of the perfect case [26]. This was shown to result from energy dissipation at the nanostrip edge, in the form of spin waves emission, preventing the apparition of localized structures (V and AV) seen in the perfect case. This is demonstrated in Figure 6.8B plotting the evolution of the maximum and minimum of the perpendicular magnetization for one case: comparison with Figure 6.5B′ and D′ shows variations of these extremum values that are much larger than the fluctuations seen when a V or AV is present, due to mesh discreteness. It should be noted that the best samples fabricated nowadays are better than those of this calculation (see Section 2.5), so that predictions for perfect samples may be not so unrealistic.

6.2.4.3 Increased Effective Damping

The micromagnetic study of the effect of "bulk" disorder, realized by a fluctuation of M_s that stands for a fluctuation of thickness, has given interesting results [31]. Indeed, it has been found by varying the drive field for a VW in a NiFe nanostrip, that the average velocity curve was very similar to that for a perfect sample, only with a larger damping constant. The enhancement was up to a factor of 2 for reasonable magnitudes of disorder. The effect is attributed to spin waves: a DW that moves in a sample in which the magnetic properties fluctuate emits some spin waves that propagate off the DW and are damped out. At this point one should note that, from Eqs. (6.1) and (6.2), one can show that under a constant applied field the total energy of the sample varies with time as:

$$\frac{\mathrm{dE}}{\mathrm{d}t} = -\alpha \frac{\mu_0 M_s}{\gamma_0} \int \left(\frac{\mathrm{d}\boldsymbol{m}}{\mathrm{d}t}\right)^2. \tag{6.9}$$

Assuming that the time variation of magnetization due to the motion of a DW creates spin waves proportionally, we get from Eq. (6.9) that this is globally equivalent to having an increased damping constant α in a medium without fluctuations.

6.2.5 Comparison to Experiments

The experimental investigation of DW dynamics by field in nanostrips has started rather recently, once samples have become available (the reader should consult Chapter 4 by T. Ono and T. Shinjo).

The main observable was, initially, the DW velocity and its dependence on field. In the first experiments [23,32] only one regime was seen, and the propagation fields were rather high. More recently, however, Walker breakdown could be experimentally seen in a few cases [27,33,34] where the propagation field was very low, 1.5 and 5 Oe, the Walker breakdown occurring between 6 and 10 Oe.

The other signature of Walker breakdown is the oscillation with time of the DW structure, velocity, etc., as shown in Figure 6.5. In one experiment [27], oscillations of DW velocity were indeed detected, however as a function of DW position rather than of time. By a careful analytical and numerical work, they could nevertheless be related to the Walker breakdown process [35]. On the other hand, oscillations of the DW width have been inferred from time-resolved measurements of the anisotropic magnetoresistance (AMR) of a nanostrip sample [36] for large enough applied fields. The variation of the resistance R of the sample due to AMR reads:

$$\Delta R = -\frac{\Delta\rho}{S^2} \int (m_y^2 + m_z^2)\mathrm{d}^3 r, \tag{6.10}$$

where $\Delta\rho$ is the change of resistivity by AMR. Thus, AMR is sensitive to a geometric DW width and not to the dynamic DW width of Thiele. Nevertheless, an oscillation of AMR due to DW dynamics above the Walker threshold is also anticipated, similarly to that of Δ_T (Figure 6.5). The nonobservation of this phenomenon in earlier experiments may be due to insufficient sample quality masking the regularity of the DW dynamics slightly over the Walker field. Indeed, as shown in the previous section, the Walker breakdown may even be suppressed at large roughness [26]. Also, the control of the Walker field predicted by the numerical simulations quoted above has been observed. Applying large transverse fields, DW velocities up to 4500 m/s were measured in NiFe [37].

Thus, we may say that the micromagnetic predictions fit with experiments (a case with a precise and detailed comparison between both was published in Ref. [38]).

6.3   Domain Wall Motion by Spin-Polarized Current

This section describes the research about spin-polarized current-induced DW motion in soft nanostrips, and, more generally, current in plane (CIP) spin transfer, from a micromagnetic point of view. At the time of the first edition of this book (2009), the way to introduce CIP STT in micromagnetics was still controversial. This is no longer the case now, as experiments have confirmed the relevance of the phenomenology that will be presented below. We start by describing the controversy before showing numerical results and comparing them to experiments. Analytic results, using schematic DW models, are presented in Chapter 5 by H. Kohno and G. Tatara.

The reader is also referred to the number of reviews about this subject that have appeared since the first edition of this book [39–42].

6.3.1   CIP Spin Transfer in Micromagnetics

In the early theoretical work by Luc Berger [43,44], spin-transfer torque on a DW was considered in a space-integrated form, from the change of conduction electrons angular momentum after crossing the DW. Bazaliy et al. [45] were the first to propose an expression for the local torque due to spin transfer inside a DW. This expression applies to the situation of wide walls, when the DW width δ (for the Bloch wall profile one has $\delta = \pi\Delta$, Δ being called the wall width parameter) is large compared to the so-called Larmor length:

$$\delta \gg \frac{\hbar v_{\mathrm{F}}}{E_{sd}}. \tag{6.11}$$

In this expression pertaining to free "s" electrons exchange-coupled to localized "d" electrons responsible for the material's magnetization, v_{F} is the Fermi velocity and E_{sd} the energy splitting of the s electrons' up and down bands. For wide walls, and in similarity to neutrons traveling inside a polarization rotation device [46], the conduction electrons' spin is expected to closely follow the local d electrons spin direction. This is called the adiabatic limit. The spin transfer torque (STT) is then obtained as the differential change of the angular momentum of all conduction electrons (we use the symbol L in these formula as J already denotes the current density, but the total angular momentum is meant here). For a sample slice of thickness $\mathrm{d}x$ (along the electron flow), the time evolution of its angular momentum can be evaluated both by the evolution of the local magnetization and the amount of angular momentum deposited by the conduction electrons:

$$\frac{\mathrm{d}}{\mathrm{d}t}(\mathrm{d}\boldsymbol{L}) = -\frac{\mu_0 M_{\mathrm{s}}}{\gamma_0}\frac{\partial \boldsymbol{m}}{\partial t}\,\mathrm{d}x$$

$$= \frac{J}{e}P\frac{\hbar}{2}[\boldsymbol{m}(x+\mathrm{d}x) - \boldsymbol{m}] \tag{6.12}$$

As the time derivative of an angular momentum is a torque, this effect is called STT. Therefore, the spin-polarized current leads to a time evolution of the magnetization that can be expressed as:

$$\left.\frac{\partial m}{\partial t}\right|_{STT} = -\frac{Jg|\mu_B|P}{2eM_s}\frac{\partial m}{\partial x} \equiv -(u \cdot \nabla)m. \tag{6.13}$$

In this equation, e is the absolute value of the electron charge and the sign of the current density J corresponds to the direction of motion of the electrons. We refer to this form of STT as CIP STT, since the current flows along the magnetic layer with nonuniform magnetization. Adiabatic CIP spin transfer can thus be represented by an equivalent velocity u. This velocity is proportional to the current density J and its polarization ratio P and, importantly, inversely proportional to the magnetization M_s. The velocity u may be compared to the classical drift velocity v_d associated to the current by $J = n_e e v_d$ where n_e is the conduction electrons density. One obtains:

$$u = v_d P \frac{n_e}{n_s}, \tag{6.14}$$

where n_s is the density of Bohr magnetons. For NiFe, the prefactor $g|\mu_B|/(2eM_s)$ is numerically equal to 7×10^{-11} m³/C. Thus, a typical current density $J = 10^{12}$ A/m² with a polarization $P = 0.7$ corresponds to $u = 50$ m/s. This form of the adiabatic CIP STT is well accepted, with u also denoted by b_J [47,48] (but the double vector product form $m \times (\partial_x m \times m)$ is unnecessarily complicated as it is always equal to $\partial_x m$ because $|m| = 1$).

A first issue of controversy appears when this torque has to be incorporated into the dynamics equation of micromagnetics. One may add this term to the right-hand side of the LLG Eq. (6.1) [47,49,50] or of its LL form Eq. (6.3) [51−53]. The first procedure, namely:

$$\dot{m} = \gamma_0 H_{eff} \times m + \alpha m \times \dot{m} - (u \cdot \nabla)m \tag{6.15}$$

(where from now on an overdot is used to denote a time derivative) has a solved form that reads:

$$\dot{m} = \frac{\gamma_0}{1+\alpha^2} H_{eff} \times m + \frac{\alpha\gamma_0}{1+\alpha^2} m \times (H_{eff} \times m)$$
$$-\frac{1}{1+\alpha^2}(u \cdot \nabla)m - \frac{\alpha}{1+\alpha^2} m \times [(u \cdot \nabla)m], \tag{6.16}$$

In this last equation, two terms derived from the spin-polarized current appear, whereas if the STT is added directly to the solved (or LL) form Eq. (6.3) one gets:

$$\dot{m} = \frac{\gamma_0}{1+\alpha^2} H_{eff} \times m + \frac{\alpha\gamma_0}{1+\alpha^2} m \times (H_{eff} \times m) - (u \cdot \nabla)m. \tag{6.17}$$

One presentation of the difference of conceptions leading to these two equations is that Eq. (6.15) considers damping as a viscosity with respect to the lattice, that creates a friction field proportional to the magnetization velocity, whatever its origin. On the other hand, a motivation behind Eq. (6.17) is that damping always decreases the energy of the sample and, as the adiabatic STT cannot be derived from an energy term, only the magnetic energy of the sample is considered for damping [53].

It should also be noted that Eq. (6.17) has a straightforward solution for a sample with translation invariance, like a wire without any defect. Indeed, if $m_0(r)$ is a solution of this equation with $u = 0$, for example a DW, then the solution for an arbitrary $u(t)$ is:

$$m(r, t) = m_0 \left(r - \int_{-\infty}^{t} u(\tau)d\tau \right). \tag{6.18}$$

Such is not the case with Eq. (6.15), as explained hereafter.

The second point of controversy concerns the presence or not of a CIP STT term beyond that of the adiabatic approximation. Domain walls, even if they are wide in nanowires made of soft materials (because the DW width is determined by the wire size and not by anisotropy), are never infinitely wide so that a deviation to adiabaticity has to occur. However, the microscopic calculation of this truly nonadiabatic contribution has shown that it was vanishingly small for the DW widths of the materials used in the experiments [54]. On the other hand, as first pointed out by Zhang et al. [55], the relaxation of the spin of the carriers, well known in the context of giant magnetoresistance, gives rise to another torque term [56]. From a mathematical point of view, the only possible other torque term linear in the spatial gradient of m is of the form $m \times [(u \cdot \nabla)m]$, as $\dot{m}$ has to be orthogonal to m because $|m| = 1$. From the micromagnetic results to be shown below, such a term has a great impact on DW dynamics. Therefore, the equation often considered in LLG form is [56,57]:

$$\begin{aligned}\dot{m} &= \gamma_0 H_{\text{eff}} \times m + \alpha m \times m \\ &\quad - (u \cdot \nabla)m + \beta m \times [(u \cdot \nabla)m],\end{aligned} \tag{6.19}$$

where $\beta \ll 1$ expresses a small deviation to adiabaticity (the denomination $c_J \equiv \beta u$ is also used). The solved form of Eq. (6.19) is:

$$\begin{aligned}\dot{m} &= \frac{\gamma_0}{1 + \alpha^2} H_{\text{eff}} \times m + \frac{\alpha \gamma_0}{1 + \alpha^2} m \times (H_{\text{eff}} \times m) \\ &\quad - \frac{1 + \alpha\beta}{1 + \alpha^2} (u \cdot \nabla)m + \frac{\beta - \alpha}{1 + \alpha^2} m \times [(u \cdot \nabla)m].\end{aligned} \tag{6.20}$$

This expression shows that, when the CIP STT terms are added to the LLG equation, the equation is equivalent to an LL form to which a slightly renormalized adiabatic term $u(1 + \alpha\beta)/(1 + \alpha^2)$ and a nonadiabatic term $u(\beta - \alpha)/(1 + \alpha^2)$ have been added. In this sense, the various points of view presented above end up into different expectations for the value of β.

One should also note that microscopic calculations based on the electronic structure of the materials have been recently developed so as to compute *ab initio* the relaxation-type nonadiabatic coefficient β [58,59]. In such a framework, the coefficient β has the same physical origin of the damping coefficient α. As the contributions from the electron relaxation to these two coefficients differ, in general one expects $\alpha \neq \beta$ [60]. In the LL formulation, this means the second torque term is also nonzero.

In this chapter, we will describe results in the framework of the LLG formulation Eq. (6.19). Results can be transformed back to the LL formulation by replacing β in LLG by $\beta - \alpha$ in LL. In order to be specific and allow for direct comparison with the field-driven case, the nanostrip of 300×5 nm^2 cross section will be kept. Starting from the simplest situation, complexity will be added step by step in the models.

6.3.1.1 Other Terms

Apart from the spin transfer effect, the large current flowing inside the nanostrip gives rise to two other effects that affect magnetization dynamics: the Œrsted field created by the current and the increase of sample temperature.

The Œrsted Field

The existence of this field is the reason for which sample width and thickness have to be small. For a nanostrip with flat rectangular cross section $h \ll w$, the field has mainly a perpendicular z component, that is maximum at the strip edges, and reads (for an infinitely long straight strip, J being the current density):

$$H_{z,\max} = \frac{J}{4\pi}\left[h \ln\left(1 + \left(\frac{2w}{h}\right)^2\right) + 4w\,\text{Atan}\frac{h}{2w} \right]. \tag{6.21}$$

This value is roughly proportional to the nanostrip thickness h. The y field component (obtained from Eq. (6.21) by exchanging h and w) is slightly smaller, and changes sign across the strip thickness so that for thin nanostrips (a few exchange lengths thick) this field should have virtually no effect. Note that a nonzero average for the y field exists if the sample consists of several metallic layers, as in spin-valve structures [61] or if different electron scattering rates exist at both interfaces of the magnetic layer [62]. For a straight wire the axial (x) field component is zero, so that there is no direct effect on the DW position.

The maximum values of the Œrsted field are not large for usual current densities. For example, in a 300×5 nm^2 nanostrip with a typical current density $J = 1 \times 10^{12}$ A/m^2, one gets $\mu_0 H_{z,\max} = 5.8$ mT. This field has to be compared to the demagnetizing field opposing an out-of-plane component, so that neglecting

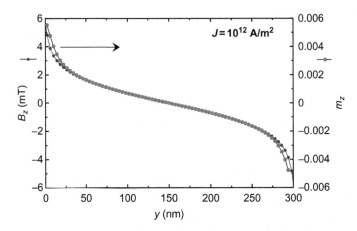

Figure 6.9 Œrsted field in a nanostrip 300×5 nm^2 under a uniform current density $J = 10^{12}$ A/m^2. The field component perpendicular to the film plane is evaluated at the center of the thickness, and has the expression.

$$H_z(y) = \frac{J}{4\pi}\left[h \ln\left(\frac{(w-y)^2 + (h/2)^2}{y^2 + (h/2)^2}\right) + 4(w-y)\text{Atan}\frac{h/2}{w-y} - 4y\,\text{Atan}\frac{h/2}{y}\right]$$

The perpendicular magnetization component m_z induced by this field in the uniformly magnetized nanostrip is also shown. The scales of both curves are such that the approximate expression $m_z \approx H_z/M_s$ can be visually tested: a larger out-of-plane magnetization component is seen close to the edges, due to the reduced local perpendicular demagnetizing field.

exchange effects one expects a local $m_z \approx H_z/M_s$, here less than 1%. A profile of this field is plotted in Figure 6.9, together with the m_z component that results from a micromagnetic calculation under this nonhomogeneous field.

The presence of this perpendicular field component may affect the processes that occur at the sample edges. As seen in the field-driven case, V and AV are injected and disappear at the edges, and their core has a perpendicular magnetization. Thus, a disymmetry of dynamic processes is to be expected, in principle, similarly to the effect of an exchange coupling of the film to a perpendicularly magnetized layer [21]. However, the smallness of this field component in comparison to the typical demagnetizing field within the core, of the order of M_s, lets anticipate weak effects. One report, however, has evidenced calculated oscillations in the transient response of a DW to a current step due to the influence of the Œrsted field [63].

The influence of this field for unperfect samples, for example, with an edge roughness or even a notch that reduces locally the nanostrip width, should be investigated in each precise case. For this purpose, the current density cannot be assumed as uniform, and requires a separate evaluation.

Sample Temperature

Due to the sample resistivity $\rho(T)$, to a current density J is associated a volumic heating power ρJ^2. For $\rho = 10\ \mu\Omega$ cm and $J = 1 \times 10^{12}$ A/m^2 this power is 1×10^{17} W/m^3. Without sample cooling, such a power would rapidly destroy the sample: the volumic heat capacity of iron is, for example, $C = 3.5 \times 10^6$ J/(m^3 K) so that sample heating would proceed at the tremendous rate of 30 K/ns. An analysis of heat flow in an infinitely long nanostrip on top of an inifinitely thick substrate was performed analytically and by finite elements solution of the heat diffusion equation [64]. The result is that, if λ_S denotes the thermal conductivity of the substrate, C_S its volumic heat capacity and hence $D_S = \lambda_S/C_S$ the heat diffusion constant in the substrate, the heating power at long times t (i.e., $\sqrt{D_S t} \gg w$) applies to a cross section of the substrate that increases by diffusion as $\approx D_S t$, so that temperature rises much more slowly than linearly, as:

$$T(t) = \frac{\rho J^2}{\lambda_S}\frac{wh}{\pi}\ln\left(\frac{4\sqrt{D_S t}}{w_G}\right) \tag{6.22}$$

(the symbol w_G is the width of the gaussian profile matched to the nanostrip width for simplicity of the formula, it is of order $w_G \approx 0.5w$). Taking silicon as substrate ($C_S = 1.66 \times 10^6$ J/(m^{3}K) and $\lambda_S = 148$W/(m K) so that $D_S = 8.92 \times 10^{-5}$ m^2/s, at room temperature), the calculated temperature increase in the same conditions as before is 2.7, 5.6, and 8.6 K for times 1 ns, 100 ns, and 10 μs, respectively. But with SiO$_2$ the values become roughly 100 times larger. The first values are very conservative estimates: the resistivity of NiFe is 2$-$3 times larger than the value used here, there is an insulating layer below the sample, the substrate may be thin, and the resistivity of metals increases with temperature. The second values correspond to some experimental situations [65,66] (as described in Chapter 4 by T. Ono and T. Shinjo, temperature can be monitored experimentally through the sample resistance). The heat diffusion analysis was extended recently to the case where an insulating layer separates the nanostrip from the substrate [67].

Thus, in order to describe all possible experiments, simulations should incorporate thermal effects. In micromagnetics, temperature has two influences.

1. All micromagnetic parameters, as they derive from the magnetic free energy of the material, are temperature dependent. One has thus to consider $M_s(T)$, $A(T)$, $\alpha(T)$ and, for the current polarization, $P(T)$. The temperature dependence of magnetization is well known experimentally. The calculation of the exchange constant from the Dirac$-$Heisenberg hamiltonian provides $A(T) \propto M_s^2(T)$. Measurements of damping versus temperature show virtually no effect [68,69]. Finally, it seems reasonable to assume P is proportional to $M_s(T)$ (direct measurements have been performed on NiFe

[70]). Thus, from the definition Eq. (6.13), we expect that the equivalent velocity u does not change upon heating (this holds true provided the current is maintained constant despite the rise of the sample resistance), as well as the exchange length Eq. (6.4). The overall effect of temperature can be better appreciated by transforming the equations to nondimensioned quantities, defined by:

$$\tau = \gamma_0 M_s t$$
$$h = H/M_s \tag{6.23}$$
$$\rho \equiv (\xi, \eta, \zeta) = r/\Lambda,$$

so that for example $h_{\text{eff}} = \partial^2 m/\partial\rho^2 + h_d$. With these definitions, the LLG equation with adiabatic term becomes:

$$\frac{\partial m}{\partial \tau} = h_{\text{eff}} \times m + \alpha m \times \frac{\partial m}{\partial \tau} - \frac{u}{\gamma_0 M_s \Lambda} \frac{\partial m}{\partial \xi}. \tag{6.24}$$

This form shows clearly that the effect of temperature rise is to increase the current efficiency, as $u/(\gamma_0 M_s \Lambda)$ varies like $1/M_s$ and M_s decreases with temperature. Note that also the real-time dynamics slows down as T increases, a phenomenon well known close to the Curie point.

No systematic analysis of the effect, on current-induced DW dynamics, of a temperature elevation-induced change of the micromagnetic parameters has been performed up to now, with the exception of one paper [71] that used the finite size scaling technique for changing the micromagnetic parameters [72] and found that indeed DW could be displaced with lower currents when T rose close to T_c.

2. Magnetization fluctuations increase when temperature rises. These can, following Brown [73], be introduced into standard micromagnetics by adding to the effective field a random field H_{th} with zero mean and whose variance obeys (beware that delta functions have dimensions):

$$< H_{\text{th}}^i(r,t) H_{\text{th}}^j(r',t') > = \frac{2\alpha k_B T}{\gamma_0 \mu_0 M_s} \delta_{ij} \delta(r-r') \delta(t-t'). \tag{6.25}$$

Calculations incorporating these fluctuations have shown a very weak effect of temperature, at least for metallic samples. In fact, the energy barriers between different structures of a DW are most of the time very high compared to the thermal energy [50,74,75] (see for example Figure 6.3B). Large thermal fluctuation effects were however observed by using in Eq. (6.25) a "micromagnetic temperature" that is typically 10 times larger than the considered temperature [29]. Note finally that when combining both aspects of temperature rise, the scaled typical thermal field is reinforced as it varies with temperature as $\sqrt{T}/M_s(T)$.

Another effect of the sample heating under current is the development of temperature gradients across the sample. In such conditions, the DW may be traversed by a flow of magnons, emitted in the hot regions and traveling toward the cold regions. Similarly to the electronic STT, a magnonic STT is also expected [76,77] on DWs, that has been observed [78,79]. In addition, within the nascent field of spin caloritronics (see a recent review [80], and Chapter 2 by J. Inoue), a direct effect of temperature gradients on domain walls is predicted [81], but not yet unambiguously observed. For strong temperature gradients, the magnonic STT-induced DW motion can be larger than the motion due to STT [78].

6.3.2   Perfect Case with Adiabatic STT

An ideal situation will be discussed first: the nanostrip is assumed to be perfectly straight, with no defect. It contains a DW at rest, and at time $t = 0$ the current is switched on instantaneously. Only the adiabatic term is included in the LLG Eq. (6.15). The Œrsted field is not included either.

6.3.2.1   Transverse Wall

Starting from an STW solution at rest, the time evolution of several quantities that allow monitoring the DW dynamics is plotted in Figure 6.10, for various values of the velocity u representing the spin-polarized current torque.

For currents below a threshold ($u_{c,TW} \approx 645$ m/s here), the DW displaces a little along the wire and finally comes to rest under current (Figure 6.10A). The DW initial velocity is equal to u, as remarked early [47]. The structure at rest under current differs from the initial one: the DW transverse moments μ_y and μ_z have changed (Figure 6.10B), similarly in fact to the steady-state solution under field, as the y moment decreases and the z moment increases. In addition, the maximum z component of the magnetization reaches higher and higher values as current becomes larger (Figure 6.10C). The characteristic time for the equilibration of the DW structure is short (≈ 0.5 ns), with some clear oscillations. These are due to the conversion during motion of the STW into the ATW structure (that is the stable form of TW at this nanostrip size, see Table 6.1 and Figure 6.3), as proved in Figure 6.10C.

The fact that the DW eventually stops implies that the CIP STT is compensated by another torque. Such a torque is provided by the deformation of the DW structure. In order to understand this deformation, we reduce the TW structure to the central moment in the DW (i.e., the moment oriented along the transverse y direction, at equilibrium). The CIP STT on this moment, starting from equilibrium, is the same as that resulting from a field applied along the perpendicular z direction. Therefore, at equilibrium under current, this moment will have precessed toward the z direction, leading to an effective field (of demagnetizing origin, mostly) also along the z direction that compensates the spin transfer. Although their full spatial form and dependence on magnetization orientation are different, the fact that initially spin transfer is equivalent to a perpendicular field helps understanding why no continuous DW motion under current should be expected. Note finally that DW transient displacement and DW structural transformation are intimately related, as shown [82] using the concept of DW angle (or momentum [83–85]), and as demonstrated experimentally [86].

For $u > u_{c,TW}$, the DW is set into continuous motion (Figure 6.10D), and the DW velocity reaches, in a small current interval, values of the order of several times 100 m/s. This motion is not a mere DW translation, but occurs under rotation

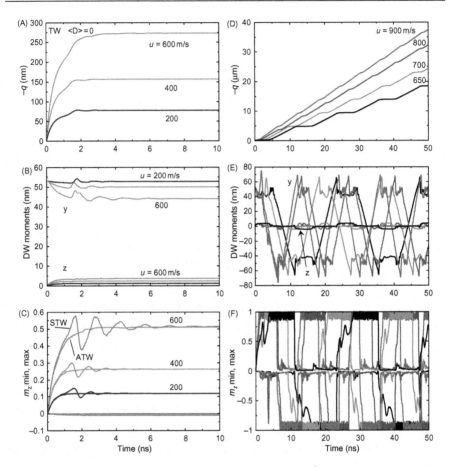

Figure 6.10 Time-resolved dynamics of a STW in a 300×5 nm^2 perfect nanostrip submitted at $t = 0$ to a constant spin-polarized current, in the adiabatic model. The low current regime $u < u_{c,\text{TW}}$ is shown on the left figures (A−C) and the high current regime on the right (D−F), with different scales at left and right. The DW position is shown in the top (A, D) panels. Panels (B, E) show the y and z moments of DW magnetization (defined in Eq. (6.6)). Finally, the values of the maximum and minimum perpendicular component m_z of magnetization are plotted in (C, F). The electric current is along the positive x direction, so that electrons move in the negative x direction. Panel (C) contains in addition the data obtained when starting from the stable ATW structure, where the oscillations seen for the STW are absent. (For interpretation of the references to color in this figure legend, the reader is referred to the web version of this book.)

of the DW moment (Figure 6.10E), realized through the injection, displacement across the nanostrip width, and expulsion of AVs, as shown by the plots of the maximum and minimum of the perpendicular magnetization component (Figure 6.10F), as well as by looking at the movie of the magnetization evolution. This is fully similar to the field-driven case. The comparison of Figure 6.10D and

Figure 6.10F reveals that DW motion occurs only when the AV is present, another illustration of the relation between DW displacement and DW structure transformation.

These phenomena have been recognized early [47,48,50]. Composite models of DW dynamics based on AV and V energetics and dynamics are being constructed presently [20,87], that provide simple and efficient explanations of this behavior. The basic ingredient of these models is the Thiele equation, that is introduced in Chapter 5 by H. Kohno and G. Tatara, and discussed later here.

6.3.2.2 Vortex Wall

The case of the VW structure is similarly illustrated in Figure 6.11. At very low currents, the VW structure is conserved. Similarly to the TW case, the DW first moves but finally stops (Figure 6.11A), adopting a deformed structure (Figure 6.11B) under current. The quantitative difference with the TW case is that the VW is displaced much more (for the same current) than the TW. This is in fact related to the much larger structural charateristic time of this structure, equal to ≈ 20 ns in this case (Table 6.1). The deformation of the VW structure is mainly apparent by the change of the y DW moment, that corresponds to an offset of the vortex core (Figure 6.11C). The other quantitative difference with the TW is the magnitude of the currents involved: the VW structure is maintained only for $u < u_{c,VW} \approx 60 \ m/s$.

For intermediate currents, the VW transforms to a TW by expulsion of the V core, as shown in Figure 6.11E and F. This transformation is accompanied by a sizable displacement of the DW position, that is in fact "quantized" [82,86]. The intermediate regime, defined by $u_{c,VW} < u < u_{c,TW}$, is such that even after conversion to a TW the wall eventually stops. The total DW displacement incurred by the current application is then the algebraic sum of the "quantized" displacement due to the structure conversion and the displacement of the TW when submitted to the current.

At large currents $u > u_{c,TW}$ we recover the DW continuous motion seen before.

6.3.2.3 Comparison to Experiments

From the study of these two cases, we conclude that in the perfect and adiabatic situation there is no continuous DW displacement under constant current, except for very large values of the current where DW motion occurs at the price of a continuous transformation of the DW structure, with a large velocity. Thus, an intrinsic current threshold for DW motion is predicted in this framework [49,50]. For other nanostrip sizes, results are similar with only a change in numerical values. Some plots of the important quantities in the (w, h) plane are provided in Ref. [82].

These conclusions are in strong disagreement with experiments. Indeed, with a polarization $P = 0.7$ for NiFe, a current equivalent velocity $u_c = 650 \ m/s$ corresponds to $J_c = 13 \times 10^{12} \ A/m^2$, a value 20 times larger than the experimental

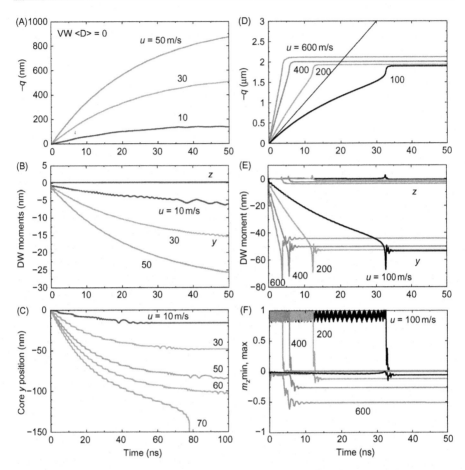

Figure 6.11 Time-resolved dynamics of a VW in a 300×5 nm^2 perfect nanostrip submitted at $t = 0$ to a constant spin-polarized current, in the adiabatic model. The low current regime $u < u_{c,VW}$ is shown on the left figures (A–C, $u = 10, 30, 50$ m/s) and the intermediate current regime on the right (D–F, $u = 100, 200, 400, 600$ m/s). The electric current is along the positive x direction, so that electrons move in the negative x direction. The DW transverse moments (B, E) are the scaled integrals of the transverse (y and z) magnetization components (Eq. 6.6). The vortex core y position is shown (C) for long times and low currents, including in addition those close to the transition to the TW ($u = 60, 70$ m/s). The straight line in (D) shows that for $u = 100$ m/s the DW initial velocity is indeed $v = u$.

results for close sizes [88–90]. Moreover, the current threshold for DW motion has been shown [91] to be proportional to the DW propagation field, and is therefore extrinsic. Several ways to lift this paradox, whereby the current appears to be much more efficient in reality than in models, have been proposed: the inclusion of a nonadiabatic term, the use of the LL equation with adiabatic term only, and the increase of sample temperature.

(i) In Section 6.3.1.1, we have shown that sample heating depends enormously on sample architecture, and varies very rapidly with current density J (faster than J^2). From the micromagnetic analysis of the temperature dependence of the material parameters, it has also been shown that the CIP spin transfer is expected to be amplified by an increase of temperature. Therefore, the rise of T up to the vicinity of T_c may give rise to DW motion in the perfect case, at values of current compatible with those of the experiments. However, first, the current window for such DW motion should be very narrow, as with just a little more current the sample reaches T_c. There are indeed many reports [65,66,92,93] that T_c could be reached in experiments, as the nucleation of structures such as pairs of domain walls, was observed. Even without reaching the Curie temperature, the transformation of the DW structure is also often observed: change of chirality of the DW [94], or change between TW and VW [95,96] (but DW structure transformations can also be ascribed to STT). Second, there exist also experiments with different sample structures and experimental conditions where heating was measured to be relatively small, and nevertheless DW motion observed, so that alternatives to heating have to be considered.

(ii) The introduction of adiabatic CIP spin transfer into the LL equation (see Eq. (6.17)) changes the situation completely. Indeed, as explained in Section 3.1, the solution for a DW under current is then trivial: a translation as a whole without any deformation, with instantaneous velocity $v(t) = u(t)$. There would therefore be no current threshold for DW motion.

From the field-driven case, however, we know that DW propagation is hindered by many sorts of defects. The existence of these defects breaks the translation invariance of the nanostrip, so that the analytic solution just discussed no longer applies and micromagnetic models with disorder have to be employed. These show the appearance of a threshold current, depending on the disorder model and magnitude, so that the very existence of a threshold current is not a sensitive test of the models.

Another prediction of the model is the velocity relation $v = u$. Velocities measured were initially much smaller, of the order of a few meters per second [88,97], but recently much higher values were measured with shorter current pulses [36,93], that do not always obey $v = u$.

The other strong prediction of the adiabatic LL formulation is the displacement without change of structure. Several experiments with a VW have shown a lateral y shift of the vortex core following current pulses [92, 94, 95] (due to defects that prevent the core from returning to the nanostrip center) that contradicts the rigid translation model. This may well be the most sensitive experimental test of the CIP spin transfer micromagnetic models. However, all conclusions of this model should be tested for robustness to disorder.

(iii) The third alternative is to stick to LLG and add a nonadiabatic term Eq. (6.19), with coefficient β. This encompasses the previous idea as a special case, namely $\beta = \alpha$ (up to a very small renormalization of u by a factor $1 + \alpha^2$). Therefore, we will in the following discuss the micromagnetic numerical results obtained in this framework. An analytic description can be found in Chapter 5 by H. Kohno and G. Tatara.

6.3.3   Perfect Case with Nonadiabatic STT

The equation governing magnetization dynamics considered now is Eq. (6.19), or its solved form Eq. (6.20). We will discuss first the results for the transverse wall before analyzing those for the VW and comparing them.

6.3.3.1 Transverse Wall

The dependence of the TW velocity on current u as a function of the nonadiabaticity parameter β is shown in Figure 6.12, that also incorporates the results of Section 6.3.2. The velocity here is the velocity measured at long times, the current being applied in a steplike way at time $t = 0$ on the TW structure at rest. In the case where DW motion is not stationary, as it is still periodical once the transients are over, the velocity is the average value over one period.

The first striking result is that, as soon as $\beta > 0$, DW motion is obtained under infinitely small currents. Thus, there is no longer an intrinsic threshold current for DW motion, and small DW velocities can be easily reached. The reason for DW motion as soon as $\beta \neq 0$ can be "read" from Eq. (6.20): in order to avoid the full cancellation of the CIP STT, the ratio of both spin transfer terms $((\alpha - \beta)/(1 + \alpha\beta)$ here) has to differ from the corresponding ratio for the effective field (α).

Steady-State Regime

The regime found at low currents does not show any curvature, at variance with the field-driven case. The 1D collective coordinates model of DW dynamics (see Chapter 5 by H. Kohno and G. Tatara) predicts indeed that:

$$v = \frac{\beta}{\alpha} u \tag{6.26}$$

(note that for $\beta = \alpha$ we recover $v = u$ as seen above). In fact, this relation has a much greater range of validity, as it has been shown to hold as soon as there is a steady-state DW motion, whatever the DW structure [57]. In stationary conditions

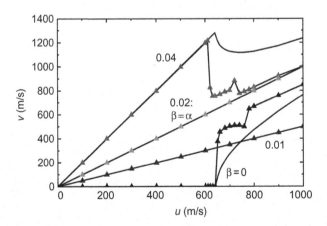

Figure 6.12 Velocities of a TW in a 300×5 nm^2 nanostrip without any defect under constant current represented by the equivalent velocity u, with the nonadiabaticity coefficient β as a parameter (material is NiFe with damping $\alpha = 0.02$). continuous curves are the prediction of the 1D model with constant DW width [57], with $u_c = 644.7$ m/s. As explained in the text, the figure applies also to the VW.

indeed, time derivatives are space derivatives and by integration general relations can be obtained, known as the Thiele equation [17,85] and its generalization to the current-driven case [57]. The slopes in Figure 6.12 are perfectly reproduced by Eq. (6.26).

However, this does not mean that the DW structure is unchanged. The evolution with time of the parameters of the TW is shown in Figure 6.13A−C for some values of the parameters β and for a large current $u = 100$ m/s. From that figure (panels B and C) the evolution of the DW transverse magnetization moment is apparent. It should be noted that the evolutions for $\beta = 0$ and 0.04 are opposite to each other, that for $\beta = 0.01$ being one half of that for $\beta = 0$ and that for

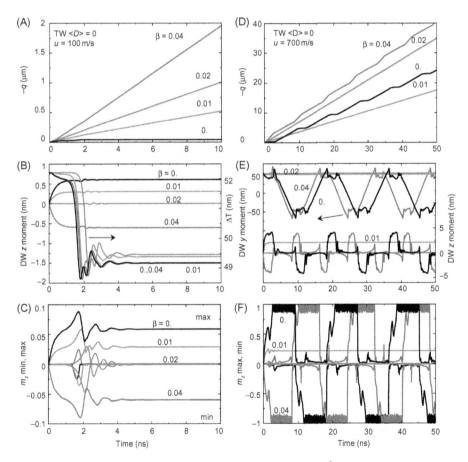

Figure 6.13 Time-dependent response of a STW in a 300×5 nm^2 nanostrip without any defect ($<D> = 0$) submitted at time 0 to a current represented by the equivalent velocity u, with the nonadiabaticity coefficient β as a parameter ($\alpha = 0.02$). Two values of current are shown, $u = 100$ m/s (A−C) in the linear regime and $u = 700$ m/s (D−F) above the Walker threshold, for some values of β.

$\beta = 0.02 \equiv \alpha$ being nil (except for the STW to ATW transition shown in Figure 6.13B by the Thiele DW width, due to the unstability of the STW).

More quantitatively, DW structures in steady-state motion under current are virtually undiscernable from those obtained under field (shown in Figure 6.6), provided they are compared at the same *relative velocity* v_R defined as $v_R = v - u$. This comes directly from the Lagrangian formulation of the LLG equation (see Chapter 5 by H. Kohno and G. Tatara), and its extension to include CIP spin transfer that is described below.

Lagrangian Formulation of Micromagnetics under CIP STT

In micromagnetics, the concept of Lagrangian was introduced by W. Döring [18], who used it to show that any DW structure has a maximum velocity, and to compute the moving DW structure to first order in damping (see the discussion in Ref. [13]; values of these velocities for TW and VW are given as a function of w and h in Ref. [82]). In the presence of CIP spin transfer the Lagrangian incorporates a term due to adiabatic spin transfer, whereas the nonadiabatic term plays no role as it belongs to the terms that do work, like damping and applied field along the strip axis. For steady-state motion, the Lagrangian is found to depend on the relative velocity v_R only. Thus, steady-state motion at velocity v under STT described by the equivalent velocity u obeys:

$$|v - u| < v_{\max}. \tag{6.27}$$

As a consequence, the regime where the DW does not move ($v = 0$) can exist only for $|u| < v_{\max}$, as seen above. In the case $\beta = 0$ where this solution exists, we have the following important relation between the field and current-driven regimes:

$$u_{c,TW} = v_{\max,TW}. \tag{6.28}$$

This equation shows that two quantities apparently unrelated are in fact equal (this is one of the motivation for describing DW dynamics by field and by current together in this chapter).

When $\beta \neq 0$ the steady-state relation (6.26) implies that the steady state ends at $u = u^*$ (with a DW velocity v^*) given by:

$$u^* = \frac{\alpha}{|\beta - \alpha|} v_{\max}$$

$$v^* = \frac{\beta}{|\beta - \alpha|} v_{\max}. \tag{6.29}$$

For the values of β considered in the figures, we obtain the following analytical predictions for $\alpha = 0.02$: $u^* = 2v_{\max,TW}$ and $v^* = v_{\max,TW}$ for $\beta = 0.01$, $u^* = v_{\max,TW}$ and $v^* = 2v_{\max,TW}$ for $\beta = 0.04$, and $u^* = \infty$ for $\beta = 0.02$ (the meaningful parameter being β/α). With the value of $v_{\max,TW}$ given in Table 6.1, we see that the analytical predictions for the end of the linear regime are well obeyed. The agreement is not perfect as the "Döring principle" is in fact a low damping approximation [13].

The dependence on v_R only of the Lagrangian in the stationary regime, together with the expression of DW velocity Eq. (6.26) in that regime, show that the modification of the DW structure in the stationary regime depends on $(\beta - \alpha)u/\alpha$. This explains the magnitudes and signs of the DW structural deformations seen in Figure 6.13.

Beyond the Walker Regime

For currents above u^*, DW motion is no longer stationary, as shown in Figure 6.13D–F in the particular case $u = 700$ m/s. The dynamics is very similar to that seen under field (Figure 6.5) with precession of the DW magnetization through the injection and displacement of isolated antivortices (moderately above u^*), followed by vortices and multiple localized structures at larger and larger currents.

A notable difference with the field case is that, depending on the value of β, the velocity can increase (if $\beta < \alpha$) or decrease (if $\beta > \alpha$) compared to the linear regime (Figure 6.12). To explain this, we may remark from Figure 6.13D that DW velocity is lower when there are no localized structures for $\beta < \alpha$, and higher when $\beta > \alpha$, generalizing the observation made in the case $\beta = 0$ (Figure 6.10). In all cases velocities converge to $v = u$ at very large currents (Figure 6.12), as predicted by the 1D collective coordinates model.

In addition, although this is not a general feature as it depends on nanostrip size, we see two regions in the above Walker regime with a velocity jump between them. Examination of the associated movies shows that this is due to the time required to inject the AV. The peaks at $u = 760$ m/s for $\beta = 0$ and $u = 720$ m/s at $\beta = 0.04$ are characterized by the injection and motion of an isolated V (instead of an AV). In fact, there appears to be a competition between injection of an AV at one edge of the nanostrip with the injection of a V at the opposite edge. This competition depends on the aspect ratio of the nanostrip [20].

6.3.3.2 Vortex Wall

The DW velocities evaluated at long times (stationary regime, or periodic average above) for a VW are identical to those found for the TW and displayed in Figure 6.12. In the stationary regime, this result is proved directly by the Thiele approach recalled earlier, that gives Eq. (6.26) independently of the DW structure. Above this regime, as the DW structural transformation starts by the expulsion of the vortex core and transformation to a TW, the similarity of TW and VW is also understood.

Linear Regime

The time evolution of the VW under two values of current is illustrated in Figure 6.14. The first value $u = 50$ m/s (panels A–C) was chosen such that for all values of β the current is in the linear region for the VW, whereas for $u = 100$ m/s

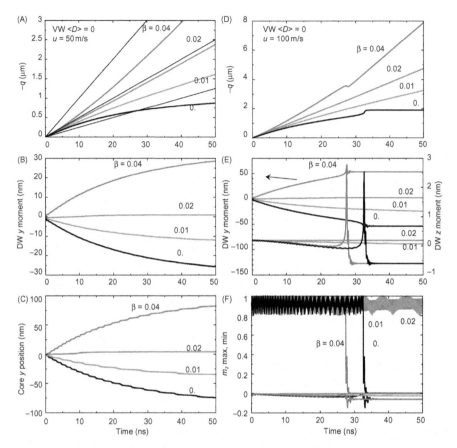

Figure 6.14 Time-dependent response of a VW in a 300×5 nm^2 nanostrip without any defect ($<D> = 0$) submitted at time 0 to a current represented by the equivalent velocity u, with the nonadiabaticity coefficient β as a parameter ($\alpha = 0.02$). Two values of current are shown, $u = 50$ m/s (A–C) in the VW linear regime and $u = 100$ m/s (D–F) above the VW Walker threshold for some values of β. The lines in (A) express the analytical relation (6.26).

only $\beta = 0.01$ and 0.02 belong to it. Similarly to the TW, the linear regime features a progressive VW structural transformation despite the perfect linearity of the $v(u)$ relation. The transformation is displayed for example by the DW y moment (Figure 6.14B and E), or by the vortex core y position (Figure 6.14C). As already seen for the TW, there is no change when $\beta = \alpha$, the change for $\beta = 0.01$ is one half of that for $\beta = 0$ and that for $\beta = 0.04$ is opposite to that for $\beta = 0$.

This was well understood with the Lagrangian formulation of micromagnetics under CIP STT. Equivalently, the study of the VW dynamics by simple analytical models rests on the application of the Thiele equation to the AV and V. As explained in Chapter 5 by H. Kohno and G. Tatara, this equation shows that AV and V are submitted to a "gyrotropic" force that is orthogonal to v_R and changes sign with the core z polarization direction and between AV or V.

The big difference between TW and VW that appears in the time-dependent results comes from the very different characteristic times of these two DW structures (see Table 6.1). The calculations show that the structural deformations under current have not yet reached their steady-state values after 50 ns (Figure 6.14B). Consequently, the DW velocities are still off from the steady-state values, as shown in Figure 6.14A where the lines corresponding to Eq. (6.26) were added. As the DW initial velocity is u, the transient DW velocities deduced from Figure 6.14A and D lie between $v = u$ and $v = (\beta/\alpha)u$.

This figure also shows that the DW displacements converge to this analytic relation with an offset, itself varying like $\beta - \alpha$. This offset is of the same nature as the transient displacement seen in the $\beta = 0$ case previously (Figure 6.11), and is directly related to the change of DW angle of the VW structure (the DW angle for a VW is defined precisely from the general definition in Ref. [82]).

The values of current at which the linear regime for the VW ends are given by the analytical prediction Eq. (6.29), with the micromagnetic value of $v_{max,VW}$ given in Table 6.1, hence $u^* = 122$ m/s for $\beta = 0.01$ and 61 m/s for $\beta = 0.04$ and 0. This is in agreement with the results shown in Figure 6.14 for $u = 50$ and 100 m/s.

Above the VW Linear Regime

The end of the linear regime for the VW is marked by the vortex expulsion (Figure 6.14F). As shown in Figure 6.14D, the DW velocity subsequently rapidly adopts the analytical value Eq. (6.26), as the characteristic time of the TW is short (see Table 6.1) and $u < u^*$ for the TW. This explains why Figure 6.12, that plots the long time velocity, is identical for TW and VW.

6.3.3.3   Comparison to Experiments

Remembering that the experimental values of u are at most 100 m/s, we see that the steady-state regime should apply, the DW velocity being thus directly linked to the β parameter, independent of the DW structure, and with no current threshold. However, all experiments show that a current threshold exists, with a value of the order of 1×10^{12} A/m^2 for NiFe [36,98]. Thus, experiments are now performed with short pulses, of nanosecond [36,93,99,100] duration. This also limits the sample heating (and its destruction).

In such a situation, the steady-state regime is not relevant at first sight: the characteristic time of the DW structure plays a role, especially for the VW where it can be very large (a map of this time as a function of nanostrip width and thickness is given in Ref. [82]). In this respect, Figure 6.11A has shown that, even with $\beta = 0$, a VW transient displacement as large as 1 μm can be obtained under a reasonable current $u = 50$ m/s. However, when the current is switched off the VW will come back to its original position in this case. Thus, transient effects have indeed to be included both at pulse risetime and falltime.

To interpret the pulsed current results, a first solution consists in performing micromagnetic simulations of the experiment in order to find out which values of u/J (hence P) and β fit the experimental results best. The possibility to follow this way depends strongly on the sample dimensions. Alternatively, it has been recently shown that, provided the DW structure is the same after the current pulse, the two transient effects cancel upon integration so that the total DW displacement is given by the steady-state velocity integrated over the pulse profile [86]. This holds true even if the current pulse is much shorter than the DW structure relaxation time (case of VW). Knowing this, systematic measurements of DW displacement [36] in NiFe are well in line with the phenomenology presented here, and show that β is distinctly larger than α

In addition to the DW displacement, the DW structure has been observed, showing in most cases off-centered vortex cores after current pulses [93,95], or even DW structure transformations [96,101]. On the one hand, this corresponds to the predictions shown above (Figure 6.14), so that these observations can be taken as a proof that $\beta \neq \alpha$, provided the sample is perfect. However, the mere observation that after a current pulse the core of the vortex in a VW does not return to the center of the nanostrip is a proof that defects hinder the DW and vortex core motion. Thus, disorder has to be taken into account. This is the object of the next section.

6.3.4 Imperfect Case

As discussed for field-driven dynamics, imperfections can be introduced in micromagnetic calculations by some roughness of the nanostrip edge. Edge roughness has the additional consequence that the current density becomes nonuniform, and has to be calculated at every point. This involves solving the Poisson equation in the rough nanostrip, a calculation to perform only once, before the micromagnetic calculations. We consider here the case where the average grain diameter is $<D> = 10$ nm only (propagation field $\mu_0 H_P \approx 1.5$ mT).

6.3.4.1 Transverse Wall

Results obtained for a TW are shown in Figure 6.15, for several values of the parameter β. This figure should be compared with Figure 6.12.

In the adiabatic case $\beta = 0$, there is little change compared to the perfect case: DW motion occurs with transformation of the DW structure, starting at $u \approx u_c$.

When $\beta \neq 0$, no DW propagation is seen unless the velocity u is large enough, similarly to field case. It can therefore be said that roughness leads to the apparition of a current threshold for DW motion. This threshold is not an intrinsic quantity, but linked to disorder (and indeed calculations with a larger roughness lead to a larger current threshold). Similarly to the field case also (Figure 6.8), DW motion in a certain range of currents above this threshold is stochastic, as the DW may move and stop at some places. The plot shows therefore two values of velocity in such range, namely 0 and the velocity derived from the period of DW motion.

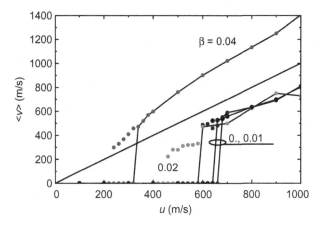

Figure 6.15 Velocities of a TW in a 300×5 nm^2 nanostrip without rough edges ($<D> = 10$ nm) under constant current represented by the equivalent velocity u, with the nonadiabaticity coefficient β as a parameter (material is NiFe with damping $\alpha = 0.02$). In the case of stochastic DW displacement two velocities are shown for the same current, and the points with lower velocity are linked together. The straight line plots the $v = u$ relation, for comparison.

The data with $\beta = 0.01$ are very similar to those at $\beta = 0$, so that this amount of adiabatic torque has no visible influence on DW motion, for the magnitude of imperfection considered in this calculation.

In the case $\beta = 0.02 = \alpha$, two regimes are seen. At the lower currents, the TW moves without notable change of structure (apart from the fluctuations of width, asymmetry, etc. due to roughness). This corresponds to what is expected in the perfect wire case. However, a relation $v \approx u - 200$ m/s is measured, instead of the expected $v = u$ rule. In this regime the wall moves and may stop. At $u > 600$ m/s (the value of u_c in the perfect case), the continuous transformation of DW structure sets in, with the successive motion of AV across the nanostrip width (at least moderately above the threshold). The DW velocity increases and becomes similar to the velocity obtained under DW structure transformation at lower values of β. The $v(u)$ relation has still the slope 1, but the offset has diminished slightly.

Finally, in the case $\beta = 0.04 > \alpha$, the low u regime with DW motion without structure transformation is also found. The perfect case relation (6.26) is obeyed, but again with an offset. In the higher current regime, the breakdown expected at $u = u_c$ (see Figure 6.12) does not occur, and the DW magnetization does not precess at least for the currents shown. Only a gradual reduction of the $v(u)$ slope occurs, toward unity (as found for a perfect nanostrip, but again with an offset).

This example shows that the influence of imperfections is (i) to mask the effect on DW motion of too small β parameters, (ii) to introduce offsets in the linear regime, (iii) to suppress velocity breakdown (as already seen in the field-driven case), and (iv) to modify the $v(u)$ slope. Note that, for this sample, the analytic slopes of the linear regime are preserved, whereas for a narrower nanostrip with the same roughness parameter $<D>$ they were reduced [57].

6.3.4.2 Vortex Wall

As we have seen in the perfect nanostrip case, the VW dynamics under current is dominated by the presence of the vortex, with the long relaxation time of its y position. This y position will be controlled by the relative DW velocity v_R.

Figure 6.16 depicts the dynamics of the VW, for two values of the current and various nonadiabaticity parameters. The data show, similarly to the results obtained

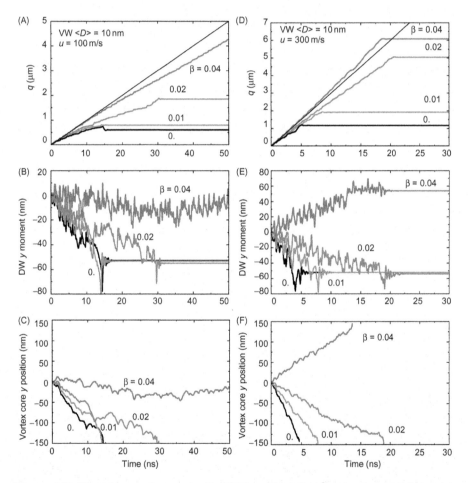

Figure 6.16 Time-dependent response of a VW in a 300×5 nm^2 nanostrip with roughness (average grain size $<D> = 10$ nm) submitted at time 0 to a current represented by the equivalent velocity u, with the nonadiabaticity coefficient β as a parameter ($\alpha = 0.02$). Two values of current are shown, $u = 100$ m/s (A–C) and $u = 300$ m/s (D–F) (note the different time scales), for some values of β. The straight lines in (A, D) display the $v = u$ relation.

for perfect nanostrips, a transient VW motion at lower currents than for the TW that lasts as long as the V is present. As the initial V core z magnetization is always the same, the V transverse motion direction is determined by v_R only.

The first value, $u = 100$ m/s (panels A−C) should be compared to the perfect nanostrip situation, shown in Figure 6.14A−C. We see that in the rough case, since $v < u$ is calculated, the DW y moment always decreases and the vortex core y position becomes negative. In the case $\beta = 0.04$ where v is close to u, the gyrotropic force on the vortex core is weak and the vortex core remains close to the nanostrip center. Thus, this calculation provides an example where microscopically $\beta = 0.04 = 2\alpha$ whereas the apparently steady-state DW motion corresponds to $v = 85.4$ m/s thus $\beta = 0.85\alpha$. This behavior exists in a certain range of currents, and the DW velocities measured in that range are plotted in Figure 6.17. Each of the measured DW velocities could be translated into a value of an effective β/α, that would vary between 0.55 and 1.13. The figure however shows that a slope of $v(u)$ can be defined in this regime (equal to 1.4 here), that lies between the values 1 and $\beta/\alpha = 2$. An offset in the $v(u)$ relation appears in addition. More generally, Figure 6.16A shows that the DW velocities in the transient regime, that in the perfect situation ranged between u and $(\beta/\alpha)u$, are reduced when roughness is present. The results for $\beta = \alpha$ are remarkable, as the DW velocity is significantly smaller than u so that the vortex is expelled and the DW stops.

For the higher value of current $u = 300$ m/s (Figure 6.16D−F), the DW velocity with $\beta = 0.04$ is now sufficiently larger than u so that the V moves upward in y and is finally expelled, bringing the DW into a TW that stops later, as the value of u is in the region of stochastic DW motion for the TW at $\beta = 0.04$. The other cases, with lower β values, show DW velocities lower than u and thus a downward expulsion of the V. The transient DW velocities are, as seen already for $u = 100$ m/s, also smaller than in the perfect case. From the figures, it can be expected that for an intermediate value $\beta \approx 0.03$, the vortex is kept and the DW continuously moves at $<v> \approx u$, that is, an apparent $\beta \approx \alpha$.

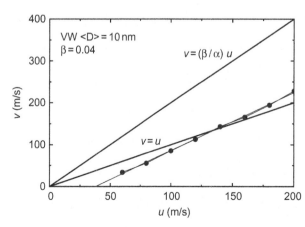

Figure 6.17 Transient velocity of a VW in a rough nanostrip (same parameters as in the previous figures) for $\beta = 0.04$, in a range of currents where the V is not expelled. The $v(u)$ slope is lower than the analytical value, with an offset, so that the relative velocity $(v - u)$ changes sign with increasing current.

Summarizing, it appears that the VW can move more easily under current than a TW, in the presence of imperfections of the sample edge [102]. This feature is clearly consistent with experiments [91,96]. The possible reason for this behavior is that the vortex core, the region with the largest magnetization gradient hence the largest CIP STT, is far from the rough edges (note that defects in the form of holes give the opposite result that the TW moves more easily than the VW [103]). In addition, a nontrivial effect of imperfections was seen, namely a general reduction of DW velocity, leading to apparent β/α ratios close to unity, and that depend on the current magnitude. This is also in qualitative agreement with experiments: many show that apparently $\beta \approx \alpha$ [33,93,104], but for similar samples other experiments require apparent $\beta/\alpha \approx 1 - 10$ [36,96,105].

These results should be taken as indicative only. As mentioned in Section 6.2, another disorder model can be considered, namely with thickness/magnetization fluctuations all over the sample. The results of such calculations [31] in the case of a VW are qualitatively similar to what has been shown above. In quantitative terms, and for the sample size considered, the overall conclusion is that an effective damping increased by disorder is the simplest way to describe the simulation results.

Altogether, these results show that the identification of the β parameter by DW velocity measurements in real samples is not straightforward. One route might be to first determine the effective α by field dynamics, and then the ratio β/α by current dynamics, so as to finally obtain a value of β.

6.4   Dynamics Under Combined Field and Current

As many experiments have considered how the CIP STT affects the field-induced DW dynamics, we will discuss this topic too.

6.4.1   Perfect Samples

The results for a perfect wire are summarized in Figure 6.18, where several values of u and β are compared. The calculations were performed for a TW.

The results for a VW differ only close to the origin, precisely speaking when $|v - u| < v_{\max,\mathrm{VW}}$.

6.4.1.1   Steady-State Regime

The linear regime has been theoretically studied in several papers, [47,48,50,56,57]. The DW velocity in steady state reads:

$$v = \frac{\gamma_0 \Delta_\mathrm{T}}{\alpha} H_x + \frac{\beta}{\alpha} u \tag{6.30}$$

with as usual Δ_T the Thiele DW width of the moving domain structure Eq. (6.8). This relation, initially derived with the 1D collective coordinates model for a TW, holds in fact for any DW structure and results only from the hypothesis of a steady

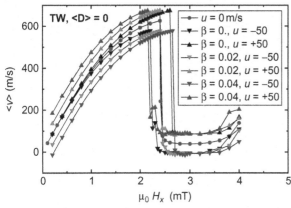

Figure 6.18 Velocity at long times of a TW in a perfect 300×5 nm^2 nanostrip under an applied field, superposed to a DC current with adiabatic and nonadiabatic STT. For each value of β, two opposite values of current $u = \pm 50$ m/s are injected.

state. Similarly to the field case, this relation has an intrinsic nonlinearity due the variation of the Thiele DW width (the accurate Döring approximation being that Δ_T depends only on the DW velocity v).

The end of the linear regime can be obtained, in the low dissipation approximation of the Döring principle, from the point where with Eq. (6.30) the maximum velocity v_{max} is reached, with a corresponding Thiele DW width $\Delta_T(v_{max})$. Thus we get a dependence of Walker field on current as:

$$\pm H_W(u) = \frac{\pm \alpha v_{max} - (\beta - \alpha)u}{\gamma_0 \Delta_T(v_{max})}. \tag{6.31}$$

This gives a relative variation $\Delta H_W / H_W$ of the Walker field at small u where the Thiele DW width has disappeared, reading:

$$\frac{\Delta H_W}{H_W} = \frac{\beta - \alpha}{\alpha} \frac{u}{v_{max}}. \tag{6.32}$$

The numerical micromagnetic results displayed in Figure 6.18 are in good agreement with Eq. (6.32), using the micromagnetic value of the maximum velocity given in Table 6.1. Therefore, at first sight, this analytic relation provides a very direct experimental measurement of β.

6.4.1.2 Above the Walker Breakdown Regime

For fields larger than $H_W(u)$, velocity decreases abruptly, similarly to the dynamics driven by field only. Figure 6.18 shows that the velocity slightly above the Walker field, in the region where $v(H)$ is flat, becomes also independent on the value of β. Moreover, the data are well described in this region by the very simple relation:

$$v = v_{min} + u \tag{6.33}$$

where the minimum velocity v_{min} was defined in Section 2.3. This result, that depends only on the relative velocity v_R, is very probably related to the Lagrangian description of the effect of CIP STT. Note that, if the steady-state relation were (erroneously) applied to such a result, a value $\beta/\alpha = 1$ would be deduced.

For higher fields where velocity increases again, however, a dependence on β reappears.

6.4.2   Imperfect Nanostrip Case

From the previous sections of this chapter, we know that imperfections may modify the conclusions reached on perfect samples. Nanostrip imperfection was mimicked here by an edge roughness as previously explained, with the same value (average diameter of the grains equal to 10 nm) used for the study of field-induced DW propagation (Section 2.4).

The results for TW motion under combined field and current obtained in such a nanostrip are shown in Figure 6.19. Similarly to the pure field case (Figure 6.8), the disappearance of the Walker breakdown in the range of fields investigated is found.

For fields much larger than the propagation field, it is found that, remarkably, DW velocity is independent of the nonadiabaticity parameter β and obeys the simple relation:

$$v = v_{max} + u. \tag{6.34}$$

This relation that mimics Eq. (6.33) corresponds to an effective ratio $\beta/\alpha = 1$. In terms of the Döring principle, relation (6.34) means simply that v_R is clipped to v_{max}. In the absence of a statistical treatment of imperfections, however, this statement should be considered just as an analogy, as there is no mathematically steady-state DW motion in a rough nanostrip. As expected, the behavior in that field range is exactly the same for a VW.

We will not comment in detail the vicinity of the propagation field, as the approach used with no thermal fluctuations and relatively short calculation times is not well suited for this purpose.

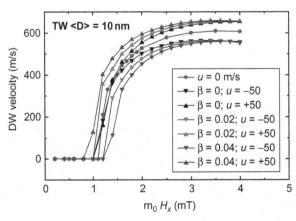

Figure 6.19 Velocity at long times of a TW, in a 300×5 nm^2 nanostrip having rough edges (mean grain diameter $<D> = 10$ nm), under an applied field superposed to a DC spin-polarized current with both adiabatic and nonadiabatic CIP STT. For each value of β, two opposite values of current $u = \pm 50$ m/s are injected.

6.4.3   Comparison to Experiments

6.4.3.1   Measurements of DW Velocity

The DW mobility under field and current in samples where the Walker breakdown could be seen, has been measured in only two cases up to now [33,104], with nanostrip cross sections equal to 300×10 and 600×20 nm^2, respectively. Both experiments have seen mainly a vertical translation of the $v(H)$ curve under current, and conclude that $\beta \approx \alpha$ in their samples, based on the expressions established in the perfect nanostrip case.

The micromagnetic calculations performed with a large roughness (Section 4.2) have however shown that a vertical translation was obtained, irrespective of β. But this roughness was so large that it wiped out the Walker breakdown. Therefore, a milder imperfection model, that gives rise to propagation fields as low as those of samples where the Walker breakdown has been observed, is very much needed.

6.4.3.2   Depinning of a DW

In addition to the measurement of DW velocity under field and current, the variation of the propagation field under current has attracted a lot of attention experimentally [94,105–110]. Samples with artificial defects (notch) pinning the DW, or just with natural defects, have been used.

For weak (natural) DW pinning, a rather linear variation of H_P with u has been measured once thermal (or thermal-like) effects were subtracted [106,111]. Keeping in mind that this field is not confidently evaluated by current micromagnetic simulations, the trend seen in Figure 6.19 proves qualitatively in agreement with these results.

The depinning results have also been analyzed either by direct simulation of the experimental situation [109], or by comparison to an Arrhenius law [110], in order to extract β values. The obtained order of magnitude is the same as that obtained by DW motion, yet no single value for NiFe emerges. As mentioned above, this may be an effect of different effective damping constants due to disorder.

6.4.4   AC Effects

Spectacular results have been obtained by setting the DW into oscillation inside a pinning potential well [105,108,112,113]. The 1D collective coordinates model was very efficient to understand these results, that were partly counterintuitive. A related experiment devised a nanostrip with small curvature under a field applied in the transverse direction so as to create a potential well for the DW [114], and resonant effects where the oscillation frequency was controlled by field were indeed observed. Briefly stated, the interest of measurements at resonance is to avoid pinning effects and to detect weaker effects (see also Chapter 3 by Y. Suzuki).

This study has been remarkably extended to the case of a disk-shaped sample containing a single vortex. The micromagnetic analysis of data obtained for the vortex displacement [115,116] has directly shown that the adiabatic CIP

STT model was quantitatively correct. The influence of the Oersted field, due to the electrical contacts to the disk, was also detected [117]. The measurement procedure and precision have been recently improved so as to obtain values of the nonadiabaticity parameter. Values larger than those obtained for a DW have been found [118,119], namely $\beta = 0.15$ in both cases, despite using different methods.

This points toward a possible dependence of the β parameter on the micromagnetic structure. The microscopic models evaluate one parameter, but in the limit of long wavelengths (zero wavevector). One way to understand this effect could be a spin diffusion effect. Indeed, the first model of β [56] takes into account diffusion effects, and only its local solution gives a single β value. The full solution of this model, coupled to micromagnetics, has recently shown the existence of a structure-related effective β [120], that depends sensitively on the size of the magnetic structure compared to the spin diffusion length.

6.5 Conclusions and Outlook

In this chapter, we have tried to present a unified view of domain wall dynamics in soft magnetic nanostrips, under the effect of field and/or spin-polarized current flowing along the nanostrip and across the domain wall. This view was based on micromagnetic simulations and their comparison to experiments.

A number of concepts have been shown to apply to both cases: the steady-state motion of the domain wall, the Walker breakdown, the maximum velocity of a domain wall structure, the Thiele domain wall width, for example. This justifies our choice to describe both driving mechanisms, even if one may appear as "old" and the other as "new." In fact, both mechanisms are only now becoming experimentally checked as better samples and experimental techniques are devised.

A few lessons can be taken from the synthesis presented here:

1. Analytic results can be obtained that apply not just to the collective coordinates (1D) model of the domain wall, but to the full micromagnetic situation. They hold true only in the limit of small damping and STT, similarly in fact to results derived recently for pillars owing to the macrospin model [121,122]. It may even prove possible to extend these results to the above breakdown case, and to imperfect nanostrips.
2. Effects of sample imperfections can be very important, up to destroying the rules obtained in the perfect case. As many models of imperfections can be thought of, it seems important now to find out which one fits best and most simply the experiments. The extension of analytic results to the imperfect situation would be a great step.
3. It seems that the value of the apparent β/α ratio is reduced toward 1 by imperfections. A number of cases where the analytical prediction for a perfect sample turned to $\beta/\alpha \sim 1$ in a rough sample were shown. Thus, in order to answer the controversies about the proper incorporation of the STT in a continuous configuration crossed by current, experimental situations where the results are not affected by imperfections should be invented.

Finally, this chapter has not covered all possible applications of micromagnetics to the effect of STT within continuous magnetic structures, partly for lack of space. We have thus deliberately left aside:

1. Nanostrips made out of hard materials with perpendicular anisotropy. The first simulations [123,124] triggered the interest for this field, that has markedly developed since. Briefly said, the advantage of such materials is twofold: they allow controlling the domain wall width by the anisotropy, and the Walker field and maximum velocity by the anisotropy and nanostrip geometry. In well-chosen cases, the intrinsic threshold for current-induced domain wall motion can be drastically reduced. Within this research area, the case of ultrathin films embedded in a structural inversion asymmetric architecture appears as extremely peculiar, with other torque terms as STT realizing the current-induced domain wall motion [125,126] and possibly a new type of DW [16].
2. Spin-valve nanostrips with two magnetic layers, where the magnetic structures are coupled and the spin-polarized current flow more complex to describe, despite many interesting results [61,107,127−129], namely lower threshold currents and higher DW velocities.
3. Nanostrips made out of magnetic semiconductors, prominently GaAs:Mn where large effects have been observed (see Chapter 7 by F. Matsukura). There are indeed major issues like the effect of temperature and carrier-mediated ferromagnetism that cast warnings on the application of standard micromagnetics to this situation.
4. The case with no domain wall, but more generally a nonuniform structure like a vortex (see however Section 4.4).

We hope nevertheless that the material presented in this chapter will be helpful to tackle these other subjects.

Acknowledgments

The work of Y.N. was partly supported by New Energy and Industrial Technology Development Organization (NEDO) of Japan. The work of A.T. was supported by the French Action Concertée Incitative NR 216 PARCOUR, the ANR projects DYNAWALL, HYFONT, and ESPERADO, and the European network MRTN-CT-2006-035327 SPINSWITCH. He also benefited from a JSPS fellowship for a stay at the Institute for Chemical Research, Kyoto University, Uji, Japan, in the summer 2006. Both thank T. Ono, S. Kasai, J. Shibata, G. Tatara, H. Kohno, M. Kläui, and J. Miltat for fruitful discussions, and J. Miltat for a critical reading of the manuscript.

References

[1] Hubert A, Schäfer R. Magnetic domains. Berlin: Springer Verlag; 1998.
[2] Aharoni A. Introduction to the theory of ferromagnetism. Oxford: Clarendon Press; 1996.
[3] Miltat J, Albuquerque G, Thiaville A. In: Hillebrands B, Ounadjela K, editors. Spin dynamics in confined magnetic structures I. Berlin: Springer; 2002. p. 1−33.
[4] Kronmüller H, Parkin S, editors. Handbook of magnetism and advanced magnetic materials, vol. 2. New York, NY: Wiley; 2007.

[5] Fidler J, Schrefl T. Micromagnetic modelling—the current state of the art. J Phys D Appl Phys 2000;33:R135.

[6] Schrefl T, Hrkac G, Bance S, Suess D, Ertl O, Fidler J. In: Kronmüller H, Parkin S, editors. Handbook of magnetism and advanced magnetic materials, vol. 2. New York, NY: Wiley; 2007. p. 765−94.

[7] McMichael RD, Donahue MJ. Head to head domain wall structures in thin magnetic strips. IEEE Trans Magn 1997;33:4167.

[8] Nakatani Y, Thiaville A, Miltat J. Head-to-head domain walls in soft nano-strips: a refined phase diagram. J Magn Magn Mater 2005;290−291:750.

[9] Tchernyshyov O, Chern GW. Fractional vortices and composite domain walls in flat nanomagnets. Phys Rev Lett 2005;95:197204.

[10] Hubert A, Rave W, Tomlinson SL. Imaging magnetic charges with magnetic force microscopy. Phys Stat Sol B 1997;204:817.

[11] Kläui M, Vaz CAF, Bland JAC, Heyderman LJ, Nolting F, Pavlovska A, et al. Head-to-head domain wall phase diagram in mesoscopic ring magnets. Appl Phys Lett 2004;85:5637.

[12] Laufenberg M, Backes D, Bührer W, Bedau D, Kläui M, Rüdiger U, et al. Observation of thermally activated domain wall transformations. Appl Phys Lett 2006;88:052507.

[13] Thiaville A, Nakatani Y. In: Hillebrands B, Thiaville A, editors. Spin dynamics in confined magnetic structures III. Berlin: Springer; 2006. p. 161−206.

[14] Hertel R, Kirschner J. Magnetization reversal dynamics in nickel nanowires. Physica B 2004;343:206.

[15] Sixtus KJ, Tonks L. Propagation of large Barkhausen discontinuities. Phys Rev 1931;37:930.

[16] Thiaville A, Rohart S, Jué É, Cros V, Fert A. Dynamics of Dzyaloshinskii domain walls in ultrathin magnetic films. Europhys Lett 2012;100:57002.

[17] Thiele AA. Steady-state motion of magnetic domains. Phys Rev Lett 1973;30:230.

[18] Döring W, Über die Trägheit der Wände zwischen Weissschen Bezirken (in german), Naturforschg Z. 1948;3a:373.

[19] Schryer NL, Walker LR. The motion of 180° domain walls in uniform dc magnetic fields. J Appl Phys 1974;45:5406.

[20] Lee J-Y, Lee K-S, Choi S, Guslienko KY, Kim S-K. Dynamic transformations of the internal structure of a moving domain wall in magnetic nanostripes. Phys Rev B 2007;76:184408.

[21] Lee J-Y, Lee K-S, Kim S-K. Remarkable enhancement of domain-wall velocity in magnetic nanostripes. Appl Phys Lett 2008;91:122513.

[22] Bryan MT, Schrefl T, Atkinson D, Allwood DA. Magnetic domain wall propagation in nanowires under transverse magnetic fields. J Appl Phys 2008;103:073906.

[23] Ono T, Miyajima H, Shigeto K, Mibu K, Hosoito N, Shinjo T. Propagation of a magnetic domain wall in a submicrometer magnetic wire. Science 1999;284:468.

[24] Gadbois J, Zhu J-G. Effect of edge roughness in nano-scale magnetic bar switching. IEEE Trans Magn 1995;31:3802.

[25] Cowburn RP, Koltsov DK, Adeyeye AO, Welland ME. Lateral interface anisotropy in nanomagnets. J Appl Phys 2000;87:7067.

[26] Nakatani Y, Thiaville A, Miltat J. Faster magnetic walls in rough wires. Nat Mater 2003;2:521.

[27] Beach GD, Nistor C, Knutson C, Tsoi M, Erskine JL. Dynamics of field-driven domain wall propagation in ferromagnetic nanowires. Nat Mater 2005;4:741.

[28] Yamaguchi A, Yano K, Tanigawa H, Kasai S, Ono T. Reduction of threshold current density for current-driven domain wall motion using shape control. Jpn J Appl Phys 2006;45:3850.

[29] Martinez E, Lopez-Diaz L, Torres L, Tristan C, Alejos O. Thermal effects in domain wall motion: micromagnetic simulations and analytical model. Phys Rev B 2007;75:174409.

[30] Ferré J. In: Hillebrands B, Ounadjela K, editors. Spin dynamics in confined magnetic structures I. Berlin: Springer; 2002. p. 127−65.

[31] Min H, McMichael RD, Donahue MJ, Miltat J, Stiles MD. Effects of disorder and internal dynamics on vortex wall propagation. Phys Rev Lett 2010;104:217201.

[32] Atkinson D, Allwood DA, Xiong G, Cooke MD, Faulkner CC, Cowburn RP. Magnetic domain-wall dynamics in a submicrometre ferromagnetic structure. Nat Mater 2003;2:85.

[33] Hayashi M, Thomas L, Bazaliy YB, Rettner C, Moriya R, Jiang X, et al. Influence of current on field-driven domain wall motion in permalloy nanowires from time resolved measurements of anisotropic magnetoresistance. Phys Rev Lett 2006;96:197207.

[34] Kondou K, Ohshima N, Kasai S, Nakatani Y, Ono T. Single shot detection of the magnetic domain wall motion by using tunnel magnetoresistance effect. Appl Phys Express 2008;1:061302.

[35] Yang J, Nistor C, Beach GSD, Erskine JL. Magnetic domain-wall velocity oscillations in permalloy nanowires. Phys Rev B 2008;77:014413.

[36] Hayashi M, Thomas L, Rettner C, Moriya R, Parkin SSP. Direct observation of the coherent precession of magnetic domain walls propagating along permalloy nanowires. Nat Phys 2007;3:21.

[37] Glathe S, Berkov I, Mikolajick T, Mattheis R. Experimental study of domain wall motion in long nanostrips under the influence of a transverse field. Appl Phys Lett 2008;93:162505.

[38] Moriya R, Hayashi M, Thomas L, Rettner C, Parkin SSP. Dependence of field driven domain wall velocity on cross-sectional area in $Ni_{65}Fe_{20}Co_{15}$ nanowires. Appl Phys Lett 2010;97:142506.

[39] Thomas L, Parkin S. In: Kronmüller H, Parkin S, editors. Handbook of magnetism and advanced magnetic materials, vol. 2. New York, NY: Wiley; 2007. p. 942−82.

[40] Beach GSD, Tsoi M, Erskine JL. Current-induced domain wall motion. J Magn Magn Mater 2008;320:1272.

[41] Boulle O, Malinowski G, Kläui M. Current-induced domain wall motion in nanoscale ferromagnetic elements. Mat Sci Eng R 2011;72:159.

[42] Malinowski G, Boulle O, Kläui M. Current-induced domain wall motion in nanoscale ferromagnetic elements. J Phys D: Appl Phys 2011;44:384005.

[43] Berger L. Low-field magnetoresistance and domain drag in ferromagnets. J Appl Phys 1978;49:2156.

[44] Berger L. Exchange interaction between ferromagnetic domain wall and electric current in very thin metallic films. J Appl Phys 1984;55:1954.

[45] Bazaliy YB, Jones BA, Zhang S-C. Modification of the Landau−Lifshitz equation in the presence of a spin-polarized current in colossal and giant-magnetoresistive materials. Phys Rev B 1998;57:R3213.

[46] Waintal X, Viret M. Current-induced distortion of a magnetic domain wall. Europhys Lett 2004;65:427.

[47] Li Z, Zhang S. Domain-wall dynamics and spin-wave excitations with spin-transfer torques. Phys Rev Lett 2004;92:207203.

[48] Li Z, Zhang S. Domain-wall dynamics driven by adiabatic spin-transfer torques. Phys Rev B 2004;70:024417.

[49] Tatara G, Kohno H. Theory of current-driven domain wall motion: spin transfer versus momentum transfer. Phys Rev Lett 2004;92:086601.

[50] Thiaville A, Nakatani Y, Miltat J, Vernier N. Domain wall motion by spin-polarized current: a micromagnetic study. J Appl Phys 2004;95:7049.

[51] Barnes SE, Maekawa S. Current-spin coupling for ferromagnetic domain walls in fine wires. Phys Rev Lett 2005;95:107204.

[52] Barnes SE, Maekawa S. In: Maekawa S, editor. Concepts in spin electronics. Oxford University Press, Oxford; 2006.

[53] Stiles MD, Saslow WM, Donahue MJ, Zangwill A. Adiabatic domain wall motion and Landau−Lifshitz damping. Phys. Rev B 2007;75:214423.

[54] Xiao J, Zangwill A, Stiles MD. Spin-transfer torque for continuously variable magnetization. Phys Rev B 2006;73:054428.

[55] Zhang S, Levy PM, Fert A. Mechanisms of spin-polarized current-driven magnetization switching. Phys Rev Lett 2002;88:236601.

[56] Zhang S, Li Z. Role of non-equilibrium conduction electrons on magnetization dynamics of ferromagnets. Phys Rev Lett 2004;93:127204.

[57] Thiaville A, Nakatani Y, Miltat J, Suzuki Y. Micromagnetic understanding of current-driven domain wall motion in patterned nanowires. Europhys Lett 2005;69:990.

[58] Garate I, Gilmore K, Stiles MD, MacDonald AH. Nonadiabatic spin-transfer torque in real materials. Phys Rev B 2009;79:104416.

[59] Gilmore K, Garate I, MacDonald AH, Stiles MD. First-principles calculation of the nonadiabatic spin transfer torque in Ni and Fe. Phys Rev B 2011;84:224412.

[60] Kohno H, Tatara G, Shibata J. Microscopic calculation of spin torques in disordered ferromagnets. J Phys Soc Jpn 2006;75:113706.

[61] Grollier J, Boulenc P, Cros V, Hamzić A, Vaurès A, Fert A. Switching a spin-valve back and forth by current-induced domain wall motion. Appl Phys Lett 2003;83:509.

[62] Thiaville A, Nakatani Y. Electrical rectification effect in single domain microstrips: A micromagnetics-based analysis. J Appl Phys 2008;104:093701.

[63] Kim WJ, Seo SM, Lee TD, Lee KJ. Oscillatory domain wall velocity of current-induced domain wall motion. J Magn Magn Mater 2007;310:2032.

[64] You C-Y, Sung IM, Joe BK. Analytic expression for the temperature of the current-heated nanowire for the current-induced domain wall motion. Appl Phys Lett 2006;89:222513.

[65] Yamaguchi A, Nasu S, Tanigawa H, Ono T, Miyake K, Mibu K, et al. Effect of Joule heating in current-driven domain wall motion. Appl Phys Lett 2005;86:012511.

[66] Togawa Y, Kimura T, Harada K, Akashi T, Matsuda T, Tonomura A, et al. Jpn J Appl Phys 2006;45:L683.

[67] You C-Y, Ha S-S. Temperature increment in a current-heated nanowire for current-induced domain wall motion with finite thickness insulator layer. Appl Phys Lett 2007;91:022507.

[68] Cochran JF, Rudd JM, Muir WB, Trayling G, Heinrich B. J Appl Phys 1991;70:6545.

[69] Counil G, Devolder T, Kim J-V, Crozat P, Chappert C, Zoll S, et al. Temperature dependences of the resistivity and the ferromagnetic resonance linewidth in permalloy thin films. IEEE Trans Magn 2006;42:3323.

[70] Zhu M, Dennis CL, McMichael RD. Temperature dependence of magnetization drift velocity and current polarization in $Ni_{80}Fe_{20}$ by spin-wave Doppler measurements. Phys Rev B 2010;81:140407(R).

[71] Seo S-M, Lee K-J, Kim W, Lee T-D. Effect of shape anisotropy on threshold current density for current-induced domain wall motion. Appl Phys Lett 2007;90:252508.

[72] Grinstein G, Koch RH. Coarse graining in micromagnetics. Phys Rev Lett 2003;90:207201.

[73] Brown Jr WF. Thermal fluctuations of a single-domain particle. Phys Rev 1963;130:1677.

[74] Duine RA, Núñez AS, MacDonald AH. Thermally-assisted current-driven domain wall motion. Phys Rev Lett 2007;98:056605.

[75] Lucassen ME, Duine RA. Fluctuations of current-driven domain walls in the nonadiabatic regime. Phys Rev B 2009;80:144221.

[76] Hinzke D, Nowak U. Domain wall motion by the magnonic spin Seebeck effect. Phys Rev Lett 2011;107:027205.

[77] Yan P, Wang XS, Wang XR. All-magnonic spin-transfer torque and domain wall propagation. Phys Rev Lett 2011;107:177207.

[78] Torrejon J, Malinowski G, Pelloux M, Weil R, Thiaville A, Curiale J, et al. Unidirectional thermal effects in current-induced domain wall motion. Phys Rev Lett 2012;109:106601.

[79] Jiang W, Upadhyaya P, Fan Y, Zhao J, Wang M, Chang L-T, et al. Direct imaging of thermally driven domain wall motion in magnetic insulators. Phys Rev Lett 2013;110:177202.

[80] Bauer GEW, Saitoh E, van Wees BJ. Spin caloritronics. Nat Mater 2012;11:391.

[81] Yuan Z, Wang S, Xia K. Thermal spin transfer torques on magnetic domain walls. Solid State Commun 2010;150:548.

[82] Thiaville A, Nakatani Y, Piéchon F, Miltat J, Ono T. Transient domain wall displacement under spin-polarized current pulses. Eur Phys J B 2007;60:15.

[83] Thiele AA. On the momentum of ferromagnetic domains. J Appl Phys 1976;47:2759.

[84] Slonczewski JC. Force, momentum and topology of a moving magnetic domain. J Magn Magn Mater 1979;12:108.

[85] Malozemoff AP, Slonczewski JC. Magnetic domain walls in bubble materials. New York, NY: Academic Press; 1979.

[86] Chauleau J-Y, Weil R, Thiaville A, Miltat J. Magnetic domain walls displacement: automotion versus spin-transfer torque. Phys Rev B 2010;82:214414.

[87] Tretiakov OA, Clarke D, Chern GW, Bazaliy YB, Tchernyshyov O. Dynamics of domain walls in magnetic nanostrips. Phys Rev Lett 2008;100:127204.

[88] Yamaguchi A, Ono T, Nasu S, Miyake K, Mibu K, Shinjo T. Real-space observation of current-driven domain wall motion in submicron magnetic wires. Phys Rev Lett 2004;92:077205.

[89] Kläui M, Vaz CAF, Bland JAC, Wernsdorfer W, Faini G, Cambril E, et al. Controlled and reproducible domain wall displacement by current pulses injected into ferromagnetic ring structures. Phys Rev Lett 2005;94:106601.

[90] Hayashi M, Thomas L, Rettner C, Moriya R, Bazaliy YB, Parkin SSP. Current driven domain wall velocities exceeding the spin angular momentum transfer rate in permalloy nanowires. Phys Rev Lett 2007;98:037204.

[91] Parkin SSP, Hayashi M, Thomas L. Magnetic domain wall racetrack memory. Science 2008;320:190.

[92] Junginger F, Kläui M, Backes D, Rüdiger U, Kasama T, Dunin-Borkowski RE, et al. Spin torque and heating effects in current-induced domain wall motion probed by transmission electron microscopy. Appl Phys Lett 2007;90:132506.

[93] Meier G, Bolte M, Eiselt R, Krüger B, Kim D-H, Fischer P. Direct imaging of stochastic domain wall motion driven by nanosecond current pulses. Phys Rev Lett 2007;98:187202.

[94] Hayashi M, Thomas L, Rettner C, Moriya R, Jiang X, Parkin SSP. Dependence of current and field driven depinning of domain walls on their structure and chirality in permalloy nanowires. Phys Rev Lett 2006;97:207205.

[95] Kläui M, Jubert P-O, Allenspach R, Bischof A, Bland JAC, Faini G, et al. Direct observation of domain wall configurations transformed by spin currents. Phys Rev Lett 2005;95:026601.

[96] Heyne L, Kläui M, Backes D, Moore TA, Krzyk S, Rüdiger U, et al. Relationship between nonadiabaticity and damping in permalloy studied by current induced spin structure transformations. Phys Rev Lett 2008;100:066603.

[97] Jubert P-O, Kläui M, Bischof A, Rüdiger U, Allenspach R. Velocity of vortex walls moved by current. J Appl Phys 2006;99:08G523.

[98] Heyne L, Kläui M, Backes D, Möhrke P, Moore TA, Kimling JG, et al. Direct imaging of current-induced domain wall motion in CoFeB structures. J Appl Phys 2008;103:07D928.

[99] Bocklage L, Krüger B, Matsuyama T, Bolte M, Merkt U, Pfannkuche D, et al. Dependence of magnetic domain-wall motion on a fast changing current. Phys Rev Lett 2009;103:197204.

[100] Heyne L, Rhensius J, Bisig A, Krzyk S, Punke P, Kläui M, et al. Direct observation of high velocity current induced domain wall motion. Appl Phys Lett 2010;96:032504.

[101] Uhlig WC, Donahue MJ, Pierce DT, Unguris J. Direct imaging of current-driven domain walls in ferromagnetic nanostripes. J Appl Phys 2009;105:103902.

[102] He J, Li Z, Zhang S. Current-driven vortex domain wall dynamics by micromagnetic simulations. Phys Rev B 2006;73:184408.

[103] Nakatani Y. ICMFS '06 Conference, Sendai, Japan (unpublished).

[104] Beach GSD, Knutson C, Nistor C, Tsoi M, Erskine JL. Nonlinear domain wall velocity enhancement by spin-polarized electric current. Phys Rev Lett 2006;97:057203.

[105] Thomas L, Hayashi M, Jiang X, Moriya R, Rettner C, Parkin S. Oscillatory dependence of current-driven magnetic domain wall motion on current pulse length. Nature 2006;443:197.

[106] Vernier N, Allwood DA, Atkinson D, Cooke MD, Cowburn RP. Domain-wall propagation in magnetic nanowires by spin-polarized current injection. Europhys Lett 2004;65:526.

[107] Grollier J, Boulenc P, Cros V, Hamzić A, Vaurès A, Fert A, et al. Spin-transfer-induced domain wall motion in a spin valve. J Appl Phys 2004;95:6777.

[108] Thomas L, Hayashi M, Jiang X, Moriya R, Rettner C, Parkin S. Resonant amplification of magnetic domain wall motion by a train of current pulses. Science 2007;315:1553.

[109] Lepadatu S, Hickey MC, Potenza A, Marchetto H, Charlton TR, Langridge S, et al. Experimental determination of spin-transfer torque nonadiabaticity parameter and spin polarization in permalloy. Phys Rev B 2009;79:094402.

[110] Eltschka M, Wötzel M, Rhensius J, Krzyk S, Nowak U, Kläui M, et al. Nonadiabatic spin torque investigated using thermally activated magnetic domain wall dynamics. Phys Rev Lett 2010;105:056601.

[111] Laufenberg M, Bührer W, Bedau D, Melchy P-E, Kläui M, Vila L, et al. Temperature dependence of the spin torque effect in current-induced domain wall motion. Phys Rev Lett 2006;97:046602.

[112] Bedau D, Kläui M, Krzyk S, Rüdiger U, Faini G, Vila L. Detection of current-induced resonance of geometrically confined domain walls. Phys Rev Lett 2007;99:146601.

[113] Moriya R, Thomas L, Hayashi M, Bazaliy YB, Rettner C, Parkin SSP. Probing vortex-core dynamics using current-induced resonant excitation of a trapped domain wall. Nat Phys 2008;4:368.

[114] Saitoh E, Miyajima H, Yamaoka T, Tatara G. Current-induced resonance and mass determination of a single magnetic domain wall. Nature 2004;432:203.

[115] Kasai S, Nakatani Y, Kobayashi K, Kohno H, Ono T. Current-driven resonant excitation of magnetic vortices. Phys Rev Lett 2006;97:107204.

[116] Yamada K, Kasai S, Nakatani Y, Kobayashi K, Kohno H, Thiaville A, et al. Electrical switching of the vortex core in a magnetic disk. Nat Mater 2007;6:270.

[117] Bolte M, Meier G, Krüger B, Drews A, Eiselt R, Bocklage L, et al. Time-resolved x-ray microscopy of spin-torque-induced magnetic vortex gyration. Phys Rev Lett 2008;100:176601.

[118] Heyne L, Rhensius J, Ilgaz D, Bisig A, Rüdiger U, Kläui M, et al. Direct determination of large spin-torque nonadiabaticity in vortex core dynamics. Phys Rev Lett 2010;105:187203.

[119] Pollard SD, Huang L, Buchanan KS, Arena DA, Zhu Y. Direct dynamic imaging of non-adiabatic spin torque effects. Nat Comm 2012;3:1028.

[120] Claudio-Gonzalez D, Thiaville A, Miltat J. Domain wall dynamics under nonlocal spin-transfer torque. Phys Rev Lett 2012;108:227208.

[121] Stiles MD, Miltat J. In: Hillebrands B, Thiaville A, editors. Spin dynamics in confined magnetic structures III. Berlin: Springer; 2006. p. 225−308.

[122] Serpico C, d'Aquino M, Bertotti G, Mayergoyz ID. Quasiperiodic magnetization dynamics in uniformly magnetized particles and films. J Appl Phys 2004;95:7052.

[123] Fukami S, Suzuki T, Ohshima N, Nagahara K, Ishiwata N. Micromagnetic analysis of current driven domain wall motion in nanostrips with perpendicular magnetic anisotropy. J Appl Phys 2008;103:07E718.

[124] Jung S-W, Kim W, Lee T-D, Lee K-J, Lee H-W. Current-induced domain wall motion in a nanowire with perpendicular magnetic anisotropy. Appl Phys Lett 2008;92:202508.

[125] Miron IM, Moore T, Szambolics H, Buda-Prejbeanu LD, Auffret S, Rodmacq B, et al. Fast current-induced domain-wall motion controlled by the Rashba effect. Nat Mater 2011;10:419.

[126] Haazen PPJ, Murè E, Franken JH, Lavrijsen R, Swagten HJM, Koopmans B. Domain wall depinnning governed by the spin Hall effect. Nat Mater 2013;12:299.

[127] Lim CK, Devolder T, Chappert C, Grollier J, Cros V, Vaurès A, et al. Domain wall displacement induced by subnanosecond pulsed current. Appl Phys Lett 2004;84:2820.

[128] Laribi S, Cros V, Muñoz M, Grollier J, Hamzić A, Deranlot C, et al. Reversible and irreversible current-induced domain wall motion in CoFeB based spin valve stripes. Appl Phys Lett 2007;90:232505.

[129] Pizzini S, Uhlíř V, Vogel J, Rougemaille N, Laribi S, Cros V, et al. High domain wall velocities at zero magnetic field induced by low current densities in spin valve nanostripes. Appl Phys Express 2009;2:023003.

7 III−V-Based Ferromagnetic Semiconductors

Fumihiro Matsukura[1,2] and Hideo Ohno[1,2,3]

[1]WPI-Advanced Institute for Materials Research, Aoba-ku, Sendai, Japan, [2]Center for Spintronics Integrated Systems, Tohoku University, Aoba-ku, Sendai, Japan, [3]Laboratory for Nanoelectronics and Spintronics, Research Institute of Electrical Communication, Tohoku University, Aoba-ku, Sendai, Japan

Contents

Nanomagnetism and Spintronics. DOI: http://dx.doi.org/10.1016/B978-0-444-63279-1.00007-1

7.1   Introduction

In magnetic semiconductors, one can make use of a variety of spin-related phenomena, not readily available in other materials. The spin-related properties are brought about by the exchange interaction among band carriers and magnetic moments. The study of magnetic semiconductors was initiated by a series of experimental and theoretical studies on compound materials like rare-earth chalcogenides and chromite spinels (e.g., $CdCr_2Se_4$). A number of exotic properties were observed in these magnetic semiconductors, such as colossal magnetoresistance (MR) and magneto-optical properties, originating from the interplay between ferromagnetism and semiconducting properties [1−3]. Although these ferromagnetic semiconductors inspired new concepts using semiconductors with utilizing the spin degree of freedom, difficulties associated with material preparation hindered further progress in experimental studies.

Today, majority of the studies on magnetic semiconductor involve alloy semiconductors, in which magnetic elements substitute a part of lattice sites in nonmagnetic semiconductor materials. Here, the host semiconductors include II−VI, III−V, IV−VI, IV, and II−VI−V$_2$. These magnetic alloy semiconductors are known as diluted magnetic semiconductors (DMSs) or semimagnetic semiconductors. Since DMSs and their heterostructures can readily be grown by modern epitaxial techniques such as molecular beam epitaxy (MBE), many of the studies conducted today on magnetic semiconductors are focusing on thin films and multilayered heterostructures prepared by epitaxy. The most extensively studied DMSs are II−VIs and III−Vs alloyed with Mn, such as (Cd,Mn)Te and (Ga,Mn)As.

Since the early stage, paramagnetic II−VI DMSs have been studied extensively [4−7]. Because Mn substitutes II-group cation in divalent state, II−VI DMS is isoelectronic and if carriers are needed they have to be provided by extrinsic means as doping or controlling defects. The d-electrons of transition metals provide localized magnetic moments. Owing to the antiferromagnetic superexchange coupling among Mn, most of II−VI DMSs show spin-glass phase or antiferromagnetic phase at low temperatures, depending on the concentration of the transition metal ions. In addition to the superexchange coupling, strong spin-dependent coupling (sp−d exchange interaction) between the band of host semiconductors (sp-state) and magnetic moments (d-state) exists in DMSs. Because of the sp−d exchange, aligning the direction of magnetic moments by external magnetic fields induces giant Zeeman splitting in the band structure, resulting in large magneto-optical effects [8]. By injecting the carriers into II−VI DMSs by optical means, bound magnetic polarons (BMPs) can be formed [9]. In BMP, electrons trapped in a shallow impurity states align locally the surrounding magnetic moments by the sp–d exchange interaction. This implies that electronic states of carriers are delocalized and extended, and when the concentration exceeds a certain value, ferromagnetism may be expected. This type of carrier-induced ferromagnetism was reported in IV–VI [10] and III–V [11,12] and later in II–VI compounds; the ferromagnetism in III–Vs is the subject of this chapter and is discussed in detail in the following sections.

In II–VI compounds, doping of delocalized *holes* in II–VI DMSs was predicted to result in the ferromagnetic order by the Ruderman–Kittel–Kasuya–Yosida (RKKY) type interaction [13] due to greater magnitude of the p–d exchange than that of the s–d exchange. This approach was taken in the experimental works and led to the observation of the ferromagnetic order in p-(Cd,Mn)Te quantum wells (QWs) [14] and in p-(Zn,Mn)Te [15]. Another approach to obtain ferromagnetic II–VI DMSs is to select Cr as magnetic elements, among which the ferromagnetic superexchange interaction is predicted by the theoretical calculation [16].

In III–V DMSs, Mn acts simultaneously as the source of localized magnetic moments and as an acceptor. Thus, without additional doping, these materials show strong p-type conduction, which provides a strong exchange interaction mediated by p–d exchange among Mn ions. Synthesis [17] and subsequent discovery of ferromagnetism in III–V-based DMSs—(In,Mn)As and (Ga,Mn)As, in 1990s [11,12]—added a new dimension to the magnetic semiconductor research, because it allowed seamless integration of ferromagnetism with established III–V heterostructures and devices. (Ga,Mn)As and (In,Mn)As are now the most well-investigated and well-understood ferromagnetic semiconductors among all and the material and their heterostructures provide an ideal test bench for demonstrating new concepts in physics as well as for spintronic device operations [18–20]. The p–d Zener model [21,22], in which hole repopulation among p–d exchange split bands stabilizes ferromagnetism, has been shown to describe qualitatively and in many cases even quantitatively a number of experimental results on these materials.

Owing to the additional spin degree of freedom, ferromagnetic semiconductors can combine semiconductor heterostructure physics and ferromagnetism, which may lead us to new spintronic functionalities [23–26]. In this chapter, we summarize the present state of the research on ferromagnetic semiconductors focusing primarily on the prototypical (Ga,Mn)As and (In,Mn)As and their device structures.

7.2 Molecular Beam Epitaxy

For the observation of a ferromagnetic phase in DMS systems, one needs to introduce a sizable amount of magnetic elements (a few percent or more), which is usually beyond the solubility limit, of the order of $10^{18} - 10^{19}$ cm^{-3} in III–V semiconductors [27,28]. For III–V DMSs which require growth far from equilibrium, MBE is so far the only established method for preparation. Under ordinary MBE growth conditions for high quality of III–Vs, such a high concentration of magnetic impurities (transition metals in most cases) results in segregation of impurity atoms and subsequent formation of unwanted compounds at the growth front, which prevents synthesis of alloy semiconductors with high concentration of transition metals. However, MBE growth can proceed under conditions far from equilibrium, especially at low growth temperatures, at which segregation and formation of unwanted compounds can be minimized or suppressed. Using low-temperature MBE (LT-MBE), preparation of uniform paramagnetic III−V DMS, (In,Mn)As, with Mn composition $x < 0.18$ was shown to be possible [17]. Here, the growth temperature T_S was reduced slightly below 300°C (typical T_S for nonmagnetic InAs is ~ 500°C). This T_S was high enough to provide metastable single-crystal epitaxial growth, yet low enough to suppress segregation of Mn and formation of thermodynamically more stable second phases such as MnAs. This initial success of (In,Mn)As was followed by the growth of uniform (Ga,Mn)As [12,29]. When growth conditions are right, majority of Mn substitutes III-group cation site [30,31], which is the reason these alloys are expressed as (III,Mn)V. Mn in cation site is an acceptor providing a hole. It also provides localized spins.

Growth of III−V DMSs is summarized in the following using (Ga,Mn)As as an example. (In,Mn)As can be prepared in a similar way [32]. LT-MBE of (Ga,Mn)As is carried out in an ultra-high vacuum MBE chamber with Ga, Mn, and As atoms impinging on GaAs (001) substrates having As-stabilized surface conditions. After removal of surface oxide and growth of buffer layer for preparation of atomically flat starting surface, growth of (Ga,Mn)As is initiated by commencing the Mn flux during the LT-GaAs growth while keeping T_S constant at ~ 250°C. Reflection high-energy electron diffraction (RHEED) pattern is a common tool to monitor the growth front. The RHEED pattern for (Ga,Mn)As is streaky (1×2) confirming growth of (Ga,Mn)As with zincblende structure. The details of this surface reconstruction are not understood; however, it may be related to the formation of Mn-related dimers as shown by combination of experimental studies (scanning tunneling microscopy (STM) and X-ray photoelectron spectroscopy) and *ab initio* calculations [33]. The Mn composition x that can be introduced in GaAs depends on T_S; the maximum x at $T_S = 250$°C is roughly 0.07 and the highest nominal x so far reported is 0.21 obtained at T_S of 150°C and only with very thin thickness less than 10 nm [34,35]. Even in such a thin (Ga,Mn)As layer with high x, uniform ferromagnetic phase is observed, confirmed by magnetization, magnetotransport, and magneto-optical signals. When T_S and/or Mn flux is too high, formation of hexagonal MnAs phase is detected by a spotty RHEED pattern [36].

We note that accurate determination of T_S is critical in growth of (Ga,Mn)As. T_S of MBE is monitored by a thermocouple placed behind the substrate holder: T_S monitored this way can be different from the actual substrate temperature by 50°C or even more, which has been shown by comparison of T_S monitored by thermocouple and by the band edge spectroscopy during (Ga,Mn)As growth. In addition, due to the high concentration of Mn acceptors in (Ga,Mn)As, the free carrier absorption in the near-infrared becomes significant and contributes to the increase of T_S after starting growth, that is, under radiation from Ga and Mn effusion cells. The effect depends on the thermal contact between the substrate and its holder as well as the distance between the substrate and the effusion cells [37]; thus it is critical to carefully establish the often machine-dependent growth temperature for (Ga, Mn)As growth.

When growth conditions are appropriate, a clear RHEED intensity oscillation is observed at the initial stage of growth [36], where surfactant effect of Mn and excess As is thought to be responsible for the two-dimensional growth at low T_S [38−40]. Magnetic and electrical properties of (Ga,Mn)As are strongly affected by the growth condition, such as V/III beam-flux ratio and T_S as well as x[41−44], which is related to the degree of compensation due to the existence of As antisites and Mn interstitials, both known as donors in GaAs.

Mn composition in the epitaxial films can be calculated from beam-equivalent pressure (BEP) measured by an ion gauge inserted in the path of the beam flux and later calibrated to relate BEP to the composition of the grown films and/or from the periods of the RHEED intensity oscillation observed at the initial stage of growth [12,45]. Conventional analyses are also employed to establish the composition, such as Auger electron spectroscopy, electron probe microanalysis, and secondary ion mass spectroscopy [12,46]. The most widely used method is the use of lattice constant a determined by X-ray diffraction (XRD) as the measure of x due to its experimental simplicity, although it is affected by the presence of species other than Mn such as As antisites [43,44,47−50]. MBE growth of (Ga,Mn)As on substrates other than GaAs (001) have also been explored, which include high-index GaAs (411), (311), and (113) substrates [51−56], GaAs (110) cleaved surface [57], Si (001) substrate [58,59], and sapphire (0001) substrate [60]. (Ga,Mn) As on sapphire (0001) substrate has a (111)-oriented zincblende structure [60].

Most of the synthesis techniques for semiconducting materials are applicable to prepare single crystals of II−VI DMSs with high quality, largely owing to the isoelectronic nature of transition metals in II−VI semiconductors [4]. Among them, MBE growth has been adopted for many heterostructures and low-dimensional structures based on II−VI materials.

7.3 Structural and Magnetic Properties

7.3.1 Structural Characterization of (Ga,Mn)As

Zincblende structure of (Ga,Mn)As is confirmed by RHEED patterns during growth as well as postgrowth XRD measurements, which show that the lattice constant a of (Ga,Mn)As is greater than that of GaAs and depends linearly on x. Asymmetric XRD showed that pseudomorphic growth of (Ga,Mn)As on GaAs substrate takes place without strain relaxation at least up to 2 μm thickness and even more [61,62]. Pseudomorphic growth is also confirmed from the reciprocal space mapping on (115) and (224) planes as shown in Figure 7.1. The freestanding lattice constant a of (Ga,Mn)As is calculated under the assumption that (Ga,Mn)As has the same elastic constants as those of GaAs (Poisson ratio = 0.311) [12]. Thus, obtained x dependence of a follows the Vegard's law, $a = a_{\text{LT-GaAs}}(1 - x) + a_{\text{ZB-MnAs}}x$, where

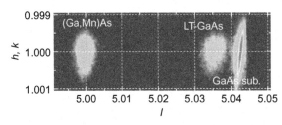

Figure 7.1 (115) Asymmetric reciprocal space map for $Ga_{0.93}Mn_{0.07}As$ (150 nm)/ LT-GaAs (50 nm)/GaAs (001) substrate measured at room temperature.

$a_{\text{LT-GaAs}}$ and $a_{\text{ZB-MnAs}}$ are the lattice constants of low-temperature-grown GaAs (LT-GaAs) ~ 0.566 nm and hypothetical zincblende MnAs ~ 0.598 nm [12]; this value is close to the theoretically predicted value of 0.59 nm [63]. Note that this way of determining a is known to suffer from uncertainties due to the degree of incorporation of defects like As antisites, As_{Ga}, and Mn interstitials, Mn_{int}, and gives inaccuracy of $x \pm 0.01$ and sometimes even more [43,44,47−50]. Theoretical calculation also shows that As antisites and/or Mn interstitials increase a of (Ga, Mn)As [64].

For Mn-doped GaAs with the doping level less than 10^{19} cm^{-3}, Mn substitutes Ga, which was confirmed by electron paramagnetic resonance and cross-sectional scanning tunneling microscopy (XSTM) [65−67]. For (Ga,Mn)As with a few percent of Mn, extended X-ray-absorption fine structure (EXAFS) study revealed that the majority of Mn was at the substitutional site, that is, on the Ga sublattice of the zincblende lattice. The Mn−As bond length was determined to be 0.249 and 0.250 nm consistent with the XRD results, longer than 0.244 nm of the host Ga−As bond length [31]. Similarly, EXAFS showed that in (In,Mn)As, majority of Mn sits on the In sublattice [30]. The ion channeling measurements, Rutherford backscattering and particle-induced X-ray emission, later revealed that a part of Mn atoms (a few tens of percent of the introduced Mn atoms) were incorporated at interstitial sites [68]. Postgrowth, low-temperature annealing (LT annealing) at or below the growth temperature [47] was found to decrease the number of Mn_{int} and simultaneously increases hole concentration p[69−71], indicating that Mn_{int} is unstable and mobile donor. The double-donor nature of Mn_{int} was also shown from *ab initio* calculations [72]. Further annealing above 350°C results in the formation of MnAs nanoclusters having NiAs structure, which has Curie temperature T_C of ~ 310 K [29]. Disappearance of (Ga,Mn)As phase after annealing was indicated by the absence of corresponding XRD peaks [36]. Double-donor Mn_{int} was calculated not to participate in the ferromagnetic order and may form pairs with substitutional Mn, Mn_{Ga}. This results in superexchange antiferromagnetic coupling between Mn_{int} and Mn_{Ga}, which compensates the Mn moments of the two participating Mn atoms and reduces the net Mn moment [73].

Electron spin resonance measurements [66] showed that neutral Mn_{Ga} in GaAs doped with Mn ($\sim 10^{17}$ cm^{-3}) is in the A^0(d^5 + h) state, that is, the neutral acceptor Mn state (A^0) has an electronic structure of five d-electrons (Mn^{2+}) and a loosely bound hole. The charge state of a single Mn in GaAs was visualized by STM images, where switching between A^0(d^5 + h) state and negatively ionized A^-(d^5)

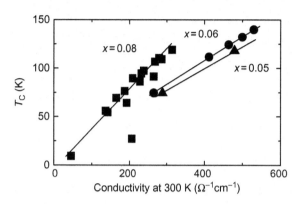

Figure 7.2 Room temperature electrical conductivity dependence of the Curie temperature of (Ga,Mn)As. Straight lines are guide for the eye [71].
Source: Reused by permission of American Institute of Physics.

state was observed by changing the bias polarity of the STM tip [74]. The hole binding energy of Mn acceptor is 0.11 eV [75,76], and this moderate ionization energy results in the metal-insulator transition (MIT) in (Ga,Mn)As, which affects magnetic and transport properties. By measuring the STM spectra, the magnetic interaction between two Mn ions was shown to depend on the separation of the two; the positions of Mn ions were controlled by an atom by atom substitution using an STM chip [77]. The shape of the acceptor wave function was also shown to be distorted by local lattice strain, when the GaAs host lattice symmetry was broken by embedded InAs quantum dots [78]. These STM measurements showed that the nature of a single Mn ion is closely related to the magnetic and electrical properties of (Ga,Mn)As. Also they showed the possibility of manipulating single magnetic spins in semiconductors.

7.3.2 Magnetic Properties of (Ga,Mn)As

Magnetization M of (Ga,Mn)As is measured using a standard magnetometer such as superconducting quantum interference device (SQUID) magnetometer. The diamagnetic response of thick GaAs substrate determined by a separate measurement must be subtracted from the measured magnetization curve in order to obtain the response from thin epitaxial layers. This subtraction process together with the removal of the residual magnetic fields of the superconducting magnet becomes critical when the magnetic layer is very thin and/or the Mn concentration is low, because substrates themselves occasionally show paramagnetic response. When H is applied in the direction of magnetic easy axis, M of (Ga,Mn)As shows a sharp and clear hysteresis in its H dependence at low temperatures, which is one of the evidences of the presence of extended ferromagnetic order in the film [12]. T_C can be determined by measuring the temperature dependence of the remanent magnetization, from the Arrott plots and/or from the Curie—Weiss plot. The value of T_C has a strong correlation with the electrical conductivity, that is, for a given x higher T_C is observed for more metallic samples as shown in Figure 7.2 [71]. This relationship can be traced using a single sample, because the conductivity of (Ga,Mn) As can be altered by annealing after sample preparation. Insulating samples, the resistance of which increases as temperature is reduced, often show a pronounced paramagnetic response in the $M-H$ curve at high magnetic fields after closure of hysteresis loop. This also indicates correlation between magnetic and electrical properties [79]. The highest T_C reported so far is 185 K for a (Ga,Mn)As film with $x = 0.09$ after annealing [80]. The p−d Zener model discussed in Section 7.6 is capable of explaining the observed magnetic properties, including the magnitude of T_C and the temperature as well as the strain dependence of the magnetic easy axis [21,22].

7.3.2.1 Magnetic Domain

Extended magnetic domain structures, which is an evidence of long-range magnetic interaction, was observed in (Ga,Mn)As samples with magnetic easy axis in-plane as well as those with easy axis perpendicular-to-plane by scanning Hall microscope, scanning SQUID microscope, magneto-optical microscope, and Lorenz microscope as shown in Figure 7.3, where the size of the domain is shown to range from a few microns to a millimeter [81−85]. The computed value of the domain width in (Ga,Mn)As by the p−d Zener model combined with micromagnetic theory [86] is in reasonable agreement with the experimental ones [81].

7.3.2.2 Magnetic Anisotropy

The direction as well as the magnitude of the magnetocrystalline anisotropy of (Ga, Mn)As can be tuned by the combination of direction and magnitude of lattice strain, and the hole concentration p. A number of methods have been employed to determine the magnetic anisotropy of (Ga,Mn)As, for example, direct magnetization measurement, AC susceptibility, ferromagnetic resonance (FMR), anomalous Hall effect (AHE), planar Hall effect (PHE), anisotropic magnetoresistance (AMR), tunnel magnetoresistance (TMR), and magneto-optical microscope [36,82,87−95]. Two measurements, FMR and transport, on the same sample, were shown to give nearly the same magnetic anisotropy constants as shown in Figure 7.4[97]. (Ga, Mn)As layers under compressive strain, for example, those grown directly on GaAs substrates, show in-plane magnetic easy axis as long as its carrier concentration is

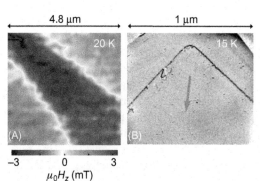

Figure 7.3 Magnetic domain structure observed for (Ga,Mn)As (A) with perpendicular easy axis by scanning Hall probe and (B) with in-plan easy axis by magneto-optical microscope. *Source*: Reproduced by permission of American Physical Society. (A) from [81] and (B) from [82].

high, while the layers under tensile strain, like (Ga,Mn)As on (In,Ga)As, show magnetic easy axis perpendicular to the plane [36]. The single ion anisotropy of Mn in GaAs is confirmed to be too small to account for this sizable magnetic anisotropy [98]. The origin of this strain-dependent magnetic anisotropy can be explained in terms of the warped anisotropic valence band due to combined effect of spin−orbit interaction (SOI) and the lattice strain in the framework of the p−d Zener model, which also predicts that the magnetic anisotropy is p dependent [21,22,99]. The p-dependent part has later been experimentally established in (Al, Ga,Mn)As, where Al reduces hole concentration; (Al,Ga,Mn)As shows perpendicular-to-plane magnetic easy axis under compressive strain as opposed to in-plane easy axis of usual (Ga,Mn)As with compressive strain [100]. In addition, the direction of magnetic easy axis was found to be temperature dependent; it was perpendicular at low temperatures and become in-plane as temperature increased as

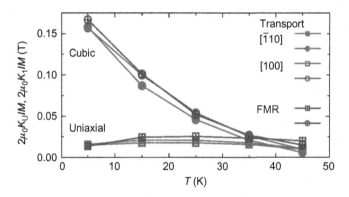

Figure 7.4 Temperature dependence of in-plane magnetic anisotropy fields determined from both magnetotransport (PHE in Hall bar along [$\bar{1}$10] channel and AMR along [100] channel) and FMR measurements [97].

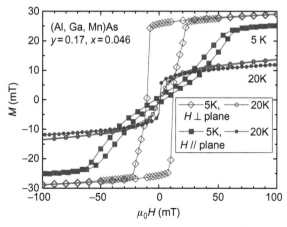

Figure 7.5 Magnetization curves of (Al,Ga,Mn)As with Al composition $y = 0.17$ and Mn composition $x = 0.046$ at 5 and 20 K. Magnetic field is applied in two different directions, showing that the easy axis of magnetization of the sample is perpendicular to the plane at 5 K and in-plane at 20 K [100].

shown in Figure 7.5[87,96,100]. This peculiar temperature dependence is also in accordance with what has been predicted by the p−d Zener model for the low p case; this is due to the carrier redistribution among different spin-split states as spin-splitting is M dependent which in turn is temperature dependent.

Compressive (Ga,Mn)As shows fourfold in-plane easy axis along $\langle 1\ 0\ 0 \rangle$ and two uniaxial easy axes along [110] or [$\bar{1}\ 1\ 0$] [88,101] and [1 0 0] or [0 1 0] [102−104]. The relative strength of these three anisotropic energies determined from planar Hall measurements is 100:10:1 [102]. The in-plane uniaxial anisotropy cannot be explained by the p−d Zener model without assuming the presence of shear strain, which has not yet been confirmed experimentally. Although a number of suggestions have been made, the origin of this in-plane uniaxial anisotropy remains unclear: suggestions include that it may be related to the anisotropic (1×2) surface reconstruction during MBE growth, the lack of top−bottom symmetry in epilayers, the existence of a trigonal distortion, the anisotropy of the strength of p−d hybridization, or anisotropic formation of Mn dimers [82,87,88,105,106]. The magnetostriction measurements by resonant nanoelectromechanical system revealed that the coupling between the magnetoelastic constants and in-plane magnetic anisotropic constants in compressive (Ga,Mn)As [107]. This makes it possible to control magnetization direction by the application of external stress, which was demonstrated by the combination of (Ga,Mn)As with piezoelectric material; the direction of magnetization was manipulated through the change of stress by applying voltage to piezoelectric material [108−112].

The magnetic overlayers on (Ga,Mn)As can induce an additional anisotropy to (Ga,Mn)As. The pinning of the M direction or exchange bias of (Ga,Mn)As is shown to be possible with overlayers of spin-glass (Zn,Mn)Se, antiferromagnetic MnO, and IrMn [113−115] as well as embedded nanosize-MnAs precipitates [116]. Another method to control the uniaxial anisotropy is lithographical nanopatterning of the (Ga,Mn)As layers [112,117−119]. From the temperature dependence of the magnetic anisotropy, this lithography-induced anisotropy may be related to the strain relaxation, as also suggested by XRD [117], and not to the classical shape magnetic anisotropy [120]. It was shown that the strain in (Ga,Mn)As is released by lifting it from the substrate, and that magnetic easy axis direction is changed by redepositing of the lifted-off (Ga,Mn)As on other substrates [121].

7.4   Electrical and Optical Properties

7.4.1   Electrical Properties of (Ga,Mn)As

7.4.1.1   Hall Effect and Hole Concentration

Since it has been show that the presence of holes stabilizes ferromagnetism in (Ga,Mn)As, it is important to measure the hole concentration p of the material. In conventional semiconductors, p can readily be determined by the Hall measurement. For ferromagnetic semiconductors, however, it is not straightforward due to the presence of the anomalous Hall effect, which may often be a significant portion of the Hall voltage even at room temperature. Hall resistance R_{Hall} in ferromagnetic materials is empirically expressed as the sum of the ordinary Hall resistance and anomalous Hall resistance, as $R_{\text{Hall}} = (R_0/d)\mu_0 H + (R_S/d)M$, where R_0 and R_S are the ordinary and the anomalous Hall coefficients, respectively, and d is the thickness of the ferromagnetic layer. Here, M is the component of the magnetization perpendicular to the sample surface. R_S is usually proportional to R_{sheet}^γ (R_{sheet}: sheet resistance) with $\gamma = 1$ or 2 depending on the origin of the anomalous Hall effect; the skew scattering results in $\gamma = 1$, whereas the side-jump in $\gamma = 2$ [122]. $\gamma = 2$ can also originate from scattering-independent topological contribution [123]. Although further compilation of the experimental data is necessary, $\gamma = 1$ for insulating and $\gamma = 2$ for metallic (Ga,Mn)As and (In,Mn)As have been reported [124,125]. The dominance of the anomalous Hall term hinders straightforward determination of the carrier type and concentration from the ordinary Hall term; the carrier concentration is given as $1/(eR_0)$ (e: elementary charge) and its positive (negative) sign corresponds to p-type (n-type) conduction. Only at low temperatures and under high H, the anomalous Hall term gradually saturates, so that R_0 can be determined with a reasonable accuracy from the remaining linear change of R_{Hall} in the H dependence [70,126,127]. Note that although M saturates at relatively low H, negative MR persists to high H, and affects the H dependence of the anomalous Hall effect through the change of the anomalous Hall coefficient. Thus, this method is not applicable to insulating (Ga,Mn)As, because of the very large MR and resistance in insulating (Ga,Mn)As [79].

Another method for determining p is the electrochemical capacitance–voltage (ECV) profiling. The reliability of this method was confirmed by comparison of p obtained from the ECV and the Hall measurements for Be-doped LT-GaAs [128–130]. A gradient of p along the growth direction (higher p near surface) was also found [129]. Determination of p by other methods, such as thermoelectric power measurements [131] or the Raman scattering analysis of the coupled plasmon-LO-phonon modes, are also possible [132,133].

These measurements revealed that p is often less than 10^{21} cm^{-3} (smaller than the nominal Mn composition x) for as-grown samples, and increases significantly to $p = x$, consistent with the acceptor nature of Mn, after appropriate annealing as long as x is below 0.07 and the film is thin enough [134].

7.4.1.2 Low-Temperature Annealing

LT annealing [47] is now a standard way to remove Mn interstitials in (Ga,Mn)As. An appropriate LT-annealing process increases the electrical conductivity (increase of p) and at the same time increases T_C. LT annealing allows us to measure the correlation between the electrical conductivity and T_C using a set of samples cleaved from the same wafer, where a monotonic positive dependence of T_C on the conductivity was observed (Figure 7.2). This is one of the evidences of the hole-induced ferromagnetism [71]. The effect was initially attributed to the reduction of As antisites, which act as a double donor in GaAs. A series of subsequent studies established that the annealing effect is coming from the reduction of Mn_{int}. As shown by the EXAFS and the channeling measurements [31,68], Mn occupies two distinct positions, Ga and interstitial sites, in zincblende host GaAs lattice. Mn_{Ga} acts as a single acceptor, while Mn_{int} as a double donor. Under the assumption that Mn is the only relevant dopant in the system, p can be expressed by the concentrations of Mn_{Ga} and Mn_{int}, as $p = [Mn_{Ga}] - 2[Mn_{int}]$. This relationship was confirmed experimentally by a combination of channeling and ECV measurements [135]. In addition, Mn_{Ga} and Mn_{int} can form antiferromagnetic pairs, in which the Mn moments are canceled [73]. Thus, when all Mn_{int} form a pair with Mn_{Ga}, the effective Mn composition contributing to ferromagnetic order is the difference of compositions of Mn_{Ga} and Mn_{int}, $x_{eff} = ([Mn_{Ga}] - [Mn_{int}])/N_0$, where N_0 is the density of the cation sites. Assuming this relationship, magnetization data were shown to yield $\sim 5\mu_B$ per Mn atom in a wide range of x from 0.02 to 0.09 [134], consistent with the expected value for Mn^{2+}. Analysis of both as-grown and annealed samples also showed that the increase of T_C after LT annealing is a combined result of increased p and x_{eff} due to the reduction of $[Mn_{int}]$. Self-compensation of Mn is enhanced at higher x, and results in almost the same p for as-grown (Ga,Mn)As with x greater than 0.02. However, overcompensation has never been observed. Thus, the self-compensation is believed to be related to the Fermi level pinning [136]. This suggests that once the Fermi level reaches a certain position in the valence band, it becomes energetically favorable to form Mn_{int} as a counter dopant [137], which was explained theoretically by calculating the formation energy of Mn_{Ga} and Mn_{int} as a function of x and the concentration of various donors [138]. The calculated x dependence of $[Mn_{Ga}]$ and $[Mn_{int}]$ based on the argument of the formation energies is in good agreement with the experimental result as shown in Figure 7.6 [139].

The effect of LT annealing is strongly structure dependent. The increase of T_C by LT annealing is greater for thinner (Ga,Mn)As and is observed only for the (Ga, Mn)As layers near the surface, that is, LT-annealing effect is suppressed with a cap GaAs layer with thickness beyond 10 monolayers [140,141]. The surface Mn

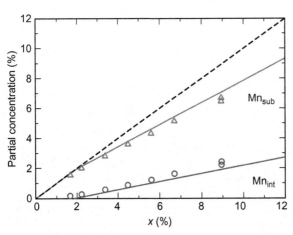

Figure 7.6 Solid lines show the partial compositions of substitutional Mn, Mn_{Ga}, and interstitial Mn, Mn_{int}, from the theoretical calculation. Symbols show the experimental results [20]. *Source*: Reproduced by permission of American Physical Society.

Auger signal increases after annealing, indicating that Mn out-diffuses and accumulates on the surface during the process [142]. This is shown to be modeled by a one-dimensional out-diffusion model as shown in Figure 7.7[142]. It has also been suggested that the diffusion of Mn to the substrate side may be limited electrostatically by a p–n junction formed by fast diffusing Mn_{int} (donor) and slowly, if any, diffusing Mn_{Ga} (acceptor). Mn_{int} at surface may become electrically inactive as a result of oxidation and p–n junction is perhaps not formed on the surface side. Thus, the LT annealing is expected to depend on the atmosphere and the layer on top of the magnetic layer. Efficient LT annealing is observed for (Ga,Mn)As annealed in O_2 and (Ga,Mn)As with an amorphous As cap layer, where the formation of Mn–O and Mn–As at the surface may have worked to reduce the number of Mn_{int}[143,144]. The LT annealing was shown to produce a homogeneous magnetization depth profile than that of an as-grown sample, revealed by polarized neutron reflectometry measurements [145]. Postgrowth hydrogenation of (Ga,Mn)As results in reduction of p from 10^{21} cm^{-3} to 10^{17} cm^{-3} without changing Mn composition and is shown to lead to suppression of ferromagnetism [146]. The reduction of p by hydrogenation changes the direction of magnetic easy axis from in-plane to out-plane at low temperatures [147], consistent again with the prediction by the p–d Zener model [21,22,99].

7.4.1.3 Spin Polarization

Spin polarization of holes P in (Ga,Mn)As at low temperatures has been measured directly by the Andreev reflection on Ga/$Ga_{0.95}Mn_{0.05}$As junctions [148] and Sn/$Ga_{0.92}Mn_{0.08}$As point contact [149] to be as high as ~80%. Similar effective P value of 77% was obtained from the magnitude of TMR (290%) at low temperatures [150]. These results are consistent with the theoretical calculations of (Ga, Mn)As [22,151]. The point contact Andreev reflection was measured in (Ga,Mn)Sb [152–154] and (In,Mn)Sb [155,156], where $P \sim 50$–60% was obtained [157,158].

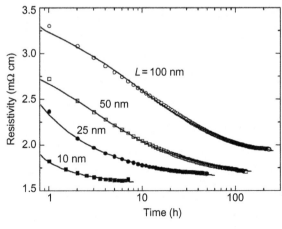

Figure 7.7 Time evolution of the resistivity change of (Ga,Mn) As with $x = 0.067$ and thickness L during annealing at 190°C. Solid lines show the fitting by a one-dimensional diffusion model [142].
Source: Reproduced by permission of American Physical Society.

7.4.1.4 Magnetotransport Properties

The SOI together with the p−d exchange interaction manifests itself in magneto-transport phenomena including the anomalous Hall effect, AMR, and the planar Hall effect, which provide valuable information on the magnetism of (Ga,Mn)As. Because of the high sensitivity, the determination of magnetization behavior from transport is an important technique for thin films of DMSs, where the total magnetic moments can be too small to be easily detected by other means. In addition, the sensitivity of transport measurements does not depend on the lateral size of the device with the fixed lateral aspect ratio.

Because R_{Hall} is dominated by the anomalous Hall effect, the temperature and magnetic field dependence of R_{Hall} reflects those of magnetization [159,160]. From the same procedure as that for magnetization measurements, T_C can be determined by the Arrott plots and the Curie−Weiss plots [161]. Since (Ga,Mn)As layers grown directly on GaAs usually have in-plane magnetic easy axis, R_{Hall} measured in perpendicular H probes the magnetization process along the hard axis of the magnetization. One can also measure the planar Hall resistance, which is the transverse resistance measured by the same probes as R_{Hall} in an in-plane H[162], to obtain information about magnetization. For (Ga,Mn)As, the signal of the planar Hall resistance is greater than that of magnetic metals by several orders and can be used to detect the in-plane M reversal as well as the domain wall (DW) dynamics [92,163,164].

The temperature and magnetic field dependence of magnetization probed by the anomalous Hall effect; however, does not show exactly the same behavior as those measured by direct magnetization measurements. The anomalous Hall effect scales with spin polarization of carriers, which is nearly proportional to the magnetization only when the spin-splitting of the bands is smaller than the Fermi energy and the contribution of the hole magnetization to the total magnetization is negligible. Furthermore, transport measurements do not probe the magnetization of the entire sample but only the region visited by carriers, where carrier-mediated magnetic interaction is strong. Thus, near MIT in particular due to the nonuniform distribution of holes in (Ga,Mn)As [21,22], transport and direct magnetization measurements may provide different information on magnetization.

7.4.1.5 Anisotropic Magnetoresistance

It has been found that (Ga,Mn)As exhibits AMR, in which MR depends on the relative orientation of the current direction with respect to the magnetic field direction, and their direction with respect to crystal axes [101]. For metallic (Ga,Mn)As, the lowest resistance is observed when H is parallel to the current as shown in Figure 7.8 [160,165]. The direction of H for the highest resistance was predicted theoretically to be different for compressive and tensile lattice strain [166]. This is corroborated by experiments; for the sample with compressive strain, the highest resistance is measured when H is perpendicular to the surface, and the sample with tensile strain it is observed when H is in in-plane perpendicular to the current [160].

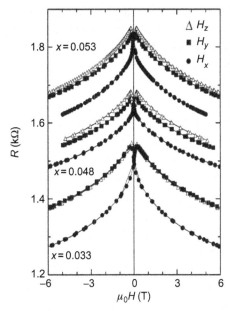

Figure 7.8 AMR of (Ga,Mn)As with different values of x. Magnetic field is applied at three different directions, x is along current, y is in-plane and perpendicular to current, and z is perpendicular to the plane [165].
Source: Reproduced by permission of American Institute of Physics.

7.4.1.6   Critical Scattering

Resistance maximum observed at around T_C together with the negative MR associated with the maximum has been attributed to the critical scattering. The resistance maximum can be interpreted in terms of a spin-dependent scattering by magnetization fluctuation via exchange interaction [167,168]. The negative MR occurs because the H-induced spin-alignment reduces the spin disorder scattering [159]. From the numerical fit to the data of the temperature dependence and H dependence of the resistance reproduces the data well and gives the magnitude of the p–d exchange interaction as $|N_0\beta| = 1.5 \pm 0.2$ eV [126], which compares favorably with that determined by photoemission experiments, $N_0\beta = -1.2$ eV [169].

7.4.1.7   Weak Localization and MIT

At low temperatures, the negative MR extends to high magnetic fields, beyond the field at which magnetic spins are fully ordered ferromagnetically according to the anomalous Hall effect response. This cannot be accounted for by the suppression of spin-disorder scattering. Here, the giant splitting of the valence bands makes both spin-disorder and spin–orbit scattering ineffective. Under such conditions, weak localization which manifests itself in negative MR is expected to show up at low temperatures, where phase breaking scattering ceases to operate [170]. The weak localization MR formula [170] was shown to explain the observed negative MR data at 2 K quite well for samples under compressive as well as those under tensile strain [160]. This three-dimensional weak localization model predicts the temperature T dependence of the conductivity σ, as $\sigma \propto T^{1/2}$. In addition to the weak localization scenario, the possibility of the Kondo effect like temperature dependence, $\sigma \propto \ln T$, was suggested [171]. The $\sigma \propto T^{1/3}$ dependence observed in a metallic (Ga,Mn)As layer down to 30 mK was explained by the Aronov–Altshular three-dimensional scaling theory with spin scattering [172].

The universal conductance fluctuation in metallic (Ga,Mn)As submicron wires and rings was observed at low temperatures (<200 mK), indicating that the phase coherence length of electrons in (Ga,Mn)As is about 100 nm at 10 mK and at even higher temperature of 100 mK [173,174]. The MIT-associated anisotropic features can also be observed for insulating (Ga,Mn)As, which shows two orders of magnitude difference in resistance between the two directions along [110] and along [$\bar{1}$ 1 0] below 1 K [175]. From the anisotropic tunnel conductance in (Ga,Mn)As/GaAs/(Ga,Mn)As tunnel junctions, the MIT behavior is shown to be M direction dependent, which may be related to the formation of an Efros–Shklovskii gap induced by the change of density of states (DOS) by M direction and the reduced screening effect of depletion layer at the interfaces [176]. As can be seen for the continued developments, III–V-based ferromagnetic semiconductors are providing new insights to the MIT, which is one of the most important topics in solid state physics [177].

7.4.1.8 DW Resistance

The Barkhausen noise caused by the scattering due to the presence of magnetic domain structure has been observed in the resistance measurement on (Ga,Mn)As [101]. The existence of magnetic DW is known to contribute to the electrical resistance, which is usually much smaller than the sample resistance and thus careful measurement is necessary. The DW resistance (DWR) of a single DW has been measured for (Ga,Mn)As; a negative sign of DWR was reported for (Ga,Mn)As with in-plane easy axis [163] and a positive one was reported for a sample with perpendicular easy axis [178]. The negative DWR may be explained by the destruction of quantum coherence of electrons at the DW [179] and/or by the AMR contribution [180]. In the latter experiment, the DW position was defined by using the coercivity H_C difference induced by the surface etching [181]. Using a series of samples having different sizes, the measured positive DWR was decomposed into extrinsic and intrinsic DWR components. Numerical evaluation of the zigzag current due to the alternating polarity of the anomalous Hall effect at DW [182] explained well the dominant extrinsic DWR. The remaining intrinsic DWR was shown to be consistent with the mistracking of the carrier spins inside DW [183]. Here, the AMR contribution was shown to be 20 times smaller than the mistracking resistance.

7.4.1.9 Thermal Transport

The Seebeck effect (the longitudinal thermopower) and the in-plane Nernst effect (the transverse thermopower) under in-plane H were measured on compressive (Ga,Mn)As. Both effects showed strong dependence on the direction of H, reflecting in-plane magnetization reversal process. These dependences indicate that the origin of the effect can be related to an AMR and the planar Hall effect, which may provide useful information for further understanding of the origin of anisotropic scattering in (Ga,Mn)As [184]. For (Ga,Mn)As grown on (In,Ga)As buffer layer with perpendicular magnetic easy axis, the anomalous Hall effect and its thermoelectric counterpart, the anomalous Nernst effect, can be observed simultaneously [185]. The relationship between the two effects can be expressed through the Mott relation [186] with $\gamma = 2$, suggesting that the both effects share a common origin related to the SOI.

7.4.2   Magneto-Optical Properties

Magneto-optical properties have been studied to elucidate the origin of ferromagnetism of (Ga,Mn)As, because they can probe the spin—split band structures induced by the sp—d exchange interaction [187]. The absorption edge of (Ga,Mn)As is not sharp due to the disorder effect [188], and the absorption coefficient is $5 \times 10^{-3} \text{ cm}^{-1}$ at 1.2 eV which is a few orders greater than GaAs owing most likely to its free carrier absorption [189]. It was shown by spectroscopic ellipsometry [105] that the critical points of (Ga,Mn)As are essentially the same as those of GaAs in the photon energy range from 0.62 to 6 eV. When the ferromagnetism is intrinsic, one expects an enhanced magneto-optical response reflecting ferromagnetism at the critical points of the host zincblende semiconductors. This is the case for (Ga,Mn)As, in which a large Faraday rotation and magnetic circular dichrosim (MCD) signals were observed at critical points E_0 and E_1, whose H dependence traces well the magnetization curves. Figure 7.9 shows the H dependence of the Faraday rotation of 2 μm thick $Ga_{0.957}Mn_{0.043}As$ at 10 and 300 K in the vicinity of the band gap, where the GaAs substrate was etched away to remove absorption of the GaAs substrate [190]. The magnitude of the Faraday rotation is proportional to

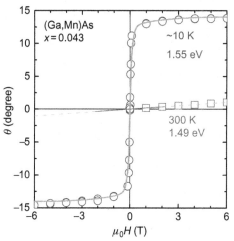

Figure 7.9 Faraday rotation as a function of magnetic field at 5 and 300 K near the absorption edge. Lines represent magnetization curves probed by anomalous Hall effect [190].

the magnetization probed by AHE. The observed value of the Faraday rotation is 6×10^4 deg/cm under 0.1 T at 10 K (ferromagnetic phase) and the Verdet constant is 9×10^{-2} deg/G cm at room temperature (paramagnetic phase). An enhancement of MCD signals in (Ga,Mn)As at GaAs critical points was observed [191,192]. The positive sign of the Faraday rotation and MCD at the band edge is opposite to most of the II−VI DMSs with negative p−d exchange, whereas theories predict negative p−d exchange for (Ga,Mn)As. This can be explained by a spin−dependent Burstein−Moss shift caused by hole redistribution [22,193]. In addition to the positive MCD, a negative contribution that increases with decreasing temperature was observed in the infrared spectrum [192]. This peculiar temperature dependence was explained as a result of a nonlinear dependence of MCD on magnetization at Burstein−Moss edge again caused by the hole redistribution on the spin-split bands [22]. To understand the experimental data, the effect of disorder on the transition probability was noted [188]. The shift of peak positions in MCD spectra was observed with (Ga,Mn)As/AlAs superlattices depending on the layer thickness, and was interpreted due to the quantum confinement in the (Ga,Mn)As QW layers [194]. The magnetic linear dichrosim is also observed, reflecting the in-plane magnetization rotation in (Ga,Mn)As, whose spectrum is considered to be related to the electronic state in the vicinity of E_F responsible for the ferromagnetism [195].

When excited by 400 nm pump pulses, (Ga,Mn)As emits terahertz radiation. The radiation occurred only in the ferromagnetic phase and its strength and H dependence showed strong correlation with the magnetization. The radiation was explained by the influence of the magnetization on the photogenerated carrier acceleration path [196].

The summary of magneto-optical properties in (III,Mn)V and related phenomena can be found elsewhere [197].

7.5   The sp−d Exchange Interaction

The most prominent feature of magnetic semiconductors is the correlation among magnetic, electrical, and optical properties, which is based on the presence of the sp−d exchange interaction in these materials [6,198]. The sp−d exchange interaction gives a sizable spin splitting of bands in the conduction (s−state) as well as in the valence (p-state) bands, when the d-states of incorporated transition metals carry magnetization. This amplifies the spin properties of semiconductors in a number of different ways; spin-dependent optical selection rules give rise to the giant magneto-optical effects and spin-polarized carriers modify the transport properties, as have seen in the previous sections. The magnitude of the sp−d interaction (expressed as $N_0\alpha$ for s−d and $N_0\beta$ for p−d exchange interaction, where N_0 is cation density) in II−VI magnetic semiconductors has been systematically investigated on various materials by photoemission and/or magneto-optical measurements. The s−d interaction is due primarily to potential interaction, which follows from the spin-dependent Coulomb interaction due to the Pauli exclusion principle between the band electrons and the localized d-electrons of transition metals. This potential interaction gives rise to ferromagnetic coupling between the delocalized s-state and the localized d-state. Because of the long-range nature of the Coulomb interaction, the magnitude of $N_0\alpha$ is nearly independent of the lattice constant a and is about 0.2 eV. The main part of the p−d exchange interaction, on the other hand, results from hybridization of the p- and the d-states. The sign of $N_0\beta$ is known to depend on the configuration of d-shell in the particular transition metal incorporated in semiconductor; for half-filled Mn 5d-state it is always negative. The magnitude of $N_0\beta$ is several times greater than $N_0\alpha$ due to hybridization. The chemical trend of $N_0\beta$ is summarized with respect to a for various II−VI magnetic semiconductors, and is found to be nearly inversely proportional to the unit cell volume as $N_0\beta \propto a^{-3}$, as shown in Figure 7.10 [22].

For (Ga,Mn)As with percent order of Mn, the value of $N_0\beta$ is determined to be -1.2 ± 0.2 eV from the photoemission measurement [169]. By adopting this value, various experimental observations, including the magnitude of T_C, magnetic anisotropy, hole polarization, and MCD spectra, can be reproduced by the p−d Zener model described in the following section. $N_0\beta$ of (In,Mn)As determined from the photoemission is -0.7 eV [199], which is consistent with the chemical trend complied for II−VI materials [22]. The paramagnetic Mn-doped GaAs with $x < 0.0005$ shows positive value of $N_0\beta$ about 2 eV from magneto-optical [200] and spin-flip Raman measurements [201]. The origin of the positive sign has not been understood.

The value of $N_0\alpha$ has been determined experimentally for paramagnetic Mn-doped GaAs. The spin-flip Raman experiments gives $N_0\alpha = 23$ meV for metal-organic vapor phase epitaxy grown Mn-doped GaAs [201]. This small value of $N_0\alpha$ has been attributed to the Coulomb repulsion between conduction band

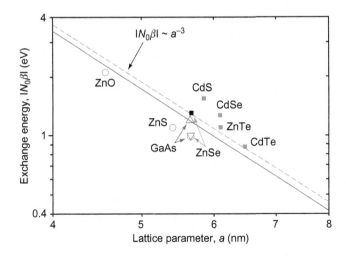

Figure 7.10 Energy of p−d exchange interaction for various materials with $x = 0.05$ as a function of lattice constant [22].

electrons and negatively charged magnetic ions [202]. For MBE-grown Mn-doped GaAs/(Al,Ga)As QWs with a QW width less than 10 nm and $x < 0.0003$, a negative $N_0\alpha$ down to -0.09 eV was observed by optical measurements [203]. This may be related to the confinement induced p-symmetry in the conduction band of Mn-doped GaAs QWs [202,204].

7.6   The p−d Zener Model of Ferromagnetism

Because T_C goes to zero when holes are compensated and because T_C of (Ga,Mn) As is higher for higher hole concentrations, it is natural to consider that the ferromagnetism of (Ga,Mn)As is hole induced. Band carrier−induced ferromagnetism was noted first by Zener in the 1950s, where the magnetic ordering is stabilized by the lowering carrier energy [205]. Because of the large p−d exchange interaction and high DOS of the valence bands compared to the s−d exchange in the conduction band, the energy gain due to the hole redistribution among spin-split subbands is greater for p-type ferromagnetic semiconductors than n-type [21,22,99]. The model is now called the p−d Zener model, in which one uses realistic band structure with the p−d exchange interaction calculated by, for example, k · p description of the valence band to obtain magnetic properties of a ferromagnetic semiconductor. The model explains surprisingly well the experimentally obtained properties qualitatively and often even semiquantitatively. For example, T_C based on the p−d Zener model shows quantitative agreement with the experimental T_C for (Ga,Mn) As and (In,Mn)As as well as p-type II−VI (Zn,Mn)Te and (Cd,Mn)Te [15,21,22]. If one considers the biaxial strain in the calculation, the strain and hole concentration-dependent magnetic anisotropy can be obtained, which compares well with the experimental observations [21,22,99]. One can also explain many other experimental results on (Ga,Mn)As, for example, MCD spectra, conductance curves of the resonant tunnel diodes with a (Ga,Mn)As emitter, AMR effect, and the magnetic domain width [19,21,22,86].

It should also be noted that the p−d Zener model has limits, for example, it may not be applicable to the regime with extremely high hole concentration, where oscillatory carrier-mediated interaction (RKKY interaction) starts to work [206−208]. This situation may lead, thorough noncollinear spin arrangement, to spin-glass phase with the increase of hole concentration. The effect of disorder was shown to enhance T_C in the model whose starting point is the impurity band of holes localized by the Mn acceptors [209].

Standard calculation methods have also been employed to explain and to predict the magnetic phases and properties of (Ga,Mn)As as well as other magnetic semiconductors. Among them, the first principle calculations have provided a number of insights to the electronic structure of the material. First principle calculations with local spin density approximation (LSDA) predicted that E_F was at the Mn d-states [210,211], which led to the suggestion that the origin of ferromagnetism was more of double exchange in nature, as opposed to the band-like picture of the p-d Zener model. Although care has to be taken to compare ground state electronic structure with excited state experiments, the calculated results do not appear to agree with the experimentally obtained DOS by photoemission, which shows very little mixture of Mn d-state at E_F. The local density approximation with the Coulomb correlation effect LDA + U[212] and self-interaction corrected LSDA [213] are known to better describe the correlation, and have been used to calculate the band structure of (Ga,Mn)As. The results suggest the presence of a sizable portion of delocalized p-state at E_F, in accordance with the experimental results.

We note that the use of band picture for the description of ferromagnetism in magnetic III—V semiconductors (the p—d Zener model) is based on the past studies of doped nonmagnetic semiconductors. A large number of studies in the past, dedicated to the understanding of the nature of the MIT in doped semiconductors, have shown that doped semiconductors near the MIT boundary exhibit duality, that is, these materials show metallic band-like nature on one type of measurement, whereas at the same time they can exhibit impurity band-like nature on another. These observations led to the proposal of the two-fluid model [214]. While whether conduction takes place in valence band or in impurity band is an unresolved issue in semiconductor physics at large [215—220], there are indications that thermodynamic properties such as specific heat can be described to a good approximation by band mass across the MIT [221]. Having these studies in mind, as pointed out in the earlier papers [21,22], the band picture was employed to describe the magnetism in magnetic III—V semiconductors.

7.7 Properties Revealed by Device Structures

To explore novel spin-dependent phenomena and their possibility for future applications, a number of ferromagnetic semiconductor-based heterostructures have been made and investigated both experimentally and theoretically. In this section, we review the results of heterostructures and device structures that exhibit properties not readily accessible in devices made only from nonmagnetic semiconductors.

7.7.1 Electrical Spin Injection into Nonmagnetic Semiconductors

Spontaneous magnetization in the ferromagnetic phase manifests itself in spin polarization of carriers below T_C. Ferromagnetic semiconductors that can be epitaxially grown on nonmagnetic semiconductor heterostructures thus are a candidate for a source of spin-polarized carriers for electrical spin injection into nonmagnetic structures in the absence of magnetic field. Injection of spin-polarized carriers (spin injection), their transport and their detection are the building blocks of semiconductor spintronic devices. The first demonstration of electrical spin injection from a ferromagnet into nonmagnetic semiconductor was done using a ferromagnetic (Ga,Mn)As semiconductor electrode as an emitter and a light-emitting diode (LED) as a spin detector (spin-LED) [222]. The employed structure was p-(Ga,Mn)

As/i-GaAs/(In,Ga)As QW/n-GaAs. Spin polarization of injected holes from (Ga,Mn)As was detected by the circular polarization of electroluminescence (EL): the change in EL polarization was $\pm 1\%$. Hysteresis loop of ΔP with respect to H and its temperature dependence traced the H dependence and temperature dependence of magnetization of (Ga,Mn)As.

For injection of spin-polarized *electrons*, an Esaki tunnel diode (ED) was employed as a spin emitter, where spin-polarized holes in the valence band of (Ga, Mn)As are injected into the conduction band of an adjacent n-GaAs by interband tunneling [223,224]. High-spin polarization of the injected electrons (80%) has been reported [225]. A three-terminal device that can bias ED and LED independently was fabricated and used to examine the bias dependence and hence energy dependence of the injected electron spin polarization as shown in Figure 7.11A [225]. EL polarization P_{EL} showed strong dependence on the biases: the highest P_{EL} of 32% (Figure 7.11B) was obtained when the valence electrons near the Fermi level of (Ga,Mn)As dominated the tunnel current. The 32% polarization corresponds to injected electron spin polarization of 85%, when spin relaxation in the QW measured by the Hanlé effect is taken into account [225,226]. The magnitude of the spin polarization as well as its bias dependence can be explained by the calculation based on Landauer−Büttiker formalism with tight-binding approximation for the ballistic transport regime [227].

The electrical spin injection from (Ga,Mn)As to n-GaAs was demonstrated also in a lateral structure by employing nonlocal spin valve effect [228], where Esaki diodes were used as both injecting and detecting contacts [229]. By using a similar structure, it was shown that dynamic nuclear spin polarization in an n-GaAs was induced by spin-polarized current through hyperfine interaction, where nuclear magnetic resonance was detected by sweeping frequency of AC magnetic field [230].

It was demonstrated that pure spin current can be injected from (Ga,Mn)As to p-GaAs by using spin pumping [231,232], and can be detected by the inverse spin Hall effect (ISHE) [233,234]. For spin pumping under FMR, galvanomagnetic effects caused by electric field of microwave generate voltage in a ferromagnetic layer in addition to the voltage produced by the ISHE in an adjacent nonmagnetic layer. Therefore, one needs to take sufficient care to distinguish voltage related to spin injection from that related to other effects [235].

7.7.2 TMR and Current-Induced Magnetization Switching

Magnetic tunnel junctions (MTJs) based on fully epitaxial (Ga,Mn)As ((Ga,Mn)As/(Al,Ga)As/(Ga,Mn)As) exhibit TMR [236,237], where parallel and antiparallel configurations of the magnetization of the two (Ga,Mn)As layers result in low and high resistance (R_P and R_{AP}) states. Note here that GaAs acts as a barrier layer for holes in (Ga,Mn)As [238]. The highest TMR ratio $[= (R_{AP} - R_P)/R_P]$ so far reported are 75% and 290% (Figure 7.12) for AlAs and GaAs intermediary barrier layers, respectively [150,239]. The latter corresponds to effective carrier spin polarization P of 77% according to the Julliere's formula [240], where TMR

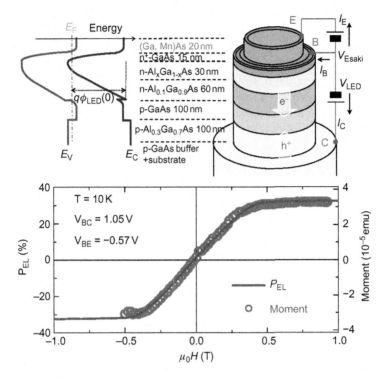

Figure 7.11 (A) Schematic band diagram and the three-terminal devices consisting of an Esaki diode and LED. (B) Magnetic field dependence of P_{EL}, which traces magnetization curve [225].

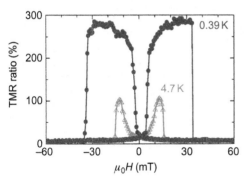

Figure 7.12 TMR of (Ga,Mn)As/GaAs/ (Ga,Mn)As under magnetic field along the [100] direction measured at low temperatures [150].

ratio $= 2P^2/(1 - P^2)$ [206]. The TMR ratio quickly decreases as the bias voltage increases, in a much faster way than seen in metallic MTJs [150]. This is understood as a result of small magnitude of band offset and the Fermi energy (~ 100 meV) in (Ga,Mn)As/GaAs [150,227]. The anisotropic TMR behavior has been observed in (Ga,Mn)As-based MTJs with in-plane magnetic easy axis, where

TMR ratio depended on the direction of in-plane H[95,150,239], which reflects the in-plane magnetocrystalline anisotropy of (Ga,Mn)As (see Section 7.3.2).

Spontaneous spin splitting in the valence bands was observed in current−voltage (I−V) characteristics of AlAs/GaAs/AlAs p-type double-barrier resonant-tunneling diodes (RTDs) having a (Ga,Mn)As emitter [241]. Here, the resonant levels in the GaAs QW act as an energy filter for hole transport, separating exchange split hole bands. In the experiment, the resonant peaks labeled HH2 and LH1 showed spontaneous splitting in the absence of magnetic field below T_C, when holes are injected from (Ga,Mn)As. This shows that RTD structure can be used to do spectroscopy of the spin splitting of the valence bands.

The interplay with the TMR effect and the quantum confinement in QWs was addressed utilizing integration of ferromagnetic semiconductor with semiconductor quantum structures. An RTD with (Ga,Mn)As emitter and collector showed the TMR effect as large as that of an MTJ with a single barrier, which was used to show the presence of spin accumulation in the QW [242]. A nonstandard TMR behavior was observed in an AlAs/(In,Ga)A/AlAs RTD with (Ga,Mn)As emitter and collector, in which an oscillatory TMR ratio was observed as a function of AlAs thickness [243]. Resonant tunneling and TMR effect through the quantized levels in a QW made of (Ga,Mn)As was reported in AlAs/(Ga,Mn)As/AlAs RTD, where oscillation of TMR ratio was observed in its bias dependence when holes are injected from the nonmagnetic p-GaAs:Be electrode as shown in Figure 7.13 [244,245].

Spin-dependent transport in a three-terminal device with a (Ga,Mn)As/AlAs/(Ga,Mn)As MTJ was also investigated. The device had three electrodes, emitter, base, and collector, formed at the top (Ga,Mn)As, the bottom (Ga,Mn)As, and p-type substrate, respectively [246]. The transistor was shown to exhibit current gain.

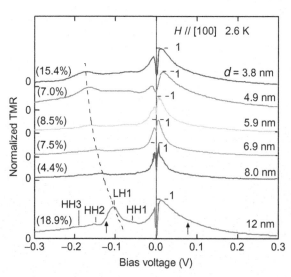

Figure 7.13 Bias dependence of normalized TMR ratio in p-type (Ga,Mn)As/AlAs/(Ga,Mn)As/AlAs/GaAs:Be RTD measured at 2.6 K with magnetic field along [100] direction. *d* represents the thickness of (Ga,Mn)As QWs. Negative bias corresponds to hole injection from GaAs:Be [244].
Source: Reproduced by permission of American Physical Society.

In submicron (Ga,Mn)As MTJs, current-induced magnetization switching (CIMS) has been observed [247,248]. Starting from the parallel magnetization state (the R_P state), current pulse from the thick layer to the thin layer of (Ga,Mn)As induces magnetization reversal of the thin (Ga,Mn)As layer, resulting in the anti-parallel state (the R_{AP} state). When the current polarity is reversed, reversal of the thinner (Ga,Mn)As layer takes place putting the MTJ into the R_P state. The critical current density J_C for switching is of the order of $10^4 - 10^5$ A/cm^2[247–249], two to three orders smaller than those observed in metal-based MTJs. Clear dependence of the switching on the current pulse direction shows that CIMS is neither due to the Oersted field nor the Joule heating. It can qualitatively be explained by the Slonczewski's spin-transfer torque model [250]. The orders of magnitude reduction of J_C compared to metal MTJs may be understood by the small magnetization of (Ga,Mn)As, although the model still predicts an order of magnitude greater J_C than the observed one. Possible effect of the large bias dependence of the TMR ratio in the CIMS process is not understood at the moment; that is, CIMS takes place at bias voltage, where the TMR ratio is almost zero. This may be related to the con-servation of the total carrier spins despite reduction of the TMR ratio [250] and/or different trend in bias dependence between spin-transfer torque vector and the TMR ratio [250]. The role of possible incoherent switching processes, such as the formation of domain structures, may also play a role in reversal [247–249,252,253].

A very large tunnel anisotropic magnetoresistance (TAMR) up to 150,000% has been reported in (Ga,Mn)As-based MTJs with vertical and lateral device structures [254–256]. The origin of TAMR [257] has been attributed to anisotropic valence band structure induced by the SOI and lattice strain in combination with the MIT [176]. Thermoelectric counterpart, tunnel anisotropic magnetothermopower, was also observed in (Ga,Mn)As/GaAs tunnel junctions [258].

CIMS in a single (Ga,Mn)As layer was demonstrated by utilizing an effective magnetic field induced by SOIs. It was shown that the lattice strain related SOI effective field generated by current density of a few milliamperes per square centi-meters is large enough to switch the magnetization direction in (Ga,Mn)As [259–261]. By applying RF current to (Ga,Mn)As, it was also shown that one can induce FMR electrically through the SOI effective field [262,263].

7.7.3 Current-Induced DW Motion

The position of a magnetic DW before and after the motion induced by current can be detected by magneto-optical microscope. The motion can also be observed by transport measurements using AHE and PHE. The DW velocity induced by external H was measured for (Ga,Mn)As with in-plane magnetic easy axis utilizing PHE. The velocity was controlled over three decades from 10^{-5} to 10^{-2} m/s by varying H. The field-driven velocity increases with the increase of temperature [164].

By the application of a current pulse across the DW, it was found that the posi-tion of DW can electrically be manipulated in the absence of magnetic fields [181]. The channel made of (Ga,Mn)As used in the experiment has magnetic easy axis

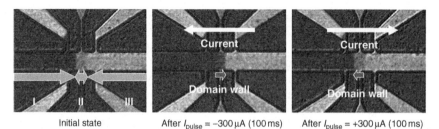

Initial state   After I_{pulse} = −300 μA (100 ms)   After I_{pulse} = +300 μA (100 ms)

Figure 7.14 Current-induced DW motion in (Ga,Mn)As in stepped structure with central thinnest region II observed by MOKE. A DW prepared at the boundary of regions I and II (left) can be moved to the opposite boundary regions of II and III by a current pulse of −300 μA (middle) and be moved back to its original position by the current with the opposite direction (left) [181].

perpendicular to the surface by inserting an (In,Ga)As or (In,Al)As buffer layer. This easy axis direction is not only useful to monitor the DW position through the anomalous Hall effect and by the magneto-optical Kerr effect (MOKE) microscope, but appears to play a role in observing the motion, as discussed later. The DW switching was observed in devices having a channel with three regions with different thicknesses, where the thinnest region was set to the center of the channel. This double-step structure allowed patterning of H_C due to nonuniformity along the growth direction in the film. In addition, each step acted as a confinement potential for DW. The device was initialized by appropriately sweeping external magnetic field in such a way that a DW was placed at one of the stepped boundaries of the thinnest region. After setting $H = 0$, a current pulse of 10^5 A/cm^2 for 100 ms at 80 K (T_C of this film was 90 K) was applied. Both the anomalous Hall signals and MOKE images indicated that the DW moved to the other step boundary, in the direction opposite to the current direction. The application of a subsequent current pulse in the opposite direction switched back the DW to its initial position as shown in Figure 7.14.

Current density j as well as the temperature dependence of the DW velocity was systematically investigated using a 5 μm wide (Ga,Mn)As channel having a single step structure [264]. DW was formed at the step by a magnetic field sweep and then was moved by current pulses. The area swept by the DW was monitored by MOKE to obtain the DW speed, v. The DW displacement was calculated by dividing the area swept by the DW by the device width. The displacement was linearly dependent on the pulse width. The change of the device temperature by the Joule heating was monitored by measuring the device resistance during the pulse application and found to be constant during the pulse. Thus obtained j dependence of v under constant temperature showed that there were at least two regimes separated by a critical current density j_C, which is a few 10^5 A/cm^2 as shown in Figure 7.15. Beyond j_C, v increased linearly with j, while a slow motion indicative of creep was observed below j_C. The quasi-linear dependence of v above j_C and its slope together with the direction of motion were found to be quantitatively consistent with the

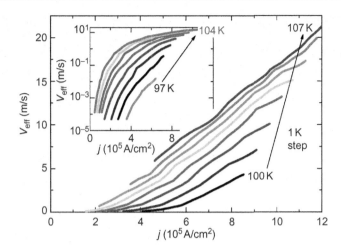

Figure 7.15 DW velocity as a function of current density at various device temperatures. The inset presents the same data but in logarithm plot to show the DW activity in subthreshold regime [264].

spin-transfer mechanism [265,266]. The value of j_C can be reproduced by the spin-transfer mechanism assuming an intrinsic DW pinning due to hard-axis anisotropy [265]. DW motion by currents has so far not been observed in (Ga,Mn)As with in-plane axis. The intrinsic pinning model may offer an explanation for this observation: j_C is expected to be at least one order of magnitude higher in this case as the relevant anisotropy, magnetocrystalline anisotropy, is an order of magnitude higher than the hard-axis anisotropy (stray field) responsible for determining j_C in samples with perpendicular easy axis. An order of magnitude higher j_C is not easily accessible experimentally because of the associated Joule heating.

Another important finding is the current-induced DW motion in the subthreshold regime, where the DW moves slowly (the inset of Figure 7.15). The j dependence of v below j_C obeys an empirical scaling law, suggesting the existence of current-induced DW creep [267]. Comparison between current-induced and field-induced creep revealed that they could be scaled using the same functional form but with different scaling exponents, indicating that the current-induced creep is fundamentally different from the field-driven one. A model based on spin-transfer torque has been put forward to explain the current-induced creep, which describes the observed scaling exponents [268]. It was also shown experimentally that the scaling exponents for field-induced creep are dependent on surface roughness [269].

On the theory side, SOI in (Ga,Mn)As was pointed out to result in spin accumulation and nonadiabaticity leading to a high mobility of the DW above j_C when transport is ballistic [270]. The importance of nonadiabaticity was also pointed out to explain the current-induced DW velocities in (Ga,Mn)(As,P) measured over a wide temperature range ($0.1T_C < T < T_C$) [271]. Spin-transfer torque assisted thermal activation process below j_C under rigid wall approximation [272,273] indicated that the DW motion should follow a specific functional form with respect to j; the predicted temperature and/or j dependence of v appears to be different from the experimental findings [268]. Further quantitative comparison between theory and experiment will reveal the rich physics involved in the DW motion of (Ga,Mn)As.

7.7.4   Control of Magnetism and Magnetization Reversal by External Means

Since the ferromagnetism in (Ga,Mn)As and (In,Mn)As is stabilized by holes, one can switch the magnetic phase without changing temperature by electrically controlling the value of p. This was shown possible using a metal-insulator-semiconductor field-effect transistor (MISFET) structure with a thin (In,Mn)As layer (≤ 5 nm) as the semiconductor channel [274]. The (In,Mn)As layer was grown on a GaAs substrate with a thick (Al,Ga)Sb buffer layer to relax the 7% lattice mismatch between (In,Mn)As and GaAs. In order to probe the magnetization through the anomalous Hall effect, Hall bar-shaped devices were prepared. Since the channel is p-type, application of positive (negative) gate electric-field E_G decreases (increases) p; for the structure under discussion $|E_G| = 1.5$ MV/cm changes several percents of total p. In the vicinity of T_C, the magnetization curves show more square shapes under negative E_G indicating enhanced ferromagnetic order, while paramagnetic-like response under positive E_G. This reversible change of T_C by $E_G = \pm 1.5$ MV/cm determined using Arrott plots can be as large as 4 K for 4 nm thick (In,Mn)As [275]. Control of T_C by electrical means is also a proof that the ferromagnetism in this material is indeed carrier induced.

Another important effect of E_G is the change of the coercive force H_C below T_C; greater (smaller) H_C for negative (positive) E_G as shown in Figure 7.16. It was shown that the temperature dependence of H_C under several E can be collapsed into a single curve, when H_C is plotted against the reduced T/T_C^*, where T_C^* is the temperature at which H_C becomes zero. The MOKE images during magnetization reversal of similar (In,Mn)As showed that the magnitude of H_C was determined by the nucleation field of domains [252]. These results suggest that the change of H_C

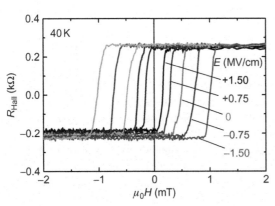

Figure 7.16 Isothermal control of coercivity of 5 nm thick (In,Mn)As with $x = 0.063$ by the application of external electric fields [275].

by E is a result of the change of the nucleation field, whose temperature dependence is determined by the magnitude of T_C^*. By using this phenomenon, a new scheme of electrical magnetization reversal, electric field-assisted magnetization reversal, was demonstrated [275] in the following way. After saturation of magnetization at positive H under $E_G = -1.5$ MV/cm, H was reduced through zero to a small negative H but still less than the coercive force, H_C. Then, E_G was switched to zero, which reduced $|H_C|$ below $|H|$ and switching took place.

For (Ga,Mn)As, control of ferromagnetism by an electrical means was initially hindered by its insulating nature at thickness t below 5 nm [276]. In order to probe magnetic properties by AHE, metallic samples with $t > 5$ nm had to be employed. On the other hand, thick metallic samples made it difficult to observe the change in T_C because of the limited ratio of the modulated sheet hole concentration to the total sheet hole concentration. The first observation of T_C by electric fields in a GaAs-based system was done in Mn-δ-doped GaAs, where holes were provided by modulation doping [277]. Later, the control of T_C by 5 K as well as H_C was shown to be possible in a 7 nm thick (Ga,Mn)As by the application of $|E_G| = 5$ MV/cm [278], where the high-quality Al_2O_3 gate-insulator deposited by atomic layer deposition (ALD) [279] was used to increase the number of modulated holes by E_G. Later, a backgate FET structure with an n-GaAs buffer layer as the backgate was also used to modulate T_C in a (Ga,Mn)As FET, where no additional gate insulator is necessary [280]. The nonvolatile T_C modulation by E_G was also demonstrated for (Ga,Mn)As MISFET by using ferroelectric as a gate insulator [281].

In many cases, the electric-field effects on magnetism on (Ga,Mn)As were detected electrically through the anomalous Hall effect. They were directly detected by magnetization measurements by using millimeter-sized field-effect structure with a 3.5 nm thick (Ga,Mn)As channel [282]. The temperature dependence of spontaneous magnetization revealed that there exist ferromagnetic and superparamagnetic-like components in thin (Ga,Mn)As. Both components show E_G dependence, and T_C of the ferromagnetic phase and the blocking temperature of the superparamagnetic-like phase decreased under positive E_G. The magnitude of spontaneous magnetic moments m_s in the ferromagnetic phase was decreased also by the application of positive E_G. The E_G dependence of T_C and m_s in the ferromagnetic phases can be explained quantitatively by the p−d Zener model taking into account hole distribution in the (Ga,Mn)As. The hole distribution is nonuniform along growth direction due to the presence of depletion layer at the interface between (Ga,Mn)As and gate insulator, and can be calculated by using one-dimensional Poisson equation. The detailed transport measurements showed that p dependence of T_C is expressed phenomenologically as $T_C \propto p^\gamma$ with $\gamma \sim 0.2$, when p was modulated by E_G[283]. The value of $\gamma \sim 0.2$, which is a few times smaller than that expected from the conventional p−d Zener model, can be reproduced by the adapted p−d Zener model with 0.5 eV of the Fermi energy pinning position at the interface above the valence band edge of (Ga,Mn)As. Since the Fermi energy pinning position is material dependent, the values of γ is also expected to be material dependent. This was confirmed in field-effect structures with a (Ga,Mn)Sb channel, which showed $\gamma \sim 1$ [284].

Because the magnitude of magnetic anisotropies of (Ga,Mn)As is p dependent [88], by applying E_G to (Ga,Mn)As in an MISFET structure, the magnetic anisotropy is expected to be manipulated, allowing electrical control of the M direction. This was demonstrated experimentally in a 4 nm thick (Ga,Mn)As layer with ALD-deposited ZrO_2 gate insulator, where the in-plane direction of M was probed by the planar Hall effect [285]. The application of E_G ranging from -4 to 4 MV/cm resulted in $\sim 10°$ of in-plane M rotation; it was shown that the change of the sign and magnitude of in-plane uniaxial anisotropy is responsible for the observed M rotation. By using AMR as a probe to monitor out-of-plane M, the modulation of perpendicular magnetic anisotropy by E_G has also been observed. Simulations showed that magnetization switching by E_G through the change of magnetic anisotropies is possible, when the set of conditions met, which provides a new scheme of magnetization switching solely by electric fields [286]. As shown in Section 7.3.2.2, because magnetic anisotropies of (Ga,Mn)As are strain dependent, by the application of acoustic pulse, magnetization switching is expected to be induced by the change of anisotropy through magnetization precession [287,288]. A picosecond acoustic pulse induced magnetization precession was observed for (Ga,Mn)As and magnetization switching for (Ga,Mn)(As,P) [289,290]. Although the mechanism is not fully understood yet, the observed switching for (Ga,Mn)(As, P) may be not due to the change of strain, because coherent behavior was not detected [290].

Peculiar temperature dependence of the anomalous Hall effect in (Ga,Mn)As was found in a thin (Ga,Mn)As channel in MISFETs. When the channel conductivity is greater than 200 S/cm, the anomalous Hall coefficient decreases with the decrease of the temperature and even changes its sign below 20 K. It was also found that the sign can be changed by the polarity of E_G. Interestingly, some samples show negative coefficient at negative E_G, while some others show the opposite trend [291]. The origin of the peculiar behavior of anomalous Hall effect has not been elucidated yet [291,292].

Photogenerated carriers can also induce similar effects on magnetic properties. Illumination turned the magnetic state of (In,Mn)As with no H_C to a state with a clear hysteresis at 5 K, which was probed by both magnetization and magnetotransport measurements [293]. The photon energy needed to see the effect was greater than 0.8 eV, which indicated that the photogenerated carriers in the GaSb buffer layer below (In,Mn)As were responsible for the observed phenomena. The state after illumination with enhanced conductivity persisted after switching the illumination was turned off, suggesting that the interface electric field at (In,Mn)As/ GaSb spatially separated the photogenerated holes and electrons leading to hole accumulation on the (In,Mn)As side of the interface. Similar photoinduced effect was observed in (In,Mn)(As,Sb)/InSb, where the effect showed up starting at low photon energy ~ 0.6 eV because of the small band gap of the buffer semiconductor [294]. The photoinduced change of H_C (photocoercivity effect, PCE) was also observed in (In,Mn)As, and the change of H_C persisted after switching the illumination off [295]. These results may partly be explained by the change of the magnetic anisotropy induced by the change of the hole concentration [296]. The PCE was also observed in (Ga,Mn)As [297], whose origin was attributed to photoinduced DW creep from the observation of Barkhausen effect [298]. The effect was used to demonstrate laser pulse triggered magnetization switching in (Ga,Mn)As [299]. The magnetization enhancement of (Ga,Mn)As by circularly polarized light illumination was observed in (Ga,Mn)As thin films [300,301] and in Mn-δ-doped GaAs [302]. (Ga,Mn)As/dye-molecules hybrid structures were shown to be effective system to increase the photoinduced effects, which showed relatively large changes in resistivity and coercivity of (Ga,Mn)As by light irradiation [303].

Time-resolved optical measurement showed the possibility of ultrafast control of magnetization in ferromagnetic semiconductors by photocarrier injection. In addition, there have been reports on softening of hysteresis loops of (In,Mn)As during a short lifetime (~ 2 ps) of the photogenerated carriers [304], demagnetization of (In,Mn)As within 100 ps [305], and the enhancement of the ferromagnetism of (Ga, Mn)As again within 100 ps [306]. It was shown that the magnetization precession in (Ga,Mn)As can be induced by photoinduced change of magnetic anisotropy [307]. This kind of photoinduced precession can be used for all-optical subnanosecond magnetization switching [308].

It was shown that the application of hydrostatic pressure enhanced ferromagnetism of (In,Mn)Sb, which was explained by the increase of exchange coupling among Mn caused by the change of lattice constant [309]. For (Ga,Mn)As, it was reported that the effect of hydrostatic pressure on T_C is relatively small (~ 1 K/GPa) and the direction of T_C variation is sample dependent [310].

The controllability of ferromagnetism by external means is not limited to III–V-based ferromagnetic semiconductors. The isothermal control of ferromagnetism was reported in II–VI (Cd,Mn)Te QW by electrical and optical means [311] and in IV GeMn by electric field [312].

Following these results obtained in ferromagnetic semiconductors, electric-field control of H_C at room temperature was demonstrated to be possible in metal ferromagnets, FePt and FePd immersed in an electrolyte [313]. The work was followed by the observation of sizable electric-field control of magnetic anisotropies in Fe and FeCo(B) at room temperature [314,315], and the demonstration of electric-field induced magnetization direction switching [316,317].

7.8 Prospects

Ferromagnetic semiconductors are providing an excellent test bench for exploring a variety of spin-dependent phenomena and to demonstrate new schemes of spintronic device operation. Although room temperature application of well-established ferromagnetic semiconductors, such as (Ga,Mn)As and (In,Mn)As, and their devices are not yet possible due to their low T_C below room temperature, the knowledge accumulated to date are of great importance for future development of spintronics in general, and in particular for semiconductor spintronics once material breakthrough in T_C is made. Initiated by theoretical predictions [21,22,318], a worldwide effort to synthesize ferromagnetic semiconductors with high T_C is currently underway.

According to the p$-$d Zener model, increase of effective Mn participating in ferromagnetic order and/or increase of hole concentration is the key to increase T_C of (Ga,Mn)As and (In,Mn)As [21,22]. Although there exists no conclusive guide to overcome the solubility limit and the self-compensation effect, some strategies are proposed, ranging from codoping of donors to suppress the self-compensation [137,319,320], optimization of postanneal condition, and use of high-index substrates other than (001) to increase Mn incorporation efficiency as well as exchange interaction [52,321]. Introduction of heterostructures to increase local Mn composition and hole concentration simultaneously was shown to be promising; T_C of 250 K was observed in a Mn-δ-doped GaAs-based structure with modulation doping to enhance the hole concentration [277].

Another direction is to synthesize ferromagnetic semiconductors based on a host semiconductor other than GaAs and InAs. Ferromagnetic semiconductors based on wide-gap semiconductors such as GaN and ZnO have extensively been investigated [322] as the chemical trend pointed out by the theoretical studies shows increase in exchange interaction. However, it has also been pointed out that high x is necessary to achieve high T_C in wide-gap materials due to the reduced range of exchange interaction among magnetic spins in these wide band-gap semiconductors [323,324]. There are many reports on observation of room temperature ferromagnetism in wide-gap materials doped with transition metals or rare earths. Some of them show correlation between magnetization and MCD signals at semiconductor critical points and/or the anomalous Hall effect [325$-$327]. To firmly establish the origin of the observed ferromagnetic order, it is now becoming increasingly clear that one needs to do a series of measurements to fully characterize the properties of the synthesized materials. Correlation among magnetization, magneto-optical, and magnetotransport properties is one of the necessary conditions of intrinsic ferromagnetism, although large magneto-optical responses from materials with a small SOI cannot be expected. The presence of carrier spin-polarization is another signature of having ferromagnetism in semiconductors, which results from spin-split states induced by the sp$-$d exchange interaction. Control of magnetism by changing carriers by external means is an important way to confirm that the ferromagnetism is carrier related [328].

Nonuniform distribution of magnetic ions in host semiconductor may result from nanoscale spinodal decomposition [329,330], which is experimentally confirmed in a number of "ferromagnetic" semiconductors [331$-$334]. Here, magnetically condensed nanometer regions have the same crystal structure as the host semiconductor. Each region behaves as a superparamagnetic cluster (even with antiferromagnetic interaction as there remain uncompensated spins) and can show ferromagnetic-like properties below blocking temperature [211,335]. Since the presence of magnetic nanoparticles fully integrated in the host lattice can enhance the magnetotransport and magneto-optical properties of semiconductors [332,336,337], it is not an easy task to separate nonuniform magnetic semiconductors from uniform ones. It should also be noted that depending on what one is looking for, this nonuniform magnetic semiconductor may provide a solution as good as the uniform ferromagnetic semiconductors. Once percolated by increasing the concentration of magnetic ions, high T_C is predicted for these materials [329]. It has been suggested that the charge state of magnetic ions during crystal growth can change the way the nanoscale phase separation takes place [335,338]. This has been experimentally verified by controllable aggregation of Cr in (Zn,Cr)Te by doping [333].

In an effort to intentionally integrate room temperature ferromagnetic materials with nonmagnetic semiconductor, ferromagnetic materials having the same crystal structure with the host semiconductor have also been investigated. Single-crystal zincblende CrAs and CrSb were grown by MBE and were confirmed to show ferro-magnetism over 400 K [339,340]. Theoretical calculation predicts that these materials are half-metallic [341].

Finally, it is important to consider how to apply the new schemes found using ferromagnetic semiconductors not only to new devices but to conventional ones using ferromagnetic metals [313−315], which will further enrich the field of spintronics.

Acknowledgment

The authors thank Ms. Noriko Sato for her help with the list of references.

References

[1] Methfessel S, Mattis DC. Magnetic semiconductors. In: Wijn HPJ, editor. Encyclopedia of physics, vol. XVIII/1, magnetism. Berlin: Springer-Verlag; 1968. pp. 389−562.
[2] Holtberg F, von Molnár S, Coey JM. Rare earth magnetic semiconductors. In: Ketler SP, editor. Handbook on semiconductors, vol. 3, Amsterdam: North-Holland; 1980. pp. 803−56.
[3] Mauger A, Godart C. Phys Rep 1986;141:51−176.
[4] Furdyna JK, Kossut J. Semiconductors and semimetals, vol. 25, diluted magnetic semi-conductors. London: Academic Press; 1988.
[5] Kossut J, Dobrowolski W. Diluted magnetic semiconductors. In: Bushchow KHJ, editor. Handbook of magnetic materials, vol. 7. Amsterdam: North-Holland; 1993. p. 231−5.
[6] Dietl T. (Diluted) Magnetic semiconductors. In: Mahajan S, editor. Handbook on semi-conductors, completely revised and enlarged edition, vol. 3B. Amsterdam: North-Holland; 1994. p. 1251−342.
[7] Dobrowolski W, Kossut J, Story T. II−VI and IV−VI diluted magnetic semiconductors —new bulk materials and low-dimensional quantum structures. In: Bushchow KHJ, edi-tor. Handbook of magnetic materials, vol. 15. Amsterdam: North-Holland; 2003. pp. 289−77.
[8] Gaj JA, Planel R, Fishman G. Solid State Commun 1979;29:435.
[9] Nawrocki M, Planel R, Fishman G, Galazka RR. Phys Rev Lett 1981;46:735.

[10] Story T, Gałazka RR, Frankel RB, Wolf PA. Phys Rev Lett 1986;56:777.

[11] Ohno H, Munekata H, Penny T, von Molnár S, Chang LL. Phys Rev Lett 1992;68:2664.

[12] Ohno H, Shen A, Matsukura F, Oiwa A, Endo A, Katsumoto S, et al. Appl Phys Lett 1996;69:363.

[13] Dietl T, Haury A, Merle d'Aubigné Y. Phys Rev B 1997;55:R3347.

[14] Haury A, Wasiela A, Arnoult A, Cibert J, Tatarenko S, Dietl T, et al. Phys Rev Lett 1997;79:511.

[15] Ferrand D, Cibert J, Wasiela A, Bourgognon C, Tatarenko S, Fishman G, et al. Phys Rev B 2001;63:085201.

[16] Blinowski J, Kacman P, Majewski JA. Phys Rev B 1996;53:9524.

[17] Munekata H, Ohno H, von Molnar S, Segmüller A, Chang LL, Esaki L. Phy Rev Lett 1989;63:1849.

[18] Ohno H. J Magn Magn Mater 1999;200:110.

[19] Matsukura F, Ohno H, Dietl T. III−V ferromagnetic semiconductors. In: Buschow KHJ, editor. Handbook of magnetic materials, vol. 14. Amsterdam: North-Holland; 2002. p. 1−87.

[20] Jungwirth T, Sinova S, Mašek J, Kučera J, MacDonald AH. Rev Mod Phys 2006;78:809.

[21] Dietl T, Ohno H, Matsukura F, Cibert J, Ferrand D. Science 2000;287:1019.

[22] Dietl T, Ohno H, Matsukura F. Phys Rev B 2001;63:195205.

[23] Awschalom DD, Loss D, Samarth N. Semiconductor spintronics and quantum computation. Berlin: Springer-Verlag; 2002.

[24] Žutić I, Fabian J, Das Sarma S. Rev Mod Phys 2004;76:323.

[25] Samarth N. Solid State Phys 2004;58:1.

[26] Awschalom DD, Flatté ME. Nat Phys 2007;3:153.

[27] DeSimone D, Wood CEC, Evans Jr. CA. J Appl Phys 1982;53:4938.

[28] Kordoš P, Janšák L, Benč V. Solid State Electron 1975;18:223.

[29] De Boeck J, Oesterholt R, Van Esch A, Bender H, Bruynseraede C, Van Hoof C, et al. Appl Phys Lett 1996;68:2744.

[30] Soo YL, Huang SW, Ming ZH, Kao YH, Munekata H, Chung LL. Phys Rev B 1996;63:195209.

[31] Shioda R, Ando K, Hayashi T, Tanaka M. Phys Rev B 1998;58:1100.

[32] Matsukura F, Ohno H. Molecular beam epitaxy of III−V ferromagnetic semiconductors. In: Henini M, editor. Molecular beam epitaxy: from research to mass production. Amsterdam: Elsevier; 2012. p. 477−86.

[33] Ohtake A, Hagiwara A, Nakamura J. Phys Rev B 2013;87:165301.

[34] Ohya S, Ohno K, Tanaka M. Appl Phys Lett 2007;90:112503.

[35] Chiba D, Nishitani Y, Matsukura F, Ohno H. Appl Phys Lett 2007;90:122503.

[36] Shen A, Ohno H, Matsukura F, Sugawara Y, Akiba N, Kuroiwa T, et al. J Cryst Growth 1997;175/176:1069.

[37] Novák V, Olejník K, Cukr M, Smrčka L, Remeš Z, Oswald J. J Appl Phys 2007;102:083536.

[38] Shen A, Horikoshi Y, Ohno H, Guo SP. Appl Phys Lett 1997;71:1540.

[39] Yasuda H, Ohno H. Appl Phys Lett 1999;74:3275.

[40] Guo SP, Shen A, Yasuda H, Ohno Y, Matsukura F, Ohno H. J Cryst Growth 2000;208:799.

[41] Ohno H. Science 1998;281:951.

[42] Matsukura F, Shen A, Sugawara Y, Omiya T, Ohno Y, Ohno H. Institute of physics, conference series No. 162, 1999. p. 547—52.
[43] Shimizu H, Hayashi H, Nishinaga Y, Tanaka M. Appl Phys Lett 1999;74:398.
[44] Myers RC, Sheu BL, Jackson AW, Gossard AC, Schiffer P, Samarth N, et al. Phys Rev B 2006;74:155203.
[45] Sadowski J, Domagała JZ, Bąk-Misiuk J, Koleński S, Sawicki M, Świątek K, et al. J Vac Sci Technol B 2000;18:1697.
[46] Sadowski J, Mathieu R, Svendlindh P, Domagała JZ, Bąk-Misiuk J, Świątek U, et al. Appl Phys Lett 2001;78:3271.
[47] Hayashi T, Hashimoto Y, Katsumoto S, Iye Y. Appl Phys Lett 2001;78:1691.
[48] Schott GM, Faschinger W, Molenkamp LW. Appl Phys Lett 2001;79:1809.
[49] Schott GM, Schmidt G, Karczewski G, Molenkamp LW, Jakiela R, Barcz A, et al. Appl Phys Lett 2003;82:4678.
[50] Sadowski J, Domagala JZ. Phys Rev B 2004;69:075206.
[51] Omiya T, Matsukura F, Shen A, Ohno Y, Ohno H. Physica E 2001;10:206.
[52] Wang KY, Edmonds KW, Zhao LX, Sawicki M, Campion RP, Gallagher BL, et al. Phys Rev B 2005;72:115207.
[53] Reinwald M, Wurstbauer U, Döppe M, Kipferl W, Wagenhuber K, Tranitz H-P, et al. J Cryst Growth 2005;278:690.
[54] Daeulbler J, Glunk M, Schoch W, Limmer W, Sauer R. Appl Phys Lett 2006;88:051904.
[55] Stefanowicz W, Śliwa C, Aleshlevych P, Dietl T, Döppe M, Wurstbauer U, et al. Phys Rev B 2010;81:155203.
[56] Dreher L, Donhauser D, Daeubler J, Glunk M, Rapp C, Schoch W, et al. Phys Rev B 2010;81:245202.
[57] Wurstbauer U, Sperl M, Soda M, Neumaier D, Schuh D, Bayreuther G, et al. Appl Phys Lett 2008;92:102506.
[58] Zhao JH, Matsukura F, Abe E, Chiba D, Ohno Y, Takamura K, et al. J Cryst Growth 2002;237—239:1349.
[59] Uchitomi N, Sato S, Jinbo Y. Appl Surf Sci 2003;216:607.
[60] Ando K, Saito H, Agarwal KC, Debnath MC, Zayets V. Phys Rev Lett 2008;100:067204.
[61] Shen A, Matsukura F, Guo SP, Sugawara Y, Ohno H, Tani M, et al. J Cryst Growth 1999;201/202:679.
[62] Welp U, Vlasko-Vlasov VK, Menzel A, You HD, Liu X, Furdyna JK, et al. Appl Phys Lett 2004;85:260.
[63] Shirai M, Ogawa T, Kitagawa I, Suzuki N. J Magn Magn Mater 1998;177—181:1383.
[64] Mašek J, Kudrnovský J, Máca F. Phys Rev B 2003;67:153203.
[65] Ilegrams M, Dingle R, Rupp Jr. LW. J Appl Phys 1975;46:3059.
[66] Schneider J, Kaufmann U, Wilkening W, Baeumler M, Köhl F. Phys Rev Lett 1987;59:240.
[67] Tsuruoka T, Tanimoto R, Tachikawa N, Ushioda S, Matsukura F, Ohno H. Solid State Commun 2002;121:79.
[68] Yu KM, Walukiewicz W, Wojtowicz T, Kuryliszyn I, Liu X, Sasaki Y, et al. Phys Rev B 2002;65:201303.
[69] Potashnik SJ, Ku KC, Chun SH, Berry JJ, Samarth N, Schiffer P. Appl Phys Lett 2001;79:1495.
[70] Edmonds KW, Wang KY, Campion RP, Neumann AC, Foxon CT, Gallagher BL, et al. Appl Phys Lett 2002;81:3010.

[71] Edmonds KW, Wang KW, Campion RP, Neumann AC, Farley NRS, Gallagher BL, et al. Appl Phys Lett 2002;81:4991.

[72] Máca F, Mašek J. Phys Rev B 2003;66:235209.

[73] Blinowski J, Kacman P. Phys Rev B 2003;67:121204.

[74] Yakunin AM, Silov AY, Koenraad PM, Van Roy W, De Boeck J, Tang JM, et al. Phys Rev Lett 2004;92:216806.

[75] Blakemore JS, Brown WJ, Stass ML, Woodbury DA. J Appl Phys 1973;44:3352.

[76] Linnarson M, Janzén E, Monemar B, Kleverman M, Thilderkvist A. Phys Rev B 1997;58:6938.

[77] Kitchen D, Richardella A, Tang J-M, Flatté M, Yazdani A. Nature 2006;442:436.

[78] Yakunin AM, Silov AYu, Koenraad PM, Tang J-M, Flatté ME, Primus J-L, et al. Nat Mater 2007;6:512.

[79] Oiwa A, Katsumoto S, Endo A, Hirasawa M, Iye Y, Matsukura F, et al. Solid State Commun 1997;103:209.

[80] Novák V, Olejník E, Wunderlich J, Cukr M, Výborný K, Rushforth, et al. Phys Rev Lett 2008;101:077201.

[81] Shono T, Hasegawa T, Fukumura T, Matsukura F, Ohno H. Appl Phys Lett 2000;77:1363.

[82] Welp U, Vlasko-Vlasov VK, Liu X, Furdyna JK, Wojtowicz T. Phys Rev Lett 2003;90:167206.

[83] Fukumura T, Shono T, Inaba K, Hasegawa T, Koinuma H, Matsukura F, et al. Physica E 2001;10:135.

[84] Thevenard L, Largeau L, Mauguin O, Partriarche G, Lemaître A, Vernier N, et al. Phys Rev B 2006;73:195331.

[85] Sugawara A, Akashi T, Brown PD, Campion RP, Yoshida T, Gallagher BL, et al. Phys Rev B 2007;75:241306.

[86] Dietl T, König J, MacDonald AH. Phys Rev B 2001;64:241201.

[87] Sawicki M, Matsukura F, Idziaszek A, Dietl T, Schott GM, Ruester C, et al. Phys Rev B 2004;70:245325.

[88] Sawicki M, Wang K-Y, Edmonds KW, Campion RP, Staddon CR, Farley NRS, et al. Phys Rev B 2005;71:121302.

[89] Wang K-Y, Sawicki M, Edmonds KW, Campion RP, Maat S, Foxon CT, et al. Phys Rev Lett 2005;95:217204.

[90] Liu X, Sasaki Y, Furdyna JK. Phys Rev B 2003;67:205204.

[91] Liu X, Lim WL, Dobrowolska M, Furdyna JK. Phys Rev B 2005;71:035307.

[92] Tang HX, Kawakami RK, Awschalom DD, Roukes MK. Phys Rev Lett 2003;90:107201.

[93] Hamaya K, Taniyama T, Kitamoto Y, Moriya R, Munekata H. J Appl Phys 2003;94:7657.

[94] Hamaya K, Moriya R, Oiwa A, Taniyama T, Kitamoto Y, Yamazaki Y, et al. Jpn J Appl Phys 2004;43:L306.

[95] Uemura T, Sone T, Matsuda K, Yamamoto M. Jpn J Appl Phys 2005;44:L1352.

[96] Sawicki M, Matsukura F, Dietl T, Schott GM, Ruester C, Gould C, et al. J Supercond 2004;16:7.

[97] Yamada T, Chiba D, Matsukura F, Yakata S, Ohno H. Phys Status Solidi C 2006;3:4086.

[98] Fedorych OM, Hankiewicz EM, Wilamowski Z, Sadowski J. Phys Rev B 2002;66:045201.

[99] Abolfath M, Jungwirth T, Brum J, MacDonald AH. Phys Rev B 2001;63:054418.

[100] Takamura K, Matsukura F, Chiba D, Ohno H. Appl Phys Lett 2002;81:2590.
[101] Hayashi T, Katsumoto S, Hashimoto Y, Endo A, Kawamura M, Zalalutdinov M, et al. Physica B 2000;284–288:1175.
[102] Pappert K, Hümpfner S, Wenisch J, Brunner K, Gould C, Schimidt G, et al. Appl Phys Lett 2007;90:062109.
[103] Lin DC, Bi GY, Li F, Song C, Wang YY, Cui B, et al. J Appl Phys 2013;113:043906.
[104] Won J, Shin J, Lee S, Yoo T, Lee H, Lee S, et al. Appl Phys Express 2013;6:013001.
[105] Burch KS, Stephens J, Kawakami RK, Awshalom DD, Basov DN. Phys Rev B 2004;70:205208.
[106] Birowska M, Śliwa C, Majewski JA, Dietl T. Phys Rev Lett 2012;108:237203.
[107] Masmanidis SC, Tang HX, Myers EB, Li M, De Greve K, Vermeulen G, et al. Phys Rev Lett 2005;95:187206.
[108] Overby M, Chernyshov A, Rokhinson LP, Liu X, Furdyna JK. Appl Phys Lett 2008;92:192501.
[109] Rushforth AW, De Ranieri E, Zemen J, Wunderlich J, Edmonds KW, King CS, et al. Phys Rev B 2008;78:085314.
[110] Goennenwein STB, Althammer M, Bihler C, Brandlmaier A, Geprägs S, Opel M, et al. Phys Stat Sol 2008;2:96.
[111] Bihler C, Althammer M, Brandlmaier A, Geprägs S, Weiler M, Opel M, et al. Phys Rev B 2008;78:045203.
[112] Casiraghi A, Rushforth AW, Zemen J, Haigh JA, Wang A, Edmonds KW, et al. Appl Phys Lett 2012;101:082406.
[113] Liu X, Sasaki Y, Furdyna JK. Appl Phys Lett 2001;79:2414.
[114] Fid KF, Stone MB, Ku KC, Maksimov O, Schiffer P, Samarth N, et al. Appl Phys Lett 2004;85:1556.
[115] Lin HT, Chen YF, Huang PW, Wang SH, Huang JH, Lai CH, et al. Appl Phys Lett 2006;89:262502.
[116] Wang K, Sawicki M, Edmonds KW, Campion RP, Rushforth AW, Freeman AA, et al. Appl Phys Lett 2006;88:022510.
[117] Wenisch J, Gould C, Ebel L, Storz J, Pappert K, Schmidt MJ, et al. Phys Rev Lett 2007;99:077201.
[118] Pappert K, Hümpfner S, Gould C, Wenisch J, Brunner K, Schimidt G, et al. Nat Phys 2007;3:573.
[119] Kohda M, Ogawa J, Shiogai J, Matsukura F, Ohno Y, Ohno H, et al. Physica E 2010;42:2685.
[120] Hümpfner S, Pappert K, Wenisch J, Brunner K, Gould C, Schimidt G, et al. Appl Phys Lett 2007;90:102102.
[121] Greullet F, Ebel L, Münzhuber F, Mark S, Astakhov GV, Kießling T, et al. Appl Phys Lett 2011;98:231903.
[122] Chien CL, Westgate CR. The Hall effect and its application. New York, NY: Plenum Press; 1980.
[123] Jungwirth T, Qian N, MacDonald AH. Phys Rev Lett 2002;88:207208.
[124] Chun SH, Kim YS, Choi HK, Jeong IT, Lee WO, Suh KS, et al. Phys Rev Lett 2007;98:026601.
[125] Oiwa A, Endo A, Katsumoto S, Iye Y, Ohno H, Munekata H. Phys Rev B 1999;59:5826.
[126] Omiya T, Matsukura F, Dietl T, Ohno Y, Sakon T, Motokawa M, et al. Physica E 2000;7:976.

[127] Sadowski J, Mathieu R, Svedlindh P, Karlsteen M, Kanski J, Fu Y, et al. Thin Solid Films 2002;412:122.

[128] Yu KM, Walukiewicz W, Wojotowicz T, Lim WL, Liu X, Sasaki Y, et al. Appl Phys Lett 2002;81:844.

[129] Koeder A, Frank S, Schoch W, Avrutin V, Limmer W, Thonke K, et al. Appl Phys Lett 2003;82:3278.

[130] Moriya R, Munekata H. J Appl Phys 2003;93:4603.

[131] Osinny V, Jędrzejczak A, Arciszewska M, Dobrowolski W, Story T, Sadowski J. Acta Phys Polon A 2001;100:327.

[132] Limmer W, Glunk M, Schoch W, Köder A, Kling R, Sauer R, et al. Physica E 2002;13:589.

[133] Seong MJ, Chun SH, Cheong HM, Samarth N, Mascarenhas A. Phys Rev B 2002;66:033202.

[134] Wang KY, Edmonds KW, Campion RP, Gallagher BL, Farley NRS, Sawicki M, et al. J Appl Phys 2004;95:6512.

[135] Wojtowicz T, Furdyna JK, Liu X, Yu KM, Walukiewicz W. Physica E 2004;25:171.

[136] Walukiewicz W. Phys Rev B 1988;37:4760.

[137] Yu KM, Walukiewicz W, Wojtowicz T, Lim WL, Liu X, Bindley U, et al. Phys Rev B 2003;68:041308.

[138] Mašek J, Turek I, Kudrnovský J, Máca F, Drchal V. Acta Phys Polon A 2004;105:637.

[139] Jungwirth T, Wang KY, Mašek J, Edmonds K, König J, Sinova J, et al. Phys Rev B 2005;72:165204.

[140] Chiba D, Takamura K, Matsukura F, Ohno H. Appl Phys Lett 2003;82:3020.

[141] Stone MB, Ku KC, Potashnik SJ, Sheu BL, Samarth N, Schiffer P. Appl Phys Lett 2003;83:4568.

[142] Edmonds KW, Bogusławski P, Wang KY, Campions RP, Novikov SN, N.R.S. Farley, et al. Phys Rev Lett 2004;92:037201.

[143] Malfait M, Vanacken J, Moschchalkov VV, Van Roy W, Borghs G. Appl Phys Lett 2005;86:132501.

[144] Adell M, Ilver L, Kanski J, Stanciu V, Svedlindh P, Sadowski J, et al. Appl Phys Lett 2005;86:112501.

[145] Kirby BJ, Borchers JA, Rhyne JJ, te Velthunis SGE, Hoffmann A, O'Donovan KV, et al. Phys Rev B 2004;69:081307.

[146] Gonnenwein STB, Wassner TA, Huebl H, Brandt MS, Philipp JB, Opel M, et al. Phys Rev Lett 2004;91:227202.

[147] Thevenard L, Largeau L, Mauguin O, Lemaître A, Khazen K, von Bardeleben HJ. Phys Rev B 2007;75:195218.

[148] Barden JG, Parker JS, Xiong P, Chun SH, Samarth N. Phys Rev Lett 2003;91:056602.

[149] Panguluri RP, Ku KC, Wojtowicz T, Liu X, Furdyna JK, Lyanda-Geller YB, et al. Phys Rev B 2005;72:054510.

[150] Chiba D, Matsukura F, Ohno H. Physica E 2004;21:966.

[151] Ogawa T, Shirai M, Suzuki N, Kitagawa I. J Magn Magn Mater 1999;196−197:428.

[152] Abe E, Matsukura F, Yasuda H, Ohno Y, Ohno H. Physica E 2000;7:981.

[153] Matsukura F, Abe E, Ohno H. J Appl Phys 2000;87:6442.

[154] Lim WL, Wojtowicz T, Liu X, Dobrowolska M, Furdyna JK. Physica E 2004;82:4310.

[155] Wojtowicz T, Cywinski G, Lim WL, Liu X, Dobrowolska M, Furdyna JK, et al. Appl Phys Lett 2003;82:4310.
[156] Yanagi S, Kuga K, Slupinski T, Munekata H. Physica E 2004;20:333.
[157] Panguluri RP, Nadgorny B, Wojtowicz T, Liu X, Furdyna JK. Appl Phys Lett 2007;91:252502.
[158] Panguluri RP, Nadgorny B, Wojtowicz T, Lim WL, Liu X, Furdyna JK. Appl Phys Lett 2004;84:4947.
[159] Matsukura F, Ohno H, Shen A, Sugawara Y. Phys Rev B 1998;57:R2037.
[160] Matsukura F, Sawicki M, Dietl T, Chiba D, Ohno H. Physica E 2004;21:1032.
[161] Ohno H, Matsukura F. Solid State Commun 2001;117:179.
[162] Jan J-P. Solid State Phys 1957;5:1.
[163] Tang HX, Masmanidis S, Kawakami RK, Awschalom DD, Roukes MK. Nature 2004;431:52.
[164] Tang HX, Kawakami RK, Awschalom DD, Roukes MK. Phys Rev B 2006;74:041310.
[165] Baxter DV, Ruzmetov D, Schreschiligt J, Sasaki Y, Liu, Furdyna JK, et al. Phys Rev B 2002;65:212407.
[166] Jungwirth T, Abolfath M, Sinova J, Kučera J, MacDonald AH. Appl Phys Lett 2002;81:4029.
[167] von Molnár S, Kasuya T. Phys Rev Lett 1968;21:1757.
[168] Novák V, Olejník K, Wuderlich J, Cukr M, Výborný K, Rushforth AW, et al. Phys Rev Lett 2008;101:077201.
[169] Okabayashi J, Kimura A, Rader O, Mizokawa T, Fujimori A, Hayashi T, et al. Phys Rev B 1998;58:R4211.
[170] Kawabata A. Solid State Commun 1980;34:431.
[171] He HT, Yang CL, Ge WK, Wang JN, Dai X, Wang YQ. Appl Phys Lett 2005;87:162506.
[172] Honolka J, Masmanidis S, Tang HX, Awshalom DD, Roules ML. Phys Rev B 2007;75:245310.
[173] Wagner K, Neumaier D, Reinwald M, Wescheider W, Weiss D. Phys Rev Lett 2006;97:056803.
[174] Vila L, Giraud R, Thevenard L, Lemaître A, Pierre F, Dufouleur J, et al. Phys Rev Lett 2007;98:027204.
[175] Katsumoto S, Oiwa A, Iye Y, Ohno H, Matsukura F, Shen A, et al. Phys Status Solidi B 1998;205:115.
[176] Pappert K, Schmidt MJ, Hümpfner S, Rüster C, Schott GM, Brunner K, et al. Phys Rev Lett 2006;97:186402.
[177] Dietl T. J Phys Soc Jpn 2008;77:031005.
[178] Chiba D, Yamanouchi M, Matsukura F, Dietl T, Ohno H. Phys Rev Lett 2006;96:096602.
[179] Tatara G, Fukuyama H. Phys Rev Lett 1997;78:3773.
[180] Miyake K, Shigeto K, Mibu K, Shinjo T, Ono T. J Appl Phys 2002;91:3468.
[181] Yamanouchi M, Chiba D, Matsukura F, Ohno H. Nature 2004;428:539.
[182] Partin DL, Karnezos M, deMenezes LC, Berger L. J Appl Phys 1974;45:1852.
[183] Levy PM, Zhang S. Phys Rev Lett 1997;79:5110.
[184] Pu Y, Johnston-Halperin E, Awshalom DD, Shi J. Phys Rev Lett 2006;97:036601.
[185] Pu Y, Chiba D, Matsukura F, Ohno H, Shi J. Phys Rev Lett 2008;101:117208.
[186] Mott NF, Jones H. The Theory of the Properties of Metals and Alloys. New York, NY: Dover; 1958.

[187] Ando K. Science 2006;312:1883.
[188] Szczytko J, Bardyszewski W, Twardowski A. Phys Rev B 2001;64:075306.
[189] Casey Jr. HC, Sell DD, Wecht KW. J Appl Phys 1975;46:250.
[190] Kuroiwa T, Yasuda T, Matsukura F, Shen A, Ohno Y, Segawa Y, et al. Electron Lett
       1998;34:190.
[191] Ando K, Hayashi T, Tanaka M, Twardowski A. J Appl Phys 1998;83:6548.
[192] Beshoten B, Crowell PA, Malajovich I, Awshalom DD, Matsukura F, Shen A, et al.
       Phys Rev Lett 1999;83:3073.
[193] Szczytko J, Mac W, Twardowski A, Matsukura F, Ohno H. Phys Rev B
       1999;59:12935.
[194] Hayashi T, Tanaka M, Seto K, Nishinaga T, Ando K. Appl Phys Lett 1997;71:1825.
[195] Kimel AV, Astakhov GV, Kirilyuk A, Schott GM, Karczewski G, Ossau W, et al.
       Phys Rev Lett 2005;94:227203.
[196] Héroux JB, Ino Y, Kuwata-Gonokami M, Hashimoto Y, Katsumoto S. Appl Phys Lett
       2006;88:221110.
[197] Burch KS, Awshalom DD, Basov DS. J Magn Magn Mater 2008;320:3207.
[198] Kacman P. Semicond Sci Technol 2001;16:R25.
[199] Okabayashi J, Mizokawa T, Sarma DD, Fujimori A, Slupinski T, Oiwa A, et al. Phys
       Rev B 2002;65:161203.
[200] Szczytko J, Mac W, Stachow A, Twardowski A, Becla P, Tworzydło J. Solid State
       Commun 1996;99:927.
[201] Heimbrodt W, Hartmann TH, Klar PJ, Lampalzer M, Stolz W, Volz K, et al. Physica
       E 2001;10:175.
[202] Śliwa C, Dietl T. Phys Rev B 2008;78:165205.
[203] Myers RC, Poggio M, Stern NP, Gossard AC, Awshalom DD. Phys Rev Lett
       2005;95:017204.
[204] Stern NP, Myers RC, Poggio M, Gossard AC, Awshalom DD. Phys Rev B
       2007;75:045329.
[205] Zener C. Phys Rev 1950;81:440.
[206] Schliemann J, König J, MacDonald AH. Appl Phys Lett 2001;78:1550.
[207] König J, Schilermann J, Jungwirth J, MacDonald AH. In: Singh DJ,
       Papaconstantopoulos DA, editors. Electronic structure and magnetism of complex
       materials. Berlin: Springer-Verlag; 2003. p. 163.
[208] Das Sarma S, Hwang EH, Priour Jr. DJ. Phys Rev B 2004;70:161203.
[209] Bhatt RN, Berciu M. Phys Rev Lett 2001;87:10723.
[210] Akai H. Phys Rev Lett 1998;81:3002.
[211] Katayama-Yoshida H, Sato K, Fukushima T, Toyoda M, Kizaki H, Dinh AD, et al.
       Phys Status Solidi A 2006;204:15.
[212] Park JH, Kwon SK, Min BI. Physica B 2000;281−282:703.
[213] Fillipetti A, Spaldin NA, Sanvito S. Chem Phys 2005;309:59.
[214] Paalanen MA, Bhatt RN. Physica B 1991;169:223.
[215] Nagai Y, Kunimoto T, Nagasaka K, Nojiri H, Motokawa M, Matsukura F, et al. Jpn J
       Appl Phys 2001;40:6231.
[216] Hirakawa K, Oiwa A, Munekata H. Physica E 2001;10:215.
[217] Burch KS, Shrekenhamer DB, Singley EJ, Stephens J, Sheu BL, Kawakami RK, et al.
       Phys Rev Lett 2006;97:087208.
[218] Jungwirth T, Sinova J, MacDonald AH, Gallagher BL, Novák V, Edmonds KW, et al.
       Phys Rev B 2007;76:125206.
[219] Ohya S, Takata K, Tanaka M. Nat Phys 2011;7:342.

[220] Dobrowolska M, Tivakornsasithorn K, Liu X, Furdyna JK, Berciu M, Yu KM, et al. Nat Mater 2012;11:444.
[221] Paalanen MA, Graebner JE, Bhatt RN, Sachdev S. Phys Rev Lett 1988;61:597.
[222] Ohno Y, Young DK, Beschoten B, Matsukura F, Ohno H, Awschalom DD. Nature 1999;402:790.
[223] Kohda M, Ohno Y, Takamura K, Matsukura F, Ohno H. Jpn J Appl Phys 2001;40: L1274.
[224] Jonston-Halperin E, Lofgreen D, Kawakami RK, Young DK, Coldren L, Gossard AC, et al. Phys Rev B 2002;65:041306.
[225] Kohda M, Kita T, Ohno Y, Matsukura F, Ohno H. Appl Phys Lett 2006;89:012103.
[226] Van Dorpe P, Liu Z, Van Roy W, Motsnyi VF, Sawicki M, Borghs G, et al. Appl Phys Lett 2004;84:3945.
[227] Sankowski P, Kacman P, Majewski J, Dietl T. Phys Rev B 2007;75:045306.
[228] Jedema FJ, Heersche HB, Flip AT, Baselmans JJA, van Wees BJ. Nature 2002;416:713.
[229] Ciorga M, Einwanger A, Wurstbauer U, Schuh D, Wegschneider W, Weiss D. Phys Rev B 2009;79:165321.
[230] Shiogai J, Ciorga M, Utz M, Schuh D, Arakara T, Kohda M, et al. Appl Phys Lett 2012;101:212402.
[231] Mizukami S, Ando Y, Miyazaki T. Jpn J Appl Phys 2001;40:580.
[232] Tserkovnyak Y, Brataas A, Bauer GEW. Phys Rev Lett 2002;88:117601.
[233] Azevedo A, Viela Leão LH, Rodriguez-Suarez RL, Oliveria AB, Rezende SM. Appl Phys Lett 2005;97:10C715.
[234] Saitoh E, Ueda M, Miyajima H, Tatara G. Appl Phys Lett 2006;88:182509.
[235] Chen L, Matsukura F, Ohno H. Nat Commun 2013;4:2055.
[236] Hayashi T, Shimada H, Shimizu M, Tanaka M. J Cryst Growth 1999;201/202:689.
[237] Chiba D, Akiba N, Matsukura F, Ohno Y, Ohno H. Appl Phys Lett 2000;77:1873.
[238] Ohno Y, Arata I, Matsukura F, Ohno H. Physica E 2002;13:521.
[239] Tanaka M, Higo Y. Phys Rev Lett 2001;87:026602.
[240] Julliere M. Phys Lett 1975;54:225.
[241] Ohno H, Akiba N, Matsukura F, Shen A, Ohtani K, Ohno Y. Appl Phys Lett 1998;73:363.
[242] Mattana R, George J-M, Jaffrés H, Nguyen Van Dau F, Fert A, Lépine B, et al. Phys Rev Lett 2003;90:166601.
[243] Ohya S, Hai PN, Tanaka M. Appl Phys Lett 2005;87:012105.
[244] Ohya S, Hai PN, Mizuno Y, Tanaka M. Phys Rev B 2007;75:155328.
[245] Ohya S, Hai PN, Mizuno Y, Tanaka M. Phys Status Solidi C 2006;3:4184.
[246] Mizuno Y, Ohya S, Hai PN, Tanaka M. Appl Phys Lett 2007;90:162505.
[247] Chiba D, Sato Y, Kita T, Matsukura F, Ohno H. Phys Rev Lett 2004;93:216602.
[248] Elsen M, Boulle O, George J-M, Jaffrès H, Mattana R, Cros V, et al. Phys Rev B 2006;73:035303.
[249] Chiba D, Kita T, Matsukura F, Ohno H. J Appl Phys 2006;99:08G514.
[250] Slonczewski JC. J Magn Magn Mater 1996;159:L1.
[251] Sankey JC, Cui Y-T, Sun JZ, Slonczewski JC, Buhrman RA, Ralph DC. Nat Phys 2008;4:67.
[252] Chiba D, Matsukura F, Ohno H. J Phys D 2006;39:R1.
[253] Moriya R, Hamaya K, Oiwa A, Munekata H. Jpn J Appl Phys 2004;43:L825.
[254] Rüster C, Borzenko T, Gould C, Schmidt G, Molenkamp LW, Liu X, et al. Phys Rev Lett 2003;91:216602.

[255] Rüster C, Gould C, Jungwirth J, Sinova J, Schott GM, Giraud R, et al. Phys Rev Lett 2005;94:027203.

[256] Giddings AD, Khalid MN, Jungwirth T, Wunderlich J, Yasin S, Campion RP, et al. Phys Rev Lett 2005;94:127202.

[257] Saito H, Yuasa S, Ando K. Phys Rev Lett 2005;95:086604.

[258] Naydenova T, Dürrenfeld P, Tavakoli K, Pégard N, Ebel L, Pappert K, et al. Phys Rev Lett 2011;107:197201.

[259] Chernyshov A, Overby M, Liu X, Furdyna JK, Lyanda-Geller Y, Rokhinson LP. Nat Phys 2009;5:656.

[260] Endo M, Matsukura F, Ohno H. Appl Phys Lett 2010;97:222501.

[261] Rokhinson LP, Overby M, Chernyshov A, Lyanda-Geller Y, Liu X, Furdyna JK. J Magn Magn Mater 2012;324:3379.

[262] Fang D, Kurebayashi H, Wunderlich J, Výborný K, Zârbo LP, Campion RP, et al. Nat Nanotechnol 2011;6:413.

[263] Fang D, Skinner TD, Kurebayashi H, Campion RP, Gallagher BL, Ferguson AJ. Appl Phys Lett 2012;101:182402.

[264] Yamanouchi M, Chiba D, Matsukura F, Dietl T, Ohno H. Phys Rev Lett 2006;96:096601.

[265] Tatara G, Kohno H. Phys Rev Lett 2004;92:086601.

[266] Barnes SE, Maekawa S. Phys Rev Lett 2005;95:107204.

[267] Lemerle S, Ferŕe J, Chappert C, Mathet V, Giamarchi T, Le Doussal P. Phys Rev Lett 1998;80:849.

[268] Yamanouchi M, Ieds J, Matsukura F, Barnes SE, Maekawa S, Ohno H. Science 2007;317:1726.

[269] Kanda A, Suzuki A, Matsukura F, Ohno H. Appl Phys Lett 2010;97:032504.

[270] Nguyen AK, Skadsem HJ, Brataas A. Phys Rev Lett 2007;98:146602.

[271] Curiale J, Lemaître A, Ulysse C, Faini G, Jeudy V. Phys Rev Lett 2012;108:076604.

[272] Tatara G, Vernier N, Ferré J. Appl Phys Lett 2005;88:252509.

[273] Duine RA, Núñez AS, MacDonald AH. Phys Rev Lett 2007;98:056605.

[274] Ohno H, Chiba D, Matsukura F, Omiya T, Abe E, Dietl T, et al. Nature 2000;408:944.

[275] Chiba D, Yamanouchi M, Matsukura F, Ohno H. Science 2003;301:943.

[276] Chiba D, Yamanouchi M, Matsukura F, Abe E, Ohno Y, Ohtani K, et al. J Supercond 2003;16:179.

[277] Nazmul AM, Kobayashi S, Sugahara S, Tanaka M. Jpn J Appl Phys 2004;43:L233.

[278] Chiba D, Matsukura F, Ohno H. Appl Phys Lett 2006;89:16250.

[279] Biercuk MJ, Monsma DJ, Marcus CM, Becker JS, Gordon RG. Appl Phys Lett 2003;83:2405.

[280] Olejník E, Owen MHS, Novák V, Mašek J, Irvine C, Wunderlich J, et al. Phys Rev B 2008;78:054403.

[281] Stolichnov I, Riester SWE, Trodahl HJ, Setter N, Rushforth AW, Edmonds KW, et al. Nat Mater 2008;7:464.

[282] Sawicki M, Chiba D, Korbecka A, Nisitani Y, Majewski JA, Matsukura F, et al. Nat Phys 2010;6:22.

[283] Nishitani Y, Chiba D, Endo M, Sawicki M, Matsukura F, Dietl T, et al. Phys Rev B 2010;81:045208.

[284] Ohno H. J Appl Phys 2013;113:136509.

[285] Chiba D, Sawicki M, Nisitani Y, Matsukura F, Ohno H. Nature 2008;455:515.

[286] Chiba D, Nakatani Y, Matsukura F, Ohno H. Appl Phys Lett 2010;96:192506.

[287] Iwasaki Y. J Magn Magn Mater 2002;240:395.
[288] Linnik TL, Scherbakov AV, Yakovlev DR, Liu X, Furdyna JK, Bayer M. Phys Rev B 2011;84:214432.
[289] Scherbakov AV, Salasyuk AS, Akimov AV, Liu X, Bombeck M, Brüggemann C, et al. Phys Rev Lett 2010;105:117204.
[290] Cashiraghi A, Walker P, Akimov AV, Edmonds KW, Rushforth AW, De Ranieri E, et al. Appl Phys Lett 2011;99:262503.
[291] Chiba D, Wepachowska A, Endo M, Nishitani Y, Matsukura F, Dietl T, et al. Phys Rev Lett 2010;104:106601.
[292] Werpachowska A, Dietl T. Phys Rev B 2010;81:155205.
[293] Koshihara S, Oiwa A, Hirasawa M, Katsumoto S, Iye Y, Urano C, et al. Phys Rev Lett 1997;78:4617.
[294] Kanamura M, Zhou YK, Okunuma S, Asami K, Nakajima M, Harima H, et al. Jpn J Appl Phys 2002;41:1019.
[295] Oiwa A, Słupinski T, Munekata H. Appl Phys Lett 2001;78:518.
[296] Liu X, Lim WL, Titova LV, Wojtowicz T, Kutrowski M, Yee KJ, et al. Physica E 2004;20:370.
[297] Astaknov GV, Hoffmann H, Korenev VL, Kiessling T, Schwittek J, Schott GM, et al. Phys Rev Lett 2009;102:187401.
[298] Astaknov GV, Schwittek J, Schott GM, Gould C, Ossau W, Brunner K, et al. Phys Rev Lett 2011;106:037204.
[299] Reid AHM, Astaknov GV, Kimel AV, Schott GM, Ossau W, Brunner K, et al. Appl Phys Lett 2010;97:232503.
[300] Oiwa A, Mitsumori Y, Moriya R, Słupinski T, Munekata H. Phys Rev Lett 2002;88:137202.
[301] Kondo T, Nomura K, Koizumi G, Munekata H. Phys Status Solidi C 2006;3:4263.
[302] Nazmul AM, Amemiya T, Shuto Y, Sugahara S, Tanaka M. Phys Rev Lett 2005;95:017201.
[303] Herrera Diez L, Konuma M, Placidi E, Arciprete F, Rushforth AW, Campion RP, et al. Appl Phys Lett 2011;98:022503.
[304] Wang J, Khodaparast GA, Kono J, Oiwa A, Munekata H. J Mod Opt 2004;51:2771.
[305] Wang J, Sun C, Kono J, Oiwa A, Munekata H, Cywiński Ł, et al. Phys Rev Lett 2005;95:167401.
[306] Wang J, Cotoros C, Dani KM, Liu X, Furdyna JK, Chelma DS. Phys Rev Lett 2007;98:217401.
[307] Hashimoto Y, Kobayashi S, Munekata H. Phys Rev Lett 2008;100:067202.
[308] Carpene E, Piovera C, Dallera C, Mancini E, Puppin E. Phys Rev B 2011;84:134425.
[309] Csontos M, Mihály G, Jankó B, Wojtowicz T, Liu X, Fyrdyna JK. Nat Mater 2005;4:447.
[310] Gryglas-Borysiewicz M, Kwiatkowski A, Baj M, Wasik D, Przybytek J, Sadowski J. Phys Rev B 2010;82:153204.
[311] Boukari H, Kossacki P, Bertolini M, Ferrand D, Cibert J, Tatarenko S, et al. Phys Rev Lett 2002;88:207204.
[312] Park YD, Hanbicki A, Erwin SCT, Hellberg CS, Sullivan JM, Mattson JE, et al. Science 2002;295:651.
[313] Weisheit M, Fähler S, Marty A, Souche Y, Poinsignon C, Givord D. Science 2007;315:349.
[314] Maruyama T, Shiota Y, Nozaki T, Ohta K, Toda N, Mizuguchi M, et al. Nat Nanotechnol 2009;4:158.

[315] Endo M, Kanai S, Ikeda S, Matsukura F, Ohno H. Appl Phys Lett 2010;96:212503.
[316] Shiota Y, Nozaki T, Bonell F, Murakami S, Shinjo T, Suzuki Y. Nat Mater 2011;11:39.
[317] Kanai S, Yamanouchi S, Ikeda S, Nakatani Y, Matsukura F, Ohno H. Appl Phys Lett 2010;101:122403.
[318] Sato K, Katayama-Yoshida H. Jpn J Appl Phys 2000;39:L555.
[319] Sato K, Katayama-Yoshida H. Jpn J Appl Phys 2008;46:L1120.
[320] Fujii H, Sato K, Bergqvist L, Dederichs PH, Katayama-Yoshida H. Appl Phys Express 2011;4:043003.
[321] Franceschetti A, Dudiy SV, Barbash SV, Zunger A, Xu J, van Schilfgaaede M. Phys Rev Lett 2006;97:047202.
[322] Pearton SJ, Abernathy CR, Overberg ME, Thaler GT, Norton DP, Theodoropoulou N, et al. J Appl Phys 2003;93:1.
[323] Bergvist L, Eriksson O, Kudrnocský J, Drchal V, Korzhavyi P, Turek I. Phys Rev Lett 2004;93:137202.
[324] Sato K, Scheika W, Dederichs PH, Katayama-Yoshida H. Phys Rev B 2004;70:201202.
[325] Saito H, Zayets V, Yamagata S, Ando K. Phys Rev Lett 2003;90:207202.
[326] Fukumura T, Yamada Y, Tamura K, Nakajima K, Aoyama T, Tsukazaki A, et al. Jpn J Appl Phys 2003;42:L105.
[327] Toyosaki H, Fukumura T, Yamada Y, Nakajima K, Chikyow T, Hasegawa T, et al. Nat Mater 2004;3:221.
[328] Li L, Guo Y, Cui XY, Zheng R, Ohtani K, Kong C, et al. Phys Rev B 2012;85:174430.
[329] Sato K, Katayama-Yoshida H, Dederichs PH. Jpn J Appl Phys 2005;44:L948.
[330] Fukushima T, Sato K, Katayama-Yoshida H, Dederichs PH. Jpn J Appl Phys 2006;45:L416.
[331] Singh RK, Wu SY, Liu HX, Gu L, Simith DJ, Newman N. Appl Phys Lett 2005;86:012504.
[332] Jamet M, Barski A, Devillers T, Poydenot V, Dujardin R, Bayle-Guillemaud P, et al. Nat Mater 2006;5:653.
[333] Kuroda S, Nishizawa N, Takita K, Mitome M, Bando Y, Osuch K, et al. Nat Mater 2007;6:440.
[334] Bonanni A. Semicond Sci Technol 2007;22:R41.
[335] Dietl T. Nat Mater 2006;290:1395.
[336] Ye S, Klar PJ, Hartmann T, Heimbrodt W, Lampalzer M, Nau S, et al. Appl Phys Lett 2003;83:3927.
[337] Shinde SR, Ogale SB, Higgins JS, Zheng H, Millis AJ, Kulkarni VN, et al. Phys Rev Lett 2004;92:166601.
[338] Dietl T. Physica E 2006;35:293.
[339] Akinaga H, Manago T, Shirai M. Jpn J Appl Phys 2000;39:L1118.
[340] Zhao JH, Matsukura F, Takamura K, Abe E, Chiba D, Ohno H. Appl Phys Lett 2001;79:2776.
[341] Shirai M. J Appl Phys 2003;93:6844.

Printed in the United States
By Bookmasters